Contributions from Science Education Research

Volume 14

Contributions from Science Education Research is the international, multidisciplinary book series of the European Science Education Research Association (ESERA). The aim of the series is to synthesize, for the benefit of the scholarly community, the findings of high quality, theoretically-framed research in the domain of science education as well as comprehensive explorations of specific methodological strands in science education research. The series aims to publish books that are innovative in attempting to forge new ways of representing emergent knowledge in the field. The series includes edited collections of chapters, monographs and handbooks that are evaluated on the basis of originality, scientific rigor and significance for science education research. The book series is intended to focus mainly on work carried out in Europe. However, contributions from researchers affiliated with non-European institutions and non-members of the European Science Education Research Association are welcomed. The series is designed to appeal to a wide audience of researchers and post-graduate students in science education. Book proposals for this series may be submitted to the Publishing Editor: Claudia Acuna E-mail: Claudia.Acuna@springer.com

Connie Svabo • Michael Shanks
Chunfang Zhou • Tamara Carleton
Editors

Creative Pragmatics for Active Learning in STEM Education

Springer

Editors
Connie Svabo
STEM Centre for Research in Science
Education and Communication (FNUG)
Department of Mathematics and
Computer Science
University of Southern Denmark
Odense, Denmark

Chunfang Zhou
STEM Centre for Research in Science
Education and Communication (FNUG)
Department of Mathematics and
Computer Science
University of Southern Denmark
Odense, Denmark

Michael Shanks
Archaeology Center
Stanford University
Stanford, CA, USA

Tamara Carleton
Department of Mechanical Engineering
Blekinge Institute of Technology
Karlskrona, Sweden

ISSN 2213-3623 ISSN 2213-3631 (electronic)
Contributions from Science Education Research
ISBN 978-3-031-78722-5 ISBN 978-3-031-78720-1 (eBook)
https://doi.org/10.1007/978-3-031-78720-1

This Springer imprint is published by the registered company Springer Nature Switzerland AG
The registered company address is: Gewerbestrasse 11, 6330 Cham, Switzerland

Foreword by Andrew Pickering

Thomas Kuhn (1962) described scientific revolutions as gestalt switches in which foregrounds and backgrounds swap places—as exemplified in the famous gestalt image that can be seen as either a duck or a rabbit but not both at once. It seems to me that something revolutionary like that is happening in STEM education, as traditional approaches are not so much displaced as complemented by the arrival of the new educational paradigm that Svabo, Shanks, Carleton and Zhou call Creative Pragmatics. It also makes me happy to think that my own field—science and technology studies, STS—feeds into this transformation. How does that go?

The key observation is that how we teach science (and its STEM relatives) depends on what we think science is. Science is traditionally seen as a body of positive knowledge, running, say, from the basics of geometry to quantum field theory. In this view, education necessarily takes the form of the transmission of knowledge from teacher to student. One can imagine this sort of pedagogy as passive rote learning, like reciting multiplication tables over and over again, but usually it includes an active-learning component too—asking students to figure out how to use items of knowledge to solve textbook problems. That is how I once learned physics, chemistry and maths at school and university—I really enjoyed the experience—and it is still how STEM subjects are typically taught today.

There is nothing wrong with this view of science as a body of knowledge waiting to be passed on in the classroom. Actually, it is indispensable in a technoscientific culture such as ours. But there is another way of thinking about science which derives from STS.

Andrew Pickering is a distinguished figure in science and technology studies, renowned for his influential work on the history, sociology, and philosophy of science. Pickering has reshaped our understanding of scientific practice and culture with notable books such as Constructing Quarks: A Sociological History of Particle Physics, The Mangle of Practice: Time, Agency and Science, and The Cybernetic Brain: Sketches of Another Future. Throughout his career, Pickering has held prestigious fellowships at institutions like MIT, Princeton, and Stanford. As a professor emeritus at the University of Exeter, his interdisciplinary research continues to explore the intersections of cybernetics, art, and environmental science. His latest book is Acting with the World: Agency in the Anthropocene. (Editors)

As a field, STS centres on empirical studies of science in practice—what does science in the lab actually look like?—and in his pioneering work in this area Bruno Latour (1987) made a key distinction between what he called 'ready-made science' and 'science in action'. The former refers to the finished contents of science—theories like Newton's laws or instruments like spectrometers—while the latter points to a very different picture of what science is, as something that scientists *do*, an organised way of acting in and finding out about the world. These two facets of science are intimately connected to the extent that ready-made science is both an input to and an output from science in action. But still, they are different kinds of thing: knowing Newton's laws (ready-made science), say, tells us precisely nothing about what laboratory research looks like (science in action). This is precisely what Nobel laureate Peter Medawar (1963) meant in a celebrated talk arguing that the scientific paper is a fraud.

So, a new pedagogical project comes into sight here: how would you teach STEM subjects in a way that foregrounds not finished and ready-made science but science in action? I have often wondered about that—how might the findings of STS translate into teaching?—and now it seems to me that Creative Pragmatics is the answer.

What have we learned from STS? First, that science is *social*. Scientific practice draws on and transforms the knowledge and techniques of whole communities of scientists in a collaborative and dialogical fashion. The isolated scientific genius is a myth. Second, scientific research is *performative*: it centres on doing things, acting in the laboratory. Knowledge comes along with doing; it is not something separate from action. Third, this doing is *experimental*, in the sense of aiming to find out what will happen if we do this or that—a performative trip into the unknown. Fourth, science in action is *emergent*, in the sense that it is impossible to say in advance where research will lead. Novelty and creativity are thus definitional of science-in-practice. Fifth, *feedback* is crucial. Research is a trial-and-error process, and surprises of all sorts loop back to what one does next in a process I call 'mangling' (Pickering, 1995). And, of course, these are just the aspects of science that Creative Pragmatics, continuing the pragmatic tradition of William James and John Dewey, aims to foreground in teaching.

On my reading, the great strength of *Creative Pragmatics* is that individual chapters explore real-world examples of different ways in which these five points about science in action can inform STEM pedagogy. They are interested in forms of active learning that go far beyond solving problems in textbooks (valuable though that is). This is not the place to discuss specific contributions, but a focus on *design* runs through many of them, often in projects extending over a term or semester, and that is interesting to think about. Designing a novel artwork, artefact or structure is an open-ended problem without a singular well-defined solution, so a design-oriented course immediately introduces students to research as a creative and experimental trip into the unknown, as well as to the emergent aspects of research, in that the final design cannot be known in advance but can only take form in project-work. Design projects can also, of course, be conceived to draw upon elements of ready-made science already familiar from conventional teaching, helping students gain fluency in, and see the relevance of, canonical scientific knowledge. Going beyond design, Fablabs and Maker projects also introduce students to the struggles and transformative entanglements with materials that are again definitional of science-as-practice. And feedback is crucial throughout, including feedback from teachers and fellow students as well as materials, highlighting the social dimensions of STEM.

I should recognise here that assessment is an integral part of teaching that STS does not directly engage with. Assessment is straightforward in conventional STEM pedagogy and comes at the end, in examinations that determine whether students can successfully tackle assigned problems with known solutions. Creative Pragmatics makes assessment more challenging and calls for a different sort of solution that, in fact, goes back to what has just been said, evaluating the process as well as the products of project work, integrated with continuing feedback and recognising students' engagement, creativity and fluency with STEM concepts.

From my perspective, then, Creative Pragmatics is STS in action in teaching practice, fostering enquiry and initiative as well as proficiency in ready-made knowledge. That is what recommends it and draws me strongly to it, and that is the main point I want to make about this important book. STEM pedagogy can learn from STS and vice versa in a symbiotic loop. But it might be worth taking this line of thought one step further.

We need to think about ontology and worldviews, understandings of what the world is like in general terms. Conventional science teaching encourages students to think of the world as a knowable place, a field of well-defined solvable problems. These are what students learn to tackle, and no doubt this is appropriate in many real-world situations. But the world is also full of so-called wicked problems (Rittel & Webber, 1973) that have no unique solutions seeable in advance—that is the point about emergence. Global warming, environmental disasters, the extinction of species—all the problems we associate with the Anthropocene that today's students will have to grapple with—are wicked problems, in a different class from calculating the collision of billiard balls. And then we can recognise that Creative Pragmatics is both a way of introducing students to this novel ontology of a world of wicked systems and of fostering creative ways forward in living in a world like that. That worldview and pattern of action is something our children cannot get from conventional STEM teaching but which they really need in approaching a radically uncertain future, and Creative Pragmatics is an important step in the right direction.

A final thought comes up here. As presented in this book, Creative Pragmatics is about the *form* that pedagogy can take—the gestalt shift from the transmission of knowledge to active, open-ended, project-based teaching and learning. But we could also think about the *content* of the curriculum, *what* we teach rather than *how*. As I have described it so far, Creative Pragmatics can function as an oblique introduction to an ontology of unknowability and the practices that go with it—a way of introducing students to that ontology without necessarily talking about it explicitly. But pedagogy could also come at these concerns head-on. The STS literature is full of insightful discussions of emergence and more-than-human couplings, and assigned readings directly on those topics could be another vector along which STS could inform STEM education. And if the conventional contents of ready-made science hang together with an overall vision of a calculable world, less familiar sciences like complexity science and cybernetics foster instead a picture of an emergent world of wicked problems (Pickering, 2010). Words like 'cybernetics' sound intimidating but also mysterious and inviting, and cybernetic concepts like feedback are just as teachable as calculus—why not teach them alongside conventional maths and physics? There are also real-world examples of adaptive practices as novel ways forward in tackling the wicked problems of the Anthropocene (Pickering, forthcoming). These too could feature in engineering education, alongside conventional approaches.

In the end, then, Creative Pragmatics could be even more radical than it presently is—as an alternative pedagogical paradigm in content as well as form. And maybe that might call for another book in this series, *Creative Pragmatics II*.

Professor Emeritus, University of Exeter, PhD (Physics), PhD (Science Studies)
Thorverton, UK
July 2024

Andrew Pickering

References

Kuhn, T. S. (1962). *The structure of scientific revolutions*. University of Chicago Press.

Latour, B. (1987). *Science in action: How to follow scientists and engineers through society*. Harvard University Press.

Medawar, P. (1963). Is the scientific paper a fraud? *The Listener*, *70*(12), 377–378.

Pickering, A. (1995). *The mangle of practice: Time, agency, and science*. University of Chicago Press.

Pickering, A. (2010). *The cybernetic brain: Sketches of another future*. University of Chicago Press.

Pickering, A. (forthcoming). *Acting with the world: Agency in the Anthropocene*. Duke University Press.

Rittel, H., & Webber, M. (1973). Dilemmas in a general theory of planning. *Policy Sciences, 4*, 155–169.

Foreword by Jesper Bruun

As the Head of Studies for the Danish Master in STEM Education programme and a researcher focused on educational systems as complex systems, I am thrilled to introduce *Creative Pragmatics for Active Learning in STEM Education*. This book addresses a critical need in contemporary education: moving towards a holistic and dynamic understanding of knowledge, skills, and competencies within the STEM disciplines to prepare students for the multifaceted challenges of the modern world.

We know that active learning "works" when students engage in it (see, e.g. Freeman et al., 2014; Deslaurier et al., 2019; Theobald et al., 2020). Approaches such as inquiry-based, project-based and problem-based learning can significantly enhance student motivation and yield both better and more meaningful learning outcomes (Blumenfeld et al., 1991; Hmelo-Silver, 2004). Despite this, teaching and learning in schools often remain traditional. I believe that we often overlook a fundamental aspect: teaching students how to engage in active learning. Too often, students remain entrenched in the mindset of traditional STEM education, where success is measured by how well they follow instructions and arrive at a single correct solution. This approach does not equip students for the complexities they will face in their professional and personal lives.

Furthermore, teachers, pressured by the demands of standardized testing, often prioritize teaching to the test (Au, 2007; Nichols & Berliner, 2007). This focus is reinforced by students, parents, and school leadership, creating a cycle where the primary goal is high performance on national or standardized tests. As a result, summative assessment becomes a key driver of teaching practices. This emphasis leaves little room for developing a holistic view of knowledge, skills, and competencies in STEM subjects that future generations need.

Jesper Bruun is an Associate Professor in the Department of Science Education, University of Copenhagen, and serves as the Head of Studies for the innovative STEM graduate program (2021–2024). The Department of Science Education, under the Faculty of SCIENCE, is the largest science education research unit in Denmark and one of the largest globally with approximately 80 employees (of which 62 are scientific staff). Jesper is the chair of the 2025 ESERA (European Science Education Research Association) conference, contributing significantly to the advancement of science education research and practice. (Editors)

Educational systems are dynamic and complex, necessitating frameworks that account for the diverse outcomes, social and cognitive mechanisms, and probable learning trajectories influenced by this complexity (Davis & Sumara, 2006). Complex systems are inherently non-linear; thus, universal mechanisms for learning, to the degree that they may exist, cannot be scaled linearly to achieve more learning. Furthermore, while some educational principles might be universal, others are deeply rooted in the local context. This means that we need frameworks and examples that can help us investigate, understand, and develop teaching and learning activities as they unfold in many-faceted learning spaces.

This book offers a comprehensive framework and contextual examples illustrating how to navigate and embrace the complexity of teaching and learning in STEM. Creative Pragmatics recognizes educational systems as complex, with autonomous agents (for example, students and teachers) who seek to connect their learning to the real world. This perspective fosters the holistic development of each individual within society, combining creativity and pragmatism. It will not only enhance learning outcomes but also prepare students to be adaptive, innovative thinkers capable of addressing the complex challenges of our time. I believe that such an approach is needed in order to transform teaching across educational systems.

The Master in STEM Education programme is designed to empower compulsory school teachers with cutting-edge STEM knowledge and the teaching strategies advocated in this book. Many of our students directly apply these strategies, changing their own professional practice towards providing more agency to students while acknowledging their creativity as valuable. They act locally as agents of change as they seek to contribute to a culture of continuous improvement in STEM education. As such, we welcome Creative Pragmatics; it is a contribution that can help our teachers in their future work of transforming teaching in their local context.

The themes and methodologies presented in *Creative Pragmatics for Active Learning in STEM Education* are essential for educators, policymakers, and researchers globally. By adopting the principles of Creative Pragmatics, perhaps we can finally move beyond traditional teaching models to create more engaging, relevant, and effective STEM education.

In my view, this book can become a critical resource for a broad array of stakeholders in STEM educational systems. It offers a holistic and practical approach to transforming educational systems. I highly recommend this book and look forward to seeing its impact on the global educational community.

Head of Studies, STEM graduate program, Jesper Bruun
Copenhagen University
Copenhagen, Denmark
June 2024

References

Au, W. (2007). High-stakes testing and curricular control: A qualitative metasynthesis. *Educational Researcher, 36*(5), 258–267. https://doi.org/10.3102/0013189X07306523

Blumenfeld, P. C., et al. (1991). Motivating project-based learning: Sustaining the doing, supporting the learning. *Educational Psychologist, 26*(3–4), 369–398.

Davis, B., & Sumara, D. (2006). *Complexity and education: Inquiries into learning, teaching, and research.* Routledge.

Deslauriers, L., McCarty, L. S., Miller, K., Callaghan, K., & Kestin, G. (2019). Measuring actual learning versus feeling of learning in response to being actively engaged in the classroom. *Proceedings of the National Academy of Sciences, 116*(39), 19251–19257.

Freeman, S., Eddy, S. L., McDonough, M., Smith, M. K., Okoroafor, N., Jordt, H., & Wenderoth, M. P. (2014). Active learning increases student performance in science, engineering, and mathematics. *Proceedings of the National Academy of Sciences*, 111(23), 8410–8415.

Hmelo-Silver, C. E. (2004). Problem-based learning: What and how do students learn? *Educational Psychology Review, 16*, 235–266.

Nichols, S. L., & Berliner, D. C. (2007). *Collateral damage: How high-stakes testing corrupts America's schools*. Harvard Education Press.

Prince, M. (2004). Does active learning work? A review of the research. *Journal of Engineering Education, 93*(3), 223–231.

Theobald, E. J., Hill, M. J., Tran, E., Agrawal, S., Arroyo, E. N., Behling, S., ... & Freeman, S. (2020). Active learning narrows achievement gaps for underrepresented students in undergraduate science, technology, engineering, and math. *Proceedings of the National Academy of Sciences, 117*(12), 6476–6483.

Foreword by Søren Nedergaard

We live in a time and a world with a greater need for science education than ever before. Urgent global societal challenges, such as climate change, inequity in access to healthcare, misinformation, and pressure on democracies, cannot be resolved without improvements in STEM education. It is equally urgent that STEM graduates acquire the ability to work across disciplines and with colleagues outside STEM fields to produce the changes needed by global policymakers. The complexity of societal challenges is reaching unprecedented levels, and it increasingly seems that the only certainty we have is that the problems of tomorrow will require creativity, improvisation, and changes in scientific and technological approaches to solutions. Solutions will increasingly require an understanding of the local and global political context and innovative ways of combining knowledge and ever-changing perspectives on the problems at hand. At the core is STEM education—how STEM is being taught and how learners learn in science and technology. It must be obvious that teaching and learning in STEM have to change as our common world changes, and in that regard, this book on new and creative approaches to active learning in STEM education is a welcome breath of fresh air.

Chief Operating Officer, Novo Nordisk Foundation — Søren Nedergaard
Copenhagen, Denmark
July 2024

Søren Nedergaard is the chief operating officer at the Novo Nordisk Foundation, appointed in 2021. The foundation is a leading philanthropic organisation, investing billions annually to drive scientific and technological advancements, education, and innovation. Nedergaard's extensive experience includes various leadership roles in Denmark's Ministry of Higher Education, at the University of Copenhagen, and the Association of Colleges of Applied Sciences (Danske Professionshøjskoler) in Denmark. From 2006 to 2009, he was the first research and technology attaché at Innovation Center Denmark in Silicon Valley, fostering significant academic connections between universities in the San Francisco Bay Area and Denmark. (Editors)

Preface

> We travel together, passengers on a little space ship, dependent on its vulnerable reserves of air and soil; all committed for our safety to its security and peace; preserved from annihilation only by the care, the work, and, I will say, the love we give our fragile craft. (Adlai Stevenson Speech to the Economic and Social Council of the United Nations, Geneva, Switzerland, July 9, 1965)

As American diplomat Adlai Stevenson describes in his quote above, a guiding metaphor for comprehending this book is *navigation*: making and appreciating a shared journey, finding our way with diverse perspectives, and adjusting to shifting grounds. We move from *foundations* towards *orientations*, leaving solid ground and moving into sometimes turbulent and muddy waters.

This book advances a transformative approach to STEM (science, technology, engineering, mathematics) education. It doesn't just focus on the foundational subjects of science, technology, engineering, and mathematics; it deals with the mixtures of these domains in practices of education. With this, we are engaging with post-foundational thinking and with ideas of nomadism and performativity.

This book focuses on transversal skills and competencies that connect disciplinary expertise. It provides new lines of sight, new points of orientation that are essential for preparing students for learning in and living with the complexities of the twenty-first century. It explores why such skills and competences are crucial in contemporary STEM education, and it offers pedagogical lines of direction on the basis of transdisciplinary scholarship. With this focus, we argue that education should support learners in developing their full capacity to communicate science, develop professional identity and a sense of autonomy and responsibility, as well as interpersonal sensibility.

Taking orientation from pragmatist philosophy and ontologies of performativity/performative approaches to science, this book positions creativity and pragmatics as bearings for active teaching and learning in STEM Education. Promoting STEM Education as open-ended, dynamic processes of creating, designing, and innovating within and across STEM disciplines, the book advocates for holistic and creative approaches to education, reflected in the various chapters' examples and cases.

Yet Creative Pragmatics is not a map. In a metaphorical sense, a Creative Pragmatic stance will help learners navigate through challenges. It is a belief system that aids decision making in situations of incomplete information, and it can sensitise STEM graduates to diversity and help them build tolerance for multiple ways of knowing. It equips them to bridge the gap between disciplinary knowledge and transversal skills, between science and society, and helps them develop realistic plans and visionary ambitions for handling the multifaceted challenges of our times. Mastering the highest level of understanding in a discipline is made manifest in creative agency within the specialised area, and goes hand in hand with building autonomy and a sense of agency and responsibility.

The philosophical standpoint and transformative pedagogical approach, Creative Pragmatics, empower science educators and learners to navigate the fluid currents of innovation and active learning in STEM.

We wrote this book to contribute to ongoing shifts in STEM education, encouraging dynamic and participatory approaches. In presenting and exploring Creative Pragmatics, we aim to inspire educators and researchers to embrace engaging and interactive approaches and to reconsider the balance between teaching bodies of knowledge and facilitating student-centred activities. We believe that beyond imparting knowledge, educators must empower students with transversal competencies essential for dialogue, self-leadership, and collaboration. This book provides practical examples and philosophical points of orientation.

As editors, we embarked on this journey driven by curiosity about science education and wanting to contribute our diverse research foundations and experiences of the impact of design and creativity in individual and group processes of learning and development, plus our vast experiences of working pragmatically in STEM contexts spanning four continents and multiple disciplines over several decades.

We are travellers professionally and personally too. Connie Svabo's childhood spanned three continents, four languages, and more than a handful of countries. In her academic life, she has journeyed across three major faculty areas: the social sciences, humanities, and is now located in the sciences—continuously working in areas of border-crossing and transition, bringing various disciplines together in open-ended conversation. Michael Shanks's life is stretched across two continents, 8000 kilometres, 5000 years, and uncountable cultures. Academically, he holds degrees in archaeology, anthropology, classics, and education, teaching in programmes of rhetoric, design, science technology and society, and urban studies. Tamara Carleton regularly crosses borders as she grew up internationally and works globally. In addition, she holds degrees in social sciences and engineering, plus deliberately takes an integrative approach in her teaching at multiple business schools and technical programmes worldwide—including the United States, Sweden, Switzerland, Mexico, and Japan. Chunfang Zhou has had her own journeys on creativity across engineering, science, technology and society (STS), and STEM education, transitioning from radically different educational systems between China and Northern Europe.

Drawing from our lived experiences, we can say: life is lived in multiple ways, there are multiple ways of knowing and multiple routes to learning. We hope our readers see this sentiment in our book, and we sought contributors who navigate education in complementary ways.

The book is guided by shared values in pedagogy. *For* participation, empowerment, creativity, design, and transdisciplinarity. *Against* siloed, rote learning with the purpose of passing exams, institutionalised policing of boundaries, and the lack of interest in other perspectives, traditions, and cultures. We as editors want knowledge institutions to empower learners. We want universities to reach out to various publics and engage with matters of pressing concern.

The book is informed by a deep interest in knowing, learning, pedagogy, and education—and in how institutionalised and formalised education can draw on insights from the broader field of design. The book seeks to strengthen the creativity potential of higher educational institutions by opening up and inquiring into learning processes as they take place in our institutions. By integrating insights from pragmatist philosophy, science and technology studies, and an expanded field of STEM education, we seek to contribute to new currents in science education and research.

The book is crafted for a diverse audience encompassing educators, researchers, and policymakers engaged in various educational levels—primarily focusing on tertiary education, but also touching upon teaching in primary and secondary education. The book seeks to contribute new models to think with as inspiration for iterative and visionary redesign of STEM education. It particularly extends its reach to teachers, teacher educators, institutional leaders, and research groups across disciplines, offering valuable insights and practical guidance for navigating STEM education with Creative Pragmatics as a bearing point.

The book has travelled from its original concept—across institutions, geography, and time. The story is too long to tell it all, but it is worth mentioning that the first version of the idea for the book had the title *Creativity in Complexity* and grew out of a previous book project called "Situated Design Methods", which was published by MIT and edited by an interdisciplinary group from computer science and participatory design to urban planning, communications, and performance design, all of whom were part of the Roskilde University Strategic Research Initiative Designing Human Technologies. This was directed by Professor Jesper Simonsen, and Connie was affiliated with the group during her time as Associate Professor at Roskilde University.

Michael is an Honorary Professor at Roskilde, and he welcomed the opportunity to bridge Stanford design thinking with Nordic student-centred and participatory pedagogies in the form of a book. The early editorial team consisted of Connie and Michael with Jesper Simonsen and Magnus Rotvit Perlt Hansen—the latter two from Roskilde University. In the early phase, Michael invited Tamara to join as his longtime collaborator on strategic foresight and comprehensive design from Stanford University. Building on more than a decade of collaboration in overlapping constellations and networks—plus writing and teaching together on creative, aesthetic, visual, and design-based approaches—the concept of "Creative Pragmatics" emerged as a guiding stance for how to adapt learning situations ranging from education and professional development to global consultancy and institutional politics.

As the book concept progressed, children were born, job positions opened and closed, a global COVID-19 pandemic occurred, and MIT Press changed its publishing mission, while Connie moved to the University of Southern Denmark to direct a unit on STEM Education and Science Education (Forskningscenter for Naturvidenskabelig Uddannelse og formidling, abbreviated as FNUG). In response to this change, the original book idea was recast with a view to science education and the ESERA book series. The current editorial group was established, now anchored at Connie's new research centre.

The recently established research centre FNUG focuses on educational research within natural science, technology, and mathematics. The vision is to ensure that children and young people, from early education to university, experience engaging science, technology, engineering, and mathematics teaching. The centre is rooted in the robust scientific research environments at the Faculty of Science at SDU. FNUG's ongoing projects cover all educational levels and often transcend boundaries between them, from early childhood education to lifelong learning. The centre has a particular focus on tertiary education and professional development. Educating the educators, FNUG aims to delve deeper into the teaching of individual disciplines while also exploring STEM and cross-disciplinary competence and skills, at SDU referred to as 21st Century Science Skills. FNUG focuses on developing "T-shaped professionals"—individuals with both deep disciplinary knowledge and cross-disciplinary competencies. FNUG is a greenhouse for STEM educational innovation, contributing to the national research capacity in education with a focus on the natural sciences and mathematics. This book and the transdisciplinary network that has made it possible is an important move in this direction: bridging science and society and designing engaging learning experiences through Creative Pragmatics.

We invite you onto this little spaceship!

Odense, Denmark — Connie Svabo
Stanford, CA, USA — Michael Shanks
Odense, Denmark — Chunfang Zhou
Karlskrona, Sweden — Tamara Carleton

Acknowledgements

Connie Svabo

Creative Pragmatics has become a book! A small artefact in a big world, advocating for flexible thinking and innovative practices in STEM education. Bridging science and society, the book is a singular node in an expansive network, mediating multiple ways of knowing.

Special thanks to the dedicated community at the FNUG Center for Research in Science Education and Communication, University of Southern Denmark (SDU), for joyful collaboration and co-learning. In alphabetical order, thank you to: Anja Cetti Andersen, Chunfang Zhou, Dorte Moeskær Larsen, Emanuela Marchetti, Gabriele Characiejiene, Gitte Miller Balslev, Jane Morthorst, Jens Jakob Ellebæk, Jing Qin, Jonathan Gammeltoft, Karen Bollingberg, Karina Kiær, Katrine Bergkvist Borch, Kean Najmeddini Gindesgaard, Keith Devlin, Lisbet Foged, Luís Cruz-Filipe, Maiken Westen Holm Svendsen, Magnus Wahlberg, Mads Toudal, Michael Petersen, Morten Christensen, Rie Troelsen, Sara Egemose, Simon Enderby, Søren Krogh Hansen, Stefan Jänicke, Stine Mariegaard, Thomas Albrechtsen, and Tina Maria Brinks. Special thanks also to the SDU Faculty of Science leaders and mentors: Marianne Holmer, Martin Svensson, Poul Nielsen, Gary Banta, Sebastian Mernild, Claus Michelsen, and colleagues at IMADA.

Pioneering efforts in innovating science education are taking place at several educational levels in Denmark. Connie extends her gratitude to the co-leaders and co-developers in the ambitious Novo Nordisk Foundation-funded programme for university-based science education research (USE, 2025–2030): Anne Mette Mørcke (AU), Jan Alexis, Lars Ulriksen, and Sebastian Horst (UCPH). Thank you to the Novo Nordisk Foundation for supportive funding for science education research, including Berith Bjørnholm, Bente Guldbrandsen, Stefania Xella, and Thomas Dyreborg.

Connie also thanks collaborators in the Danish Academy of Science Education (NAFA, 2022–2028): Anders Vestergård Thomsen, Birgitte Lund Nielsen, Bjørn Friis Johannesen, Claus Auning, Daniel Spikol, Ivar Lykke Ørnby, Jan Sølberg,

Jette Reuss Schmidt, Kia Wied, Lykke Brogaard Bertel, Maria Møller, Martin K. Sillasen, Morten Rask Petersen, Peer Daugbjerg, and Rune Müller Kristensen. Additional thanks to the extended network of colleagues and mentors in the science education community in Denmark: Adrienne Traxler, Henriette Holmegaard, Jens Dolin, Jesper Bruun, Lene Møller Madsen, Linda Ahrenkiel, and Marianne Achiam.

Thank you to fellow creative practitioners in the Roskilde-based art, design, and practice research collective and Experience Lab group, which Connie initiated and led for a number of years, working with Anne Julie Arnfred, Anja Lindelof, Dorte Jelstrup, Eduardo Abrantes, Eric Komlavi Hahonou, Henriette Christrup, Julie Bønnelycke, Klaus Hildebrand Frederiksen, Lisbet Frølunde, Mads Høbye, Pernille Welent Sørensen, and Susana Tosca. Thank you to Danish and international performance design colleagues, to the research environment in cultural geography and urban planning at RUC, the FabLab RUC team, as well as the Roskilde University Strategic Initiative for Designing Human Technologies: Jesper Simonsen and Magnus Rotvit Perlt Hansen, your participatory spirits are reflected in the book.

A special thanks also to the modern museum of natural history Naturama, the UNESCO Heritage site Stevns Klint, and the inspiring Danish STS collective. These research sites and communities offered formative experiences that extend into Connie's current practices spanning art, science, and technology.

Connie is honoured to advise FabLab Spinderihallerne, Fjord & Bælt, Frilandsmuseet, Geo Bio Science Center Syd, UNESCO Global GeoPark, The South Fyn Archipelago, Karen Blixen Museet, and Museum SydøstDanmark on designs for experience and learning and to consult internationally via Kimihiko Iwamura in Silicon Valley. Significant mentors, friends, and colleagues who have helped lay important foundations for Creative Pragmatics bridging practice and academia are Signe Lykke Littrup, Jens Friis Jensen, Marianne Qvick, Henrik Holt Larsen, René la Cour Sell, Chris Ford, Michael Bell, Rolf Hughes, Silje Kamille Friis, Sune Gudiksen, Martin Avila, and Susanne Kern.

A heartfelt thank you goes to Connie's co-editors—Tamara, Chunfang, and particularly Michael—for their unwavering perseverance with the project, even as its character evolved over the long process. And ending on a personal note: thank you with much love to Jasper and Luka.

Michael Shanks

Michael celebrates the colleagues and collaborators across many projects and institutions that have built the deep foundations for Creative Pragmatics.

Archaeology is little acknowledged as a field of extraordinary transdisciplinary exploration, combining genetics and art history, big data and critical theory, international heritage policy and the culture wars of local identity. This is where Michael finds the experience and praxis, and the historical roots, that can inform our learning journeys into the future. Integrated STEM and way way beyond.

Thanks therefore to Bill Rathje, sorely missed originator of the archaeological science of garbage, Tucson Arizona, before he joined Michael's lab at Stanford in 2002. Dear archaeological companions (several co-authored books) also include Bjørnar Olsen (Tromsø), Chris Witmore (Texas Tech), Gary Devore (New Hampshire), and Alfredo Gonzalez Ruibal (Madrid and Santiago Spain).

In the creative arts, Michael sends heartfelt thanks to Gabriella Giannachi (Exeter UK), Alessandra Lopez y Royo (London UK), Nick Kaye (Exeter UK), and artists Lynn Hershmann Leeson (posthumanist embodiments in California!). Above all, Michael cherishes 30 years working in the performing arts with theatre company Brith Gof. Art Director Cliff McLucas taught him the science of art. Mike Pearson was a fellow traveller in an extraordinary journey they called theatre/archaeology into border territories of performance art and world building, between pasts, presents, and futures.

Stanford Humanities Lab (2001–2009) pioneered radical student-centred pedagogy in collaborative transdisciplinary research projects before it was shut down by reactionary interests. Michael directed with Jeffrey Schnapp (now Harvard), Tim Lenoir (now Duke), and Henry Lowood (Stanford Libraries). Fred Turner (Stanford) was a key figure in seminars that took them into the wildness of information and communication tech and media ecologies, understanding which is so central to a savvy future-oriented pedagogy.

Bernie Roth (Stanford Engineering) and Bill Moggridge (founder of the design consultancy IDEO) introduced Michael to the Design Group at Stanford and their project of design thinking (now sadly so diluted by hyperbolic business puffery in Stanford d.school), but authentically represented in this book. Two key colleagues and co-teachers from whom Michael learned so much: Larry Leifer (Center for Design Research) and David Kelley (founder of IDEO and Stanford faculty). This is where Michael met Tamara. A spin-off from Stanford's design group was the Revs Program—an extraordinary experiment (2010 to 2015) in building a transdisciplinary research and learning programme focused on the design of mobility past, present, and future. Cliff Nass was the visionary instigator who invited Michael to direct; Miles Collier (Naples Florida) was the thought leader who saw into new possibilities for a culturally engaged STEM programme.

MediaX and H-Star were long-running programmes at Stanford (until 2023) that bridged education, EdTech, and innovation in their industrial affiliate programmes under the leadership of Keith Devlin (H-Star) and Martha Russell (mediaX). Michael was closely connected with both and sends thanks for the welcome they gave to him. They were a platform for his shift into business consultancy. It was as a Fellow of H-Star that Connie first met Michael. With Chunfang they worked with Janet Carlson (Stanford Graduate School of Education) in helping design a high school in China (2022–2023).

It was Bill Cockayne who introduced Michael to design foresight when he was Executive Director of Stanford Humanities Lab; Michael has learned and applied so many insights from Bill since then. With Tamara, Michael has collaborated on many projects, sometimes through Stanford mediaX, in business consultancy. This is an applied archaeology, a transdisciplinary interest in coaching leadership and in building learning communities to suit future social and cultural needs through what can easily be called an expanded STEM curriculum. Michael worked for more than a decade from 2008 with the city and port of Rotterdam (the International Advisory Board, under Sander de Ionge). Colleagues who have joined him in his Stanford lab in this effort in applied archaeology include Victor Taratukhin (SAP and Muenster), Natalia Pulyavina (Plekhanov Business School, Moscow), and Sabine Remdisch (Leuphana). Riel Miller (emeritus head of UNESCO's Futures Literacy Program) has also joined as an interested and inspiring party. With Kimihiko Iwamura (Valley Breeze Consulting), Michael is currently, as of the publication of this book, working with Japanese corporations (Biprogy, CEO Akiyoshi Hiraoka; Aisin, CEO Kenji Suzuki and General Manager Sadafumi Shirai) to build corporate capacity in what is described in this book as Creative Pragmatics.

Tamara Carleton

I second all of my fellow editors' acknowledgements above! And an extra thanks to Michael who first invited me on this book adventure.

I want to acknowledge several other major influences that have shaped my view of Creative Pragmatics in education and learning—though this is, by no means, a complete list. During high school, I attended a county magnet school for literary arts that emphasised project-based learning and contextual inquiry. This experience was empowering to me as a young student, and I thank my teachers Mrs. Jean Fripp and Mrs. Sylvia Trimble. Then my time observing, participating in, managing, and later documenting the ME310 course at Stanford University was instrumental because I saw how many alumni and corporate sponsors celebrated a project-based course for its active learning model and outcomes over multiple decades. I thank Stanford Professors Larry Leifer and Mark Cutkosky for their generous support and access to ME310. Likewise, I thank Bill Cockayne for countless experiments and variations of active learning formats, especially jointly inventing and teaching methods of strategic foresight and innovation around the world and at Stanford's Silicon Valley Innovation Academy. Dedicated readers will recognize this hands-on and collaborative philosophy reflected in our *Playbook for Strategic Foresight and Innovation*.

Chunfang Zhou

Besides all colleagues in FNUG Center (as Connie mentioned), Chunfang thanks Aalborg University (AAU) for its long tradition of Problem and Project-Based Learning (PBL). She spent more than 10 years there working on active learning, creative pedagogies, and STEM education across different research centres. Special thanks go to AAU Learning Lab where Chunfang organised university teacher education in PBL. In the Research Group on Techno-anthropology and Participation,

Chunfang carried out research and teaching using methods of participatory design and co-creation. Special mention here goes to the UNESCO Chair in PBL.

Chunfang's European network includes colleagues on the editorial board of the journal *Discover Education*. Thanks go to all contributors in her Brill book series Arts, Creativities and Learning Environments in Global Perspectives. Special thanks to Professor Taisir Subhi Yamin at the International Center for Innovation in Education (ICIE), Dr Julia von Thienen at Stanford-Potsdam Designing Thinking Research Program at the Hasso Plattner Institute, Professor Giovanni Emanuele Corazza at Marconi Institute for Creativity at the University of Bologna, and Professor Maria Mar Carrio Llach at the Polytechnic University of Catalonia.

Chunfang's Asian Network includes Professor Lingling Luo and Professor Hongbing Chen at the Research Center in Science, Technology and Society (STS), Northeastern University in China, Professor Zhiliang Zhu, Professor Xinbo Sun, and Professor Lanxia Zhang at the PBL Teaching Innovation Center. Cherished colleagues also include Professor Xiaomin Sun at Beijing Normal University, Professor Min Jou at National Taiwan Normal University, and Professor Fangqi Xu at Kindai University and in the Japan Creativity Society.

Additional Thanks, Funding, and Institutional Support

Special thanks are due to our chapter contributors, as well as to Andrew Pickering, Jesper Bruun, and Søren Nedergaard for their encouraging forewords. Thank you Gabriele for excellent editorial assistance.

Appreciation is also extended to the team at Springer and ESERA for their support.

Acknowledgement goes to the institutional support and funding that has contributed to making possible the projects that are explored in the chapters. In the United States, this includes Stanford, Columbia, and University of Nebraska. In Denmark—Roskilde and University of Southern Denmark. Funding agencies who have supplied direct support include EU Interreg, the Novo Nordisk Foundation, and Region Syddanmark.

Contents

Contributors

Thomas R. S. Albrechtsen University College of Southern Denmark, Esbjerg, Denmark

Gitte Miller Balslev STEM Centre for Research in Science Education and Communication (FNUG), Department of Mathematics and Computer Science, University of Southern Denmark, Odense, Denmark

Michael Bell Graduate School of Architecture, Planning and Preservation, Columbia University, New York, NY, USA

Katrine Bergkvist Borch STEM Centre for Research in Science Education and Communication (FNUG), Department of Mathematics and Computer Science, University of Southern Denmark, Odense, Denmark

Martin Malthe Borch Technical University of Denmark, Kongens Lyngby, Denmark

Tamara Carleton Department of Mechanical Engineering, Blekinge Institute of Technology, Karlskrona, Sweden

Gabriele Characiejiene STEM Centre for Research in Science Education and Communication (FNUG), Department of Mathematics and Computer Science, University of Southern Denmark, Odense, Denmark

Sara Daugbjerg Roskilde University, Roskilde, Denmark

Amalie Thorup Eich-Høy STEM Centre for Research in Science Education and Communication (FNUG), Department of Mathematics and Computer Science, University of Southern Denmark, Odense, Denmark

Chris Ford School of Engineering, Stanford University, Stanford, CA, USA

Michael Haldrup Roskilde University, Roskilde, Denmark

Mads Hobye Roskilde University, Roskilde, Denmark

Karina Kiær University of Southern Denmark, University College of Southern Denmark, Haderslev, Odense, Denmark

Dorte Moeskær Larsen STEM Centre for Research in Science Education and Communication (FNUG), Department of Mathematics and Computer Science, University of Southern Denmark, Odense, Denmark

Schack Lindemann Roskilde University, Roskilde, Denmark

Nikolaj Møbius Roskilde University, Roskilde, Denmark

Martin Niss Department of Science and Environment, Roskilde University, Roskilde, Denmark

Nicolas Padfield Roskilde University, Roskilde, Denmark

Sofie Pedersen Roskilde University, Roskilde, Denmark

Michael Shanks Archaeology Center, Stanford University, Stanford, CA, USA

Connie Svabo STEM Centre for Research in Science Education and Communication (FNUG), Department of Mathematics and Computer Science, University of Southern Denmark, Odense, Denmark

Maiken Westen Holm Svendsen STEM Centre for Research in Science Education and Communication (FNUG), Department of Mathematics and Computer Science, University of Southern Denmark, Odense, Denmark

Bo Thorning Roskilde University, Roskilde, Denmark

Rie Troelsen University of Southern Denmark, Odense, Denmark

Falk Uebernickel Hasso Plattner Institute, Potsdam, Germany

Chunfang Zhou STEM Centre for Research in Science Education and Communication (FNUG), Department of Mathematics and Computer Science, University of Southern Denmark, Odense, Denmark

About the Editors

Connie Svabo PhD, Professor of STEM Education and Science Communication, Director of STEM Center for Research in Science Education and Communication (FNUG), Department of Mathematics and Computer Science, University of Southern Denmark.

Connie is a transdisciplinary scholar and creative practitioner with expertise in learning environment design and experience. She is the founding Director of the STEM Center for Research in Science Education and Communication (FNUG), chair of the SDU research programme in Science and Mathematics Education and Communication, leader of the Learning Experience Design section and member of the management team of the Department of Mathematics and Computer Science. She promotes Creative Pragmatics as a model for knowing in a complex world and underpins this stance with a practice that ranges from art over education to design and consultancy. Her portfolio includes geological UNESCO heritage sites Stevns Klint and the South Fyn Archipelago, as well as the Danish National Gallery Open Air Museum and the museum of natural history, Naturama. Previously a Continuing Studies Instructor at Stanford University and Faculty at Roskilde University, where she directed the research centre Experience Lab, chaired the study programmes in Communication Performance Design 2013–2016 and served on the academic board. She has published books on STEM education, sustainability, competence development, design pedagogy and digital life. She holds a PhD in Social Science from Roskilde University's multidisciplinary programme in Society, Space and Technology.

Michael Shanks PhD Doc, Professor of Classics and Archaeology, Stanford University.

Acknowledged as one of the most original and influential contemporary archaeologists, Michael is a specialist in long-term humanistic views of design and creativity, innovation and social change, while exploring connections across the sciences and arts in a hybrid field of theatre/archaeology. He is senior faculty in Stanford University's programmes in Science Technology Society, Urban Studies, Writing and Rhetoric, Archaeology, Classics, and in the Center for Design Research, part of Stanford's d.school in the School of Engineering. He directed Stanford Humanities Lab, pioneering innovative project-based pedagogy in the arts and humanities, and helped found the Revs Program at Stanford, connecting the archaeological automobile with contemporary car design. Founding faculty in Stanford's Archaeology Center, Michael is a director of Stanford's research group Foresight and Innovation, and, for two decades until its retirement in 2023, he worked with mediaX, the affiliates' programme that connected Stanford with industry through technology and innovation. He served on the city of Rotterdam's International Advisory Board from 2008 to 2019 and works with many companies and organisations in developing cultures of innovation. In another mobilisation of Creative Pragmatics, Michael continues to pursue archaeological fieldwork in the northern borders of the Roman empire and beyond.

Chunfang Zhou PhD, Associate Professor of STEM Education and Science Communication, STEM Center for Research in Science Education and Communication (FNUG), Department of Mathematics and Computer Science, University of Southern Denmark.

A researcher and practitioner working on creativity, STEM education and teacher education. Chunfang has an interdisciplinary and cross-cultural educational background: she finished a Bachelor's degree (Industry Automation and Information Science) and Master's degree (Philosophy of Science and Technology) in China; during 2008–2012, Chunfang finished her doctoral research on *Group Creativity in Engineering Education in a Problem and Project-Based Learning (PBL) Environment* at Aalborg University, Denmark. She is a board member of Danish Development Research Network (DDRN) and works as a senior consultant in the PBL (Problem-Based-Learning) Teaching Innovation and Research Center at Northeastern University (NEU), China. Chunfang locates her research in Science, Technology and Society (STS), with a focus on creativity and STEM education. Since 2013, Chunfang has authored and co-authored over 110 peer-reviewed publications; she is the editor of *Handbook of Research on Creative Problem-Solving Skill Development in Higher Education* (IGI Global, 2016), the author of *Introducing Problem-Based Learning for Creativity and Innovation in Chinese Universities* (IGI Global, 2020), and an editor in the book series Arts, Creativities and Learning Environments in Global Perspective (Brill/Sense). Chunfang is an editorial board member of *Discovery Education* and the *European Journal of Education and Pedagogy*, as well as advisory board member for the *International Journal of Engineering Education.*

Tamara Carleton PhD, Universitetslektor in Mechanical Engineering, Blekinge Institute of Technology.

Tamara is an award-winning professor, industry advisor and expert in radical innovation and strategic foresight. Recognised as one of the Top 50 Women Leaders in Education (in 2023 and 2024) by Women We Admire, she teaches at multiple business schools and technical universities around the world. In addition, she holds the UNESCO Chair in Anticipatory Leadership for Innovative and Better Futures at Tecnológico de Monterrey in Mexico. Tamara is also the founder and CEO of the Innovation Leadership Group, which helps leaders and teams to build innovation capacity and map bold futures. She has written several seminal books in innovation, including *Building Moonshots: 50+ Ways to Turn Radical Ideas into Reality* and the *Playbook for Strategic Foresight and Innovation*. She holds a doctorate in mechanical engineering from Stanford University.

Chapter 1
Creative Pragmatics for Active Learning in STEM Education

Connie Svabo, Michael Shanks, Chunfang Zhou, Tamara Carleton, and Gabriele Characiejiene

1.1 Models for Knowing

We start with a big question. Is the world knowable? Much traditional education rests on the idea that this is the case: the world is finitely knowable and we can transmit positive knowledge of it. "Knowledge is the thing" (Pickering, 2010 p. 390). We challenge this conventional view of knowability, urging that we embrace the complex and dynamic nature of knowledge creation and its inherent uncertainties. Although questioning the stability and completeness of knowledge may feel uncomfortable, we suggest that it is important to endure the discomfort, in order to consider the implications for education, of a world which is not finitely knowable.

We argue that it is time for science, technology, engineering and mathematics (STEM) education to help learners, not only to master the laws of nature but also learn to navigate the dynamic realities of exceedingly complex systems and their unpredictable interrelations.

This book highlights creative agency, suggests that education could and should play a pivotal role in purposefully redesigning our actions in the world, and underscores that science and technology, as forces shaping society, can be reconfigured to better serve humanity and our companion species.

C. Svabo (✉) · C. Zhou · G. Characiejiene
STEM Centre for Research in Science Education and Communication (FNUG), Department of Mathematics and Computer Science, University of Southern Denmark, Odense, Denmark
e-mail: svabo@imada.sdu.dk; https://www.sdu.dk/en/forskning/stem

M. Shanks
Archaeology Center, Stanford University, Stanford, CA, USA

T. Carleton
Department of Mechanical Engineering, Blekinge Institute of Technology, Karlskrona, Sweden

C. Svabo et al. (eds.), *Creative Pragmatics for Active Learning in STEM Education*, Contributions from Science Education Research 14,
https://doi.org/10.1007/978-3-031-78720-1_1

> The university ought to take an active role in planning society, and in particular, in the planning of science and technology in the service of society. (p. 102)

> We will need a type of education which fosters judgement in complex and dynamically changing situations. (p. 102)

> The task is nothing less than to build a new society and new institutions for it. (p. 101)

These forward-thinking calls come from Erich Jantsch's, 1972 essay in an OECD document about the need to be interdisciplinary and more, to be transdisciplinary. They are still highly relevant today.

If there is only one takeaway for the reader from this book, it should be this: there is no solid ground of science for science education to stand on. Science and technology can reconfigure and can be reconfigured. Science, as traditionally taught, often emphasises predictability, stability, causality, and reduction. Yet, we inhabit a world that is complex, dynamic, and interdependent—a reality that even extends to natural systems. The traditional view of science as aspiring to objectivity and replicable certainty does not fully account for the complexities and ever-evolving nature of scientific understanding (Gilbert, 2004). While we value the validity, reliability, and generalizability that accounts of science in education offers, we also advocate for a broader perspective that includes truths that are contingent, partial and situated. We emphasise the need to foreground the role of the knower in the known (Davis & Sumara, 2006), recognise the co-agency of instruments and technologies in knowledge creation, and address openly the ecological and existential crises we face (Pickering, 2025).

We are not rejecting science. Far from it. Rather, we advocate an expanded science paradigm informed by complexity thinking, systems thinking, cybernetics, ecology and science and technology studies (STS). This expanded paradigm highlights the interdisciplinary nature of contemporary scientific inquiry. It emphasises transdisciplinarity, a holistic perspective, interconnectedness and interdependence between systems, and focuses on the role of science and technology in shaping society (Bijker & Law, 1994). This also encompasses agential realism, response-ability (Barad, 2003, 2007) and new materialist approaches (Bennett, 2010). This expanded science paradigm transforms how scientific problems are conceptualised and addressed, and can lead to unique research agendas, understandings, and methodologies.

In resonance with this assemblage of interdisciplinary subfields in an expanded science paradigm, we advocate for a science *education* that integrates personal experiences with a principle of human-non-human symmetry, collective practices and global realities.

Our book *Creative Pragmatics for Active Learning in STEM Education* proposes that science education should not only convey established facts but also teach students how to navigate and engage with the shifting and interconnected challenges of the world. Creative Pragmatics is about preparing learners for the future by embracing adaptive and responsive approaches in a science education that is informed by the latest insights from across the scientific disciplines.

By integrating holistic insights from an expanded science paradigm with pragmatism, and art and design, Creative Pragmatics promotes a performative model of knowing, emphasising the active, engaged learning that is essential for handling contemporary scientific and societal challenges. This approach promotes learners to see themselves as agents in the world, not mere spectators, echoing the sentiments of pragmatist philosopher Anne Ruth Putnam (Putnam & Putnam, 2017 p. 111) stating that "the future is not determined; human beings cooperating can make the world better," and—with inspiration from physicist, science and technology scholar, and one of the foreword authors of this book Andrew Pickering—we modify this sentiment for the Anthropocene epoch into the still mildly optimistic statement that the future is not determined; human beings cooperating with the world can make the world better.

In sum, this book aligns STS informed performative models of knowledge with pragmatic educational philosophy to emphasise creative agency in education, connecting skills and competencies with active, engaged learning under the banner of Creative Pragmatics.

1.2 A Holistic Perspective

In the rapidly evolving world of contemporary late modernity, our globalised societies are confronted with the many intricately intertwined and familiar challenges of climate change, biodiversity crisis, migration and displacement, populist autocracy, the persistence of armed conflict, deep-seated societal disparities, and the unforeseen consequences of remarkable scientific development and technological innovation (Peters, 2017; Van Berkel & Manickam, 2020; Svabo, 2021; National Intelligence Council, 2021).

Multiple crises have emerged through the cumulative effects of human activities on the planet. Complex phenomena reveal the limitations of orthodox models of research and knowledge building which hinge on reductionism and predictability.

Simple solutions that treat problems as isolated and controllable are inadequate in this context. Each of these phenomena is multifaceted and consists of interconnected elements, meaning that changing one aspect will inevitably trigger cascading effects throughout larger systems (Byrne & Callaghan, 2022; Sawyer, 2005). The interconnected nature of contemporary challenges requires a holistic perspective that accounts for the complex interactions between various issues. This holistic approach recognises the necessity of integrating diverse disciplines and perspectives to fully understand and address these challenges. This new reality underscores the need for education, research and innovation to develop new competence and new models for knowing.

We have to shift to models for knowing that are open, iterative, systemic, holistic and more-than-representational. We have to shift towards creative and pragmatic agency. This is also a shift from purely representational ways of understanding knowledge towards what Andrew Pickering calls performative models of knowing, underscoring practice and performance (1995, 2010).

Performative models of knowing emphasise the active, dynamic process of engaging with the world, where knowledge is constructed through interaction and practice rather than simply represented. Such models value the role of human creativity working with the world, continuously learning and dealing with new information and challenges. This approach highlights the importance of experimentation, adaptation, and the co-construction of meaning between humans, technologies, and environments, recognising that understanding emerges through doing and experiencing. By embracing performative models, we empower learners and researchers to become active participants in shaping the world, leading to more sustainable situations.

1.3 Learning in a Complex World

Complexity is a central feature of our world and learning organisations and educational institutions need to understand and engage with this (Nicholls, 2018; Gallagher & Savage, 2023). Systems thinking, holistic perspectives, and recognising societies as complex dynamic systems underscore the necessity to understand the concept of emergence.

Studies in complexity offer a radical and innovative frame for understanding the nature of professional practice, particularly away from the idea that professionals, leaders, managers, or other employees are in control. This highlights that we are living in a complex world (Davis & Sumara, 2006; Davis et al., 2008; Svabo, 2021).

Funtowicz and Ravetz's (2018) concept of post-normal science deals with high systems uncertainty and high decision stakes, while normal applied science operates in low systems uncertainty and low decision stakes. The concept of post-normal science addresses epistemic uncertainty, rethinking the nature of science and seeks to democratise science education by including students and teachers as stakeholders (Agustian, 2023). Educational institutions and organisations are confronted daily by continuous changes that motivate them to develop new strategies to respond to the emerging challenges.

The complex issues and heterogeneous and interlocking processes we deal with in contemporary life require skills and competencies to navigate complexity. As such, competencies and a skill set extending beyond conventional content learning or problem-solving is imperative (Wong, 2020; Davis & Sumara, 2006; Davis et al., 2008). Proficiency in critical thinking, creative ideation, and transdisciplinary systems analysis becomes essential.

Creative Pragmatics is a synthetic concept intended to help attend to this necessity. Creative Pragmatics underscores the importance of integrating creativity, critical thinking, problem-solving, communication and emotional intelligence within active STEM learning settings, enabling students to learn to deal with real-world challenges effectively.

1.4 Policy Highlights the Importance of Handling Complexity

Educational policies emphasise the need for knowledge, skills and competencies that allow graduates to navigate and act in complex situations. The importance of developing the capability to act in situations and circumstances of complexity is supported by policy in both the US (National Research Council, 2013) and Europe (European Qualifications Framework, 2017), as well as by global organisations, such as World Economic Forum (2023) and OECD (2019).

Policy highlights the understanding of systems, system modelling, and handling complexity as crucial learning goals. Policy emphasises the need for developing competence to work in complex situations and to communicate with non-experts, as well as the ability to come up with new solutions, to work with ill-defined problems.

The qualifications expected of students in the EU "second cycle qualification framework" are, that students "have the ability to integrate knowledge and handle complexity, and formulate judgments with incomplete or limited information, but that include reflecting on social and ethical responsibilities linked to the application of their knowledge and judgments", and third cycle qualifications expect that students: "are capable of critical analysis, evaluation and synthesis of new and complex ideas." (European Qualifications Framework, 2017).

In such a framework, students are expected to develop competence in handling complex and unpredictable work situations. It is expected that they can assume professional responsibility and with autonomy can initiate and carry out collaboration across disciplines and organisational levels. Students must be able to structure their own work, and the objective is that they develop into self-directed and autonomous individuals. An important component of this is not only to be reactive to externally motivated goals, but also to be able to formulate goals, interests and concerns in self-directed ways and to assess and make informed decisions based on incomplete information and in situations of insecurity, change and risk. This makes it necessary to develop creative competencies in assessing a situation and what is at stake, to draw on experience, communicate with other participants and stakeholders and understand complexity.

Aligned with policies like the European Qualifications Framework, Creative Pragmatics offers a frame that supports learners in developing competences to navigate complexity and to deal with unpredictability. These competences are very important for addressing the multifaceted issues of our contemporary world. There is strong advocacy for STEM education that integrates multiple disciplines to prepare students for the complexities of modern challenges (Bybee, 2013, 2018). By emphasising active learning methods such as project-based learning (Zhou, 2016), problem-based learning (Tan, 2003), studio-based learning (Zairul, 2020), scholartistry (Shanks & Svabo, 2018), experiential learning (Coker et al., 2017) and participatory design (Bang & Vossoughi, 2016), Creative Pragmatics empowers students to go beyond the acquisition of bodies of knowledge, enabling them to develop twenty-first century skills such as critical thinking, creativity, collaboration, and communication, as well as transformative competencies (OECD, 2019), fostering adaptability and innovation. This prepares them to tackle real-world problems with innovative and strategic approaches, making a meaningful impact on society's most pressing issues (Davis et al., 2008).

From the inherent necessity of engaging with world complexities to the acknowledgment by policymakers, a need can be outlined to develop analytic and synthetic competencies, and to work on integrating theory and practice. This need is also emphasised by Friedman and Stolterman (2014) and Norman (2010).

1.5 Creative Pragmatics Introduced

There is a widespread concern to promote the learning of certain skills and competencies, bodies of knowledge and ways of knowing, deemed essential to navigating, and not getting lost and overwhelmed, in a world that is experienced to be growing in complexity.

This book proposes Creative Pragmatics as a synthetic concept that summarises a particular set of ways to navigate complexity. These performative strategies and tactics include skills and competencies (Grugulis & Stoyanova, 2011; Hoffmann, 1999), bodies of knowledge and ways of knowing (Svabo, 2021), as well as pragmatic means of managing institutions and projects, formulating and implementing strategy and policy in governmental and non-governmental agencies and in businesses, designing and delivering curricula and training programs.

For the purpose of this book, Creative Pragmatics is a dynamic framework for active learning in STEM education. It is not a fixed or static set of theories or methodologies. It is an approach, a stance, a mindset, a scaffold or framework which involves iterative learning through practice, by doing, by designing, by making, by making art, by engineering knowledge, by collaborating, by observing and reflecting, through dialogue, participation, and imagination.

Why do we propose this new synthetic concept? The components of Creative Pragmatics are time-served and well-founded, all around us. We propose the concept as a prompt, a nudge to (re)connect ways of knowing and acting that have tended to drift apart and even become siloed. We suggest the concept supplies a useful articulating focus in the contemporary debates around ways of engaging with complexity in policy and practice, particularly with reference to education, pedagogy, and professional development (Davis & Sumara, 2006; Davis et al., 2008).

The concept of Creative Pragmatics connects certain performative modes, strategies, tactics, to which we give priority, by grounding them in a transdisciplinary field summarised as a 'pragmatics'. By pragmatics we mean 'doings' and 'things done'. We associate pragmatics with 'things done', through the roots in the old Greek *pragma*, which is a deed, an act, something that has been done, while retaining the active verbal forms of *to do, to act and perform, doing*. In this, Creative Pragmatics is oriented on *praxis*—thoughtful practice that is situated (Lave & Wenger, 1991), and includes within its systems outlook materialities, material conditions (Hughes, 2000), artefacts (Svabo & Shanks, 2013; see Olsen, 2012 for a detailed unpacking of this materialism). The designation pragmatics also points to foundations in philosophical pragmatism, especially as manifested in educational philosophy and science and technology studies (Dewey, 1938; Putnam & Putnam, 2017; Rorty, 1982), and in linguistics, where pragmatics is the study of how language is best understood in social context.

Furthermore, the concept of Creative Pragmatics grounds and relates its pragmatic and performative strategies and tactics in terms of their creative capacities. Creativity is about coming up with new ideas, methods, and ways of thinking and doing things (Sternberg, 1998). We follow Bloom's revised taxonomy (Anderson & Krathwohl, 2001), where 'create' is the highest level of thinking and learning goals. This includes activities like designing, building, developing, and exploring.

We take creativity not to be a separate kind of activity or practice, but an aspect of pragmatics. This associates creativity with agency—capacities to act in the world, with efficacy—capacities to make a difference within worldbuilding and sense making (Putnam & Putnam, 2017). Creative agency influences all aspects of how we build our world. Agency refers to the ability of individuals or entities to shape the world around them. Society and culture are not merely backdrops in which people exist but emerge from our interactions (Latour, 1999, 2007). We do not simply belong to society; we actively create and transform it through our engagements with each other and our environment.

This transformative capacity is what we refer to as creativity—the ongoing reconstruction of our worlds through the agency of both humans and non-human participants (Sayes, 2013). Reality is a continuous process of formation and reformation (Pickering, 2010, referencing William James, 1975; Whitehead, 1978. For further exploration of these ideas, see Chap. 2 in this volume).

In summary, Creative Pragmatics is a synthetic concept addressing the pressing need to navigate the complexity of the world. It encompasses performative strategies and tactics, prioritising skills, competencies, knowledge, and pragmatic means across various domains, grounded in a transdisciplinary field referred to as 'pragmatics'. The concept emphasises thoughtful, situated practice (praxis) and integrates material conditions, artefacts, and a dynamic outlook on adaptive living systems. Rooted in philosophical pragmatism, science and technology studies, and art and design (Fig. 1.1), Creative Pragmatics associates creativity with pragmatics, aligning it with agency and efficacy in worldbuilding and sense-making.

We are very conscious that the deep roots of these three fields lie in long traditions, going back to antiquity. It is neither practical nor necessary to trace here the genealogy of Creative Pragmatics through Heidegger, Hegel, Vico, Spinosa and back to pre-Socratics, never mind eastern equivalents. Nevertheless it is useful to indicate that Creative Pragmatics builds on three overarching process-philosophical propositions that involve a constellation of six concepts in dynamic articulation.

- *Process and relations.* In knowing, understanding, acting-in-the-world, one does well to focus upon *processes and relationships*, the holistic dynamics of interconnected living systems — processes as well as products, verbs as well as nouns.

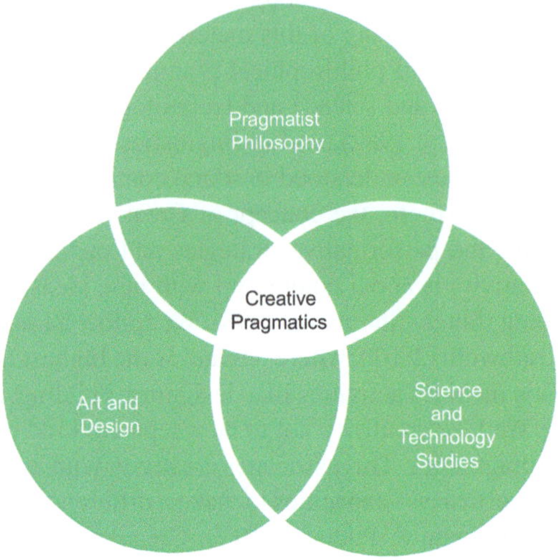

Fig. 1.1 Creative pragmatics: inspiration and sources of influence. Creative Pragmatics draws inspiration from pragmatist philosophy, science and technology studies, and art and design

- *Design and performance*. Design and performance are emergent, adaptive processes that evolve through ongoing interaction with the world. They are helpful models to think with in education where the emphasis is not on fixed outcomes but on the dynamic unfolding of action. Together, *design and performance* generate new possibilities for action and competence development, deeply entwined with the embodied, situated nature of knowing and doing.
- *Experience and agency.* Process-relations and design-performance are grounded in *experience*, another critical concept that connects the cognitive, sensory and affective in fields of embodied and creative *agency* — capacities to make a difference in the (re)productive dynamics of living adaptive systems.

We will outline this grounding for the concept Creative Pragmatics further in this chapter, especially in Sect. 1.8, and more in the next two chapters of this book, showing how these foundations offer critical insights into the ways that the skills and competencies, knowledges and ways of knowing of Creative Pragmatics may be learned and taught, in a recasting of pedagogy for the modi operandi of Creative Pragmatics.

1.6 Active Learning in STEM Education

How might we design learning to support the development of the knowledge, skills and competencies promoted by policy as essential for navigating a complex world? We need active learning approaches in STEM education, and these approaches cannot be developed through discipline-focused teaching of body of knowledge alone.

STEM is defined through the acronyms' constituent fields (science, technology, engineering, and mathematics). For many, STEM simply means science schooling.

However, in education research it is also an approach to education that focuses on the integration of these four areas, helping students learn by actively engaging with real-world problems and applying their knowledge. STEM education goes beyond content learning via memorisation and encourages practical problem-solving and critical thinking in science, technology, engineering, and mathematics. Stemming from a holistic and systemic approach to comprehend the world's complexity and interdisciplinary challenges, the goal is to prepare individuals with the literacy, skills and competences needed for professional practice in these fields (Sanders, 2009; Bybee, 2018; Larsen et al., 2022; Kristensen et al., 2024). STEM education, as an international educational phenomenon, is championed by entities like the Northamerican "National Science Teachers Association," advocating for educational responses to future workforce needs and the development of cross-disciplinary competencies (Bybee, 2013, 2018).

Despite STEM's role as a catalyst for societal investment in education, growth, and development, critics caution against potential pitfalls. They argue that an exclusive focus on STEM may narrow the educational purpose to problem-solving and technological development, relegating mathematics and natural sciences to auxiliary disciplines that need to be redesigned and technologised for relevance (Schmidt, 2019).

Active learning methods are essential in addressing these concerns, as they provide students with the opportunity to engage directly with subject themes through interdisciplinary approaches. Design principles for STEM education stress its interdisciplinary nature, support for individual STEM domains, and meaningful integration based on engagement with complexity, and attention to specific context (Bybee, 2018; Svabo et al., 2024). Whether seen through the lens of real-world connections, problem-solving, or an instructional approach, STEM education underscores the importance of authentic, problem-based learning rooted in science, mathematics, technology, and engineering (Kelley & Knowles, 2016; Shaughnessy, 2013; Johnson, 2013).

The importance of active learning methods has been highlighted in the pursuit of STEM education innovation (Felder & Brent, 2016; Penprase, 2020). These methods call for a broad range of pedagogical strategies that encourage and engage students as active participants in their learning experience (Cattaneo, 2017).

As Lee et al. (2018) have described, student active learning refers to instructional approaches that actively engage students in their learning process through collaboration and discussions rather than having them passively receive information from their instructors.

The basic meaning of the term active learning is that teachers should use teaching practices that actively engage students. An important part of active learning is that students explore and develop their own thoughts, feelings, and standpoints (Silva et al., 2022), as well as communicate and collaborate with others (McGill & Beaty, 2013).

According to Bonwell and Eison (1991), active learning is "anything that involves students in doing things and thinking about the things they are doing" (p. 2). By this, active learning highlights that students engage in reading, discussing, and writing, exploring their own attitudes and values. Emphasis is placed on developing students' skills and competencies instead of transmitting, as well as assimilating, data and information. A key point in active learning is that students engage in higher-order thinking such as analysis, synthesis, and evaluation (Børte et al., 2023).

Active learning gives students more opportunities to learn subject matter and work with creativity, critical thinking, hands-on skills, collaboration, and communication (Silberman & Biech, 2015; Revans, 2017). This helps students develop deeper understanding and it helps them apply theoretical concepts in practice.

Pedagogical methods such as design thinking (Wrigley & Straker, 2017; Goldman & Zielezinski, 2022), learning through making (Halverson & Sheridan, 2014), aesthetic learning (Uhrmacher, 2009; Shanks & Svabo, 2018) and playful learning (Honey, 2013) have been applied with similar purposes of active student participation.

Through such methods, students' engagement, critical thinking, problem-solving, and collaboration skills are enhanced (Tytler, 2020). These methods complement knowledge acquisition through rote learning with active, authentic learning experiences that better prepare students for the complexities of the modern world (Davis & Sumara, 2006; Zhou, 2016).

Integrated STEM approaches (Kelley & Knowles, 2016), SSI-approaches (Socio-Scientific Issues) (Zeidler, 2016), Inquiry-Based Learning (Sutaphan & Yuenyong, 2019), Design-Based Learning (Basham & Marino, 2013), Project and Problem-Based Learning (PBL) (Zhou, 2016), Place-Based Education (Gruenewald & Smith, 2014) and STEAM—incorporating the arts with STEM subjects (Bertrand & Namukasa, 2020) are all pedagogies that promote interdisciplinary collaboration and problem-solving.

These pedagogies typically seek to bridge the traditional boundaries between the science, technology, engineering, and mathematics disciplines through interdisciplinarity (Davis et al., 2019). They emphasise real-world orientation and promote holistic understanding (Tytler, 2020), underscoring the importance of skills development in critical thinking, decision-making and participation in discussions about issues such as climate change, ethics in the use of artificial intelligence or genetic engineering (Zeidler, 2016; Sutaphan & Yuenyong, 2019).

Furthermore, these approaches often integrate hands-on or practical aspects in educational situations (Kennedy & Odell, 2014; Kelley & Knowles, 2016), illustrating the overlap between active learning and integrated STEM education.

The next section on Creative Pragmatics will expand on these ideas, demonstrating how this concept can enhance both active learning and integrated STEM education. Creative Pragmatics offers a framework that incorporates creativity and agency into the learning process. This equips students with the competence to navigate complexity.

1.7 Challenges for Active Learning in STEM Education

STEM education often highlights the need to move beyond the boundaries of individual disciplines and embrace a more holistic approach in inquiry and problem-solving. However, traditional educational models frequently limit students to low-level cognitive tasks such as recalling facts in isolation, which can hinder their development of higher-order thinking skills necessary for tackling complex and ambiguous problems (Kennedy & Odell, 2014). Without having developed higher-order thinking skills and metacognitive capability, students may struggle to apply their learning in competent action (Thibaut et al., 2018).

A significant challenge in STEM education is the integration of interdisciplinary learning, which is essential for holistic understanding, but also often underdeveloped in current practices. This calls for more attention to interdisciplinarity in STEM and STEM integration (Nadelson & Seifert, 2017). As Anderson and Li (2020) have suggested, combining inquiry-based learning with an integrated STEM approach can provide opportunities for students to develop a range of general capabilities, such as critical thinking, self-direction, creativity and communication. However, without a strong interdisciplinary framework, students may struggle to connect ideas across disciplines and gain meaningful insights when working on real-world problems (Zhou, 2012, 2016; Holmes et al., 2018). This resonates with ongoing scholarly discussions about discipline-based science education and STEM integration (Larsen et al., 2024; Svabo et al., 2024), and indicates the need for change towards a paradigm of co-creation, with ideas including collective forms of practice and knowledge (Dollinger et al., 2018).

STEM education takes us beyond traditional disciplinary boundaries, requiring both teachers and students to develop theoretical, epistemological, methodological, and practical skills (Block et al., 2022). However, current educational models and systems often lack the flexibility needed to adapt to rapidly changing technologies and societal needs (Jantsch, 1972).

Active learning advocates a shift toward student-centred learning models and experiential and participatory pedagogies (Bonwell & Eison, 1991; Honey, 2013). Despite the benefits of these pedagogical methods, existing educational systems and teaching practices often resist change due to entrenched traditions and beliefs that 'real science teaching' should be content-focused and lecture-based. Both teachers and students may be accustomed to traditional methods and sceptical about deviating from established practices, believing that maintaining traditional forms of teaching and disciplinary boundaries is necessary (Kossybayeva et al., 2022).

In active learning practices, teachers become education designers, and there is a demand for them to teach in ways that are dialogical, open-ended, and exploratory. The teacher's role shifts from being an expert who imparts knowledge to students to one who facilitates learning and provides opportunities for students to talk, experiment, and actively engage. This requires teachers to develop creative, reflective, and research-based methods to improve their teaching (Li et al., 2019) and necessitates ongoing support and training.

Moreover, existing assessment and evaluation systems can hinder active learning approaches. Innovative STEM methods often focus on process-oriented learning, critical thinking, and collaborative skills, which may not align with traditional assessment methods (Sinatra et al., 2015). Designing assessments that foster skills like critical thinking, creativity, communication, and collaboration while meeting educational standards can be challenging (Margot & Kettler, 2019).

To support changes in educational systems, it is essential to connect students and educators with STEM fields and professionals (Dollinger et al., 2018), provide global and multi-perspective viewpoints (Kelley & Knowles, 2016), encourage outreach activities in both formal and informal learning settings (Bell et al., 2016), and select appropriate technologies to enhance teaching and learning (Kossybayeva et al., 2022). Exploring transdisciplinary settings and interactions between academic and non-academic actors is also necessary, which involves efforts in coordination, communication, and logistics (Balsiger, 2015).

These challenges indicate the need for Creative Pragmatics, which can enhance active learning in STEM by promoting innovative problem-solving, interdisciplinary integration, and adaptable learning models. The chapters of the book explore how Creative Pragmatics can be integrated into STEM education, showing examples of educators and students developing competencies to navigate a complex and dynamic world.

1.8 Creative Pragmatics and Tackling the Challenges of STEM Learning

How can we overcome these challenges in STEM education? We bring together various proven educational philosophies and sources of inspiration to urge a shift towards active learning. In this we propose Creative Pragmatics as a synthetic concept that will contribute to a transformation of how STEM disciplines are taught.

As mentioned, it is widely acknowledged that problem-solving, critical thinking, creativity, communication skills, emotional intelligence and transformative competencies are essential in addressing contemporary challenges. This book argues that such skills and competencies can be unified under the umbrella concept of 'Creative Pragmatics'. An underpinning rationale is that our present day problems can not be handled by doing more of the same. We need new, or rather refreshed approaches to how to live in, perceive and come to know the world, and these approaches have to foster skills and sensibilities which make it possible to address human and environmental needs in production, consumption and all aspects of a circular economy.

Science and technology studies (STS) provides a strong inspiration for Creative Pragmatics. STS provides nuanced understanding of scientific and technological developments, highlighting the role of instrumentality in science, the agency of materiality and the enrolment of heterogeneous forces in the making of facts. We follow those trajectories in science and technology studies which suggest that knowledge is something that we perform and achieve, and we consider such productive work to be creative. By calling knowledge creative, we don't mean it invents the world like a fictional story. Instead, we mean that knowledge and skills are accomplishments—creative accomplishments. Knowledge and skills are crafted and developed by people through their interactions and activities in the world. This process involves not only humans but also non-human elements and agents (Latour, 1999, 2007; Rathje et al., 2012 for a particular interdisciplinary science; for more on this, see Chap. 2 of this volume).

Creative Pragmatics is a dynamic framework for active learning in STEM education. It is not a fixed or static set of theories or methodologies. It is an approach, a stance, a mindset, a scaffold that involves iterative learning through practice, by doing, by designing, by making, by making art, by engineering knowledge, by collaborating, by observing and reflecting, through dialogue, participation, and imagination.

Inspired by pragmatism and pragmatist educational philosophy (Dewey, 1938; Putnam & Putnam, 2017; Rorty, 1982), Creative Pragmatics emphasises the active aspects of learning and the potential of education for societal change. This perspective underscores the importance of experiential education and the integration of theory and practice, as learners engage with the world through concrete experience, reflective observation, abstract conceptualisation, and active experimentation (Kolb, 1984).

In resonance with Pickering's (2010) prompt to sketch another future for education, Creative Pragmatics seeks to intervene in our very approach to modelling knowledge. Following Pickering's call for performative, open-ended models that go beyond mere representation, Creative Pragmatics echoes a 'more-than-representational' ethos. The concept embraces the notion of active, creative engagements with the world, epitomised by the act of making (Honey, 2013; Kennedy & Odell, 2014; McGill & Beaty, 2013).

That active agency is inherent in learning and understanding, reinforces the tenet that learners are not passive recipients but active creators of knowledge (McGill & Beaty, 2013; Vaughn, 2021). This approach to modelling knowledge places active engagements with the world as pivotal in knowledge making, learning and understanding (Svabo, 2021). Creative processes of making, designing, producing are key in developing knowledge, skills and competencies (Risopoulos-Pichler et al., 2020; Zhou, 2016).

Working creatively and with design and aesthetics, the latter broadly defined as perceptual and evaluative aspects of making, is a way of engaging in world-making, be it in the form of production of knowledge artefacts, systems or process design (Eisner, 2002). Design is purposeful practice, well conceived as the planning and management of making, of production and indeed distribution and consumption, all as creative engagements with world and environment (Honey, 2013; Wrigley & Straker, 2017). The field of design thus injects a critical dimension into Creative Pragmatics, spotlighting the interconnections between meaning and materiality (Hughes, 2000), and the aesthetic, processual and material mediation of innovation, problem-solving, ideation, and the iterative character of creative processes.

Creative Pragmatics is associated with concepts of creativity and agency, draws inspiration from studies on aesthetics, art and design, and highlights the situated performance of knowing. This relates to notions of situated learning (Engeström & Cole, 2021; Miner & Nicodemus, 2021). As Lave and Wenger (1991) described, learning, thinking, and knowing are relations among people in activity, within, and arising from the socially and culturally structured world. Creative Pragmatic approaches encourage learners to actively engage in inquiry, experimentation, making, and collaboration, inviting them to become active participants in their own learning journey.

The concept of Creative Pragmatics is rooted in the pragmatic theory of education (Dewey, 1938; Rorty, 1982), which believes that learning is active and that education is an instrument of change. Education should pay attention to learners' experiences with a process that can help integrate theory and practice. This emphasises the values of experiential education and pedagogical ways of learning by doing (Dewey, 1938). As Kolb (1984) suggested, learning occurs in a cycle through concrete experience, reflective observation, abstract conceptualisation, and active experimentation.

The convergence of these diverse fields involves their shared emphasis on active engagement (Kennedy & Odell, 2014), adaptability (Pickering, 2010), contextuality (Engeström & Cole, 2021), material mediation and the intricate interconnectedness of human experience. Together they are important threads in Creative Pragmatics as a multidimensional and dynamic concept that honours the performative, experiential, mediated, situated and transformative nature of human interaction with knowledge and the world (Svabo, 2021).

In summary, Creative Pragmatics emerges as a concept that synthesises insights from science and technology studies, pragmatist educational philosophy, art and design studies. This transdisciplinary foundation highlights the interactive, adaptive, aesthetic and contextually grounded nature of human endeavours that underscores the continuous and reciprocal relationship between individuals and their environment in the process of learning and meaning-making. By both shaping the world and constructing meaning within it, Creative Pragmatics embodies Vico's profound adage "verum ipsum factum" (Morrison, 1978), emphasising that truth is in what is made, that truth itself is not merely discovered but actively forged through acts of making and engagement.

By embracing Creative Pragmatics, STEM education can transcend traditional teaching methods and foster a holistic learning environment that empowers learners. We highlight Creative Pragmatics as a practical and philosophically grounded approach to how educators can design learning. Design informed by Creative Pragmatics inspires teachers to create learning situations where students actively build skills and competencies in knowledge work, including synthesis and integration between disciplines, the capacity to carry out self-leadership, critical analysis, creative problem-solving and communication with experts and non-experts.

Creative Pragmatics is a dynamic and adaptive approach to knowledge-building, sense-making, and problem-solving. This encompassing philosophy—or even paradigm—is characterised by an array of dimensions or vectors, with Creative

Pragmatics serving as an embodiment of infrastructuring premises or starting points, frames or mindsets within which one works, and modi operandi or orientations guiding creative practice (praxis). To use a metaphor of making or building: one establishes infrastructures and starting conditions, erects a scaffold or frame, a mindset or design philosophy, and then gets on with the making (see Table. 1.1).

Our summary overview is not prescriptive but suggestive. With knowledge intricately intertwined with perception, purpose, and imagination, a core tenet of Creative Pragmatics is its performative understanding of knowledge generation, bolstered by a pragmatic outlook (Bell, 2006). This underscores that knowledge extends beyond mere representation of a world-out-there separate from the knower and their discourses. Hands-on making, tinkering, and bricolage find home within Creative Pragmatics, synergising with an embedded, contextual, and relational disposition and orientation on practice as engagement and intervention. Creative Pragmatics prompts iterative and probing engagements and interventions, akin to riding wind and waves, sailing and surfing in a dynamically changing ocean, fostering adaptability and open-ended exploration (Svabo, 2021; Shanks & Svabo, 2018). As with engineering and design, Creative Pragmatics is not a methodology per se, though it might employ specific methodologies to serve its adaptive, located, opportunistic disposition and orientation. It is well conceived as finding way or ways (the old Greek root of *meta-hodos* is road or pathway). Consider how a design team may use quantitative and qualitative methodologies to come to know and gain empathy with a user community, use various techniques of creative ideation in exploring possible interventions, employ materials science and electrical engineering in realising a product or service, delivered through appropriate processes of manufacturing and distribution, to be assessed for use and worth to a user community. In such ways

Table 1.1 Understanding Creative Pragmatics. This table provides an overview of Creative Pragmatics, deriving the concept from deep roots in philosophical principles, influenced by science and technology studies and pragmatism, as well as by art and design. Overall, this table highlights how Creative Pragmatics integrates various dimensions to provide a holistic and transformative educational framework and practical orientations for pedagogy and policy

Infrastructures/premises	Frames within which one works/mindsets	Modi operandi/orientations on practice
Focus on process and dynamics in embracing openness and contingency	Holistic systems thinking	Iterative making as specific and located temporary interventions/solutions
Transactional relationships and contextual sensitivity	Cybernetic and adaptive sensory engagement	Participation, in distributed responsibility and autonomy
Performative understanding of knowledge	Accounting for participant context through ethnographic sensitivity in cultural diversity	Collaboration and co-creativity in teamwork and shared agency
Transdisciplinary engagement through (learner) experience	Design thinking and learning-by-doing, active experiential learning	Foresight, imagination and future worldbuilding
	Sources of influence	
Pragmatist philosophy	Science and technology studies	Art and design practice

Creative Pragmatics integrates problem-solving and decision-making processes through flexible, adaptive, experimental, and open-ended problem engagement.

So one can mix and match the components of Creative Pragmatics outlined in Table 1.1 according to need and experience, opportunity and potential. If we shift attention to pedagogy and STEM learning, Creative Pragmatics might prompt teaching strategies such as active learning, problem-based learning (PBL), and inquiry-based learning, by underscoring their practical orientations: iterative making as specific and located temporary interventions/solutions; participation, in distributed responsibility and autonomy; collaboration and co-creativity in teamwork and shared agency; foresight, imagination and future worldbuilding. These practical orientations can guide on many levels: as a teaching strategy and also on systemic, educational policy level.

Creative Pragmatics emphasises a learner-centred focus, underlying the importance of agency. Thus one of the key components is the focus on participant learning experience. Through teaching strategies embedded in Creative Pragmatics, participants (learners, as well as teachers) pursue hands-on experience: materially, physically embedded, tangible learning; adaptive variation of methods, spaces, and interactions; autotelic and playful learning; all related to specific educational content. There is a balance between cognitive and affective learning; the process is no less important than the result, where experiences of learning complement bodies of knowledge in the iterative processes that are Creative Pragmatics. Creative Pragmatics in active STEM learning thus equips learners with the capacity to engage with complexity, embrace uncertainty, and employ innovative strategies to address multifaceted challenges in STEM fields and beyond.

We seem to be the first to introduce in a scholarly milieu the concept 'Creative Pragmatics'. But neither the concept nor its dimensions and dynamic components stand isolated. It is quite appropriate to assert that Creative Pragmatics is nothing new. We have indicated in Sect. 1.5 of this chapter the diverse and deep roots of the concept. For this book particularly we draw on earlier work in pragmatist educational philosophy as well as learner-centred, production-oriented and constructivist pedagogies and teaching approaches. Creative Pragmatics adds a fresh action-oriented outlook by *integrating* various educational philosophies and pedagogies. It contributes to discussions in science education research by aiming to make it tangible for science educators in achieving learning goals for skills and competencies outlined in frameworks like the European Qualifications Framework or relevant national versions. By embracing Creative Pragmatics, STEM education is positioned to move beyond traditional teaching methods, cultivating a holistic learning environment that empowers learners to navigate a complex and uncertain world.

1.9 The Contents of This Book

This book provides research-based foundations for Creative Pragmatics in active learning and illustrates meaningful studies through various theoretical perspectives, research methods, and pedagogical models across different contexts and cultures. It also incorporates lessons learned and proven practices focused on the interaction between active learning, educational design, and participatory pedagogy in STEM education.

The collection of 13 chapters explores discussions on creative learning, teacher-student interactions, professional learning, support for instructors, assessment, science literacy, hands-on skills, problem-based learning, project-based learning, student agency, twenty-first-century skills, and national science education innovation. The collection of chapters includes voices from various educational levels and disciplines, a polyphonic format that suits the Creative Pragmatics approach.

This chapter, the book's Chap. 1, "Creative Pragmatics for Active Learning in STEM Education" launches the book by discussing the need for new knowledge models and the importance of learning to navigate in complexity. The chapter defines Creative Pragmatics as a synthetic and integrative concept for active learning in STEM education. Creative Pragmatics integrates performative models of knowledge with pragmatic educational philosophy, emphasising creative agency and active engagement.

Chapter 2, "Creative Pragmatics—Situated Performances of Knowing" provides theoretical and philosophical points of orientation, drawing inspiration from pragmatism and science and technology studies and connecting with art and design practice. Creative Pragmatics emphasises situated performances of knowing, agency and creativity—essentially as iterative processes of learning through doing and dynamic sensemaking. Art and design contribute to the principles and concepts underpinning Creative Pragmatics.

Chapter 3, "Creative Pragmatics—Knowing and Competence in a Complex World" lays the groundwork for Creative Pragmatics by exploring the idea of complexity. Complexity makes it challenging to use traditional ways of predicting and simplifying things, since it cannot be fully understood or predicted. The chapter compares how we usually think about the relationship between science and the world with a performative understanding. This new perspective requires us to rethink what we know, the skills we have, and the competencies we develop. While endorsing a competence-based approach, Creative Pragmatics aims to overcome its limitations. It emphasises that competence-based learning, although valuable for practical skills, should not succumb to reductionism but instead uphold a holistic view that acknowledges the situated and dynamic nature of competencies. This chapter concludes the three introductory chapters and sets the stage for the following case chapters, demonstrating fresh approaches to competently dealing with the complexities of our world.

Chapter 4, "Active and Agentic: Participant-Directed Inquiry in an Introductory Science Programme," demonstrates the critical importance of creative agency within the framework of Creative Pragmatics, where learning is co-designed, and agency is integral to participatory learning in the design process. Through a qualitative case study of two student projects at Roskilde University, it highlights how an open inquiry approach enables students to navigate and simplify complex situations, fostering agency within project groups. The research illustrates how students leverage natural science to address real-world challenges, such as those found in the shipping industry. The chapter underscores the transdisciplinary nature of the program, as self-directed inquiry extends beyond traditional disciplinary boundaries to engage with complex societal issues.

Chapter 5, “Artefacts and Creative Learning in Mathematics Education”, examines the artefact and material components of creative and playful learning environments, highlighting the importance of user-centred design through situated experiences and a connection with arts practice—key elements of Creative Pragmatics. The chapter challenges the isolation of learners by activating agency not only in the learners but also in the artefacts within the learning environment. It bridges the fields of artefact design, creative learning, and mathematics education, focusing on a concrete case of a booklet design process guided by Creative Pragmatics principles. Through the first author’s reflective autoethnography, the chapter demonstrates the pragmatist idea that we are part of what we seek to understand. Insights from three rounds of user testing inform both the booklet’s further development and broader considerations of artefact-mediated creative learning environments in mathematics and STEM education. By discussing the values of art, the chapter also offers inspiration for the future of mathematics and STEM education, emphasising the role of artefacts in fostering creative, playful learning environments.

Chapter 6, “Summative STEM Competence Assessment: Affect-Sensitive Learning,” explores the integration of competencies and affect in designing holistic summative assessments for integrated STEM education. By leveraging a theoretical framework that combines cognitive learning aspects with affect theory, the chapter introduces the concept of affect-sensitive learning to offer a comprehensive perspective on assessing STEM competencies. A case from the LabSTEM+ project highlights the significance of considering affect in both STEM teaching and assessment. The study examines how affect influences the analysis of summative assessments, focusing on students’ STEM competencies and their affective experiences. The case study reveals a dynamic interplay between STEM competencies and the affective domain, showing how materials like pipes and water enhance students’ engagement, motivation, and autonomy. This research contributes to a deeper understanding of the role of affect in STEM education assessment and highlights the importance of integrating affective aspects to promote a more holistic approach to learning and assessing STEM competencies, which is essential for making Creative Pragmatics work in practice.

Chapter 7, “How to Start Strong: Different Approaches for Crafting Project Briefs with Corporate Sponsors for Global Innovation and STEM-Based Team Projects,” highlights the importance of teaching tools like project briefs in putting Creative Pragmatics into action. The study marks the significance of the starting point in project-based active learning, emphasising its crucial role in the iterative, never-ending learning process that defines Creative Pragmatics. By analysing a rich data set of 68 project briefs from 19 partner universities over a decade, drawn from a long-running project course taught at Stanford University and the SUGAR Network, the chapter explores different approaches to framing these briefs, which are key to facilitating creative interventions in complex scenarios. The research found that corporate sponsors sought seven different types of project outcomes, with varied phrasing strategies such as "how might we" and "we dream..." to spark imaginative thinking. Additionally, over a third of the briefs focused on mid-term innovation horizons. The chapter concludes with a two-question guide for crafting project briefs with corporate sponsors, helping to ensure that student projects start from a stronger position.

Chapter 8, "FabLabs and the Maker Movement: Learning through Making," establishes the programmatic coherence of the maker movement as expressed through Creative Pragmatics. The chapter explores deeper associations of design processes within "epistemologies of the hand," highlighting the transdisciplinary reach of design-based methodologies. It gives shape and substance to the principle of embodiment, emphasising that active learning is not just cognitive but also affective and located in corporeality. Starting with the question, "Is 'Making' hype?", the chapter examines the ethos of the Maker movement as a new way of empowering innovation, learning, and communities. Drawing on seven years of experience from running the FabLab RUC at Roskilde University, the chapter proposes a maker mindset through six propositions, positioning "making" as a complementary methodology to design thinking.

Chapter 9, "Use of Spaces for Active Learning in University STEM Education", emphasises that space shapes the people who inhabit it, highlighting the importance of how individuals interact with their learning environment. This chapter argues that teachers' use of physical learning spaces in higher education can support active learning if they engage in a Creative Pragmatic process of reflecting and experimenting. The chapter highlights that spatial competence—understanding how to effectively use and adapt learning environments—plays a crucial role in facilitating active learning. By presenting cases where spatial competence is demonstrated, the chapter provides examples of how spaces can be manipulated to enhance learning experiences. It also offers strategies for helping teachers develop this competence by creating opportunities for dialogue about space in learning. The chapter ultimately seeks to inspire educators to engage in Creative Pragmatics by planning, conducting, and reflecting on the use of classroom spaces to promote relational and situated learning in STEM.

Chapter 10, "Design Studio for Active Learning", explores how design studios serve as dynamic learning environments that facilitate the development of competence by engaging students in active, situated learning processes. Through case studies from programs at Columbia, Stanford, and Nebraska, the chapter outlines key principles of pedagogical scaffolding that guide students through a design process. This process includes defining and briefing a problem, researching and assessing contextual issues, ideating, prototyping, and participating in studio-based critique, all while anticipating the real-world application of solutions. By focusing on how students interact with and adapt to their learning environments, the chapter demonstrates how competence extends beyond knowledge and skills to include the ability to use them in context-sensitive and collaborative ways. The concept of Creative Pragmatics offers a perspective on architecture, with further expansion of STEM through the experiential lifeworlds of physical structures, engineering, building, design. In this chapter, this wide scope is grounded in a case-study in studio-based learning focused on the design of future urban infrastructures, an epitome of complexity.

Chapter 11, "A Pedagogy of Enactment: Creative Rehearsals and Feedback Literacy in Science Teacher Education", explores the critical role of 'learning about learning' within the framework of Creative Pragmatics, emphasising the importance of practice and experimentation for those who shape classroom experiences. This chapter investigates how science teachers benefit from ongoing processes of rehearsal and reflection, highlighting the concept of performance as a fertile ground for bridging theory and practice. It demonstrates how rehearsal, enactment, scenario-building, and improvisation can be powerful tools for practice-based learning. The concept of 'creative rehearsal' illustrates the role of imagination in reconciling the tensions between theory and practice, while 'feedback literacy' is presented as a key pragmatic competency in the dynamic, iterative process of learning by rehearsing. Through this exploration, the chapter offers valuable insights into how pre-service science teachers can develop their teaching practices through active engagement and reflective practice.

Chapter 12, "Innovating Science Education at a National Level", presents a case study in navigating complex adaptive systems through the lens of Creative Pragmatics. The chapter examines the challenges of innovating science education in Denmark, focusing on policy, leadership, team dynamics, and large-scale interventions. It advocates for a holistic, ecosystemic perspective that works across administrative silos, applying abductive sensemaking iteratively, adaptively, and collaboratively. This approach is particularly emphasised in bridging large-scale initiatives with local efforts to better connect science and society. Through an auto-ethnographic assessment of the Danish Science Academy's efforts, the chapter explores how collective leadership, boundary-spanning, and building informal learning communities can drive long-term, sustainable change in science education. This analysis contributes to the broader Scandinavian discourse on large-scale educational innovation, highlighting the interconnectedness of systems in the process of educational reform. This shows how a Creative Pragmatics approach is not restricted to the classroom but can be applied broadly at all levels of learning.

Chapter 13, "A Call to Action for Embedding Creative Pragmatics in STEM Education", serves as a call to action for STEM educators, policymakers, and educational institutions to adopt Creative Pragmatics as a means to enrich current learning models with more participatory, hands-on, and student-directed approaches. This final chapter synthesises the wide range of case studies and methodologies presented throughout the book, drawing out implications for educational policy, funding models, and pedagogical practice. It argues that by integrating Creative

Pragmatics into STEM education, all stakeholders can help cultivate a pipeline of scientific and technical talent equipped to navigate the complexities of the real world. By bridging thinking and doing, this chapter explores where the dialogue on Creative Pragmatics can go next and how different stakeholders can more actively embrace and implement active learning in STEM education.

This book is about Creative Pragmatics and itself is a creative collaboration work. It indicates efforts of a group of experts from the US and Europe, including Stanford University and Columbia University in the US, University of Southern Denmark, Technical University of Denmark, Roskilde University, University College of Southern Denmark, and Blekinge Institute of Technology in Sweden.

Researchers, academics, STEM educators, advanced-level students, practitioners, policy makers, and government officials will find this book useful in furthering their research and rethinking how to develop the practice of STEM education in the future.

References

Agustian, H. Y. (2023). The critical role of understanding epistemic practices in science teaching using wicked problems. *Science & Education, 33*(1). https://doi.org/10.1007/s11191-023-00471-2

Anderson, L. W., & Krathwohl, D. R. (Eds.). (2001). *A taxonomy for learning, teaching, and assessing: A revision of Bloom's taxonomy of educational objectives*. Longman.

Anderson, J., & Li, Y. (2020). Investigating the potential of integrated STEM education from an international perspective. In J. Anderson & Y. Li (Eds.), *Integrated approaches to STEM education: An international perspective* (pp. 1–12). Springer. https://doi.org/10.1007/978-3-030-52229-2_1

Balsiger, J. (2015). Transdisciplinarity in the classroom? Simulating the co-production of sustainability knowledge. *Futures, 65*, 185–194. Elsevier. https://doi.org/10.1016/j.futures.2014.08.005

Bang, M., & Vossoughi, S. (2016). Participatory design research and educational justice: Studying learning and relations within social change making. *Cognition and Instruction, 34*(3), 173–193. Taylor and Francis. https://doi.org/10.1080/07370008.2016.1181879

Barad, K. (2003). Posthumanist performativity: Toward an understanding of how matter comes to matter. *Signs: Journal of Women in Culture and Society, 28*(3), 801–831.

Barad, K. (2007). *Meeting the universe halfway: Quantum physics and the entanglement of matter and meaning*. Duke University Press.

Basham, J. D., & Marino, M. T. (2013). Understanding STEM education and supporting students through universal design for learning. *Teaching Exceptional Children, 45*(4), 8–15. Sage. https://doi.org/10.1177/004005991304500401

Bell, V. (2006). Performative knowledge. *Theory, Culture & Society, 23*(2–3), 214–217. Sage. https://doi.org/10.1177/026327640602300245

Bell, J., Falk, J., Hughes, R., Hunt, G., Parrish, J., Ruffin, M., et al. (2016). Informal STEM education: Resources for outreach, engagement, and broader impacts. *Science Education (CAISE)*. https://doi.org/10.13140/RG.2.1.4736.3446

Bennett, J. (2010). *Vibrant matter: A political ecology of things*. Duke University Press.

Bertrand, M. G., & Namukasa, I. K. (2020). STEAM education: Student learning and transferable skills. *Journal of Research in Innovative Teaching & Learning, 13*(1), 43–56. https://doi.org/10.1108/JRIT-01-2020-0003

Bijker, W. E., & Law, J. (Eds.). (1994). *Shaping technology/building society: Studies in sociotechnical change*. The MIT Press.

Block, T., Provė, C., Dehaene, M., Abeele, P. V., & Beeckmans, L. (2022). Understanding urban sustainability from mode 2 science and transdisciplinary education: How master thesis ateliers of the Ghent *Stadsacademie* tackle wicked issues. *Environment, Development and Sustainability, 24*(8). https://doi.org/10.1007/s10668-022-02657-0

Bonwell, C. C., & Eison, J. A. (1991). *Active learning: Creating excitement in the classroom* (ASHE-ERIC Higher Education Report No. 1). School of Education and Human Development, George Washington University. https://eric.ed.gov/?id=ED336049

Børte, K., Nesje, K., & Lillejord, S. (2023). Barriers to student active learning in higher education. *Teaching in Higher Education, 28*(3), 597–615. Taylor and Francis. https://doi.org/10.1080/13562517.2020.1839746

Bybee, R. W. (2013). *The case for STEM education challenges and opportunities*. NSTA Press.

Bybee, R. W. (2018). *STEM education now more than ever*. NSTA Press.

Byrne, D., & Callaghan, G. (2022). *Complexity theory and the social sciences: The state of the art* (2nd ed.). Routledge. https://doi.org/10.4324/9781003213574

Cattaneo, K. H. (2017). Telling active learning pedagogies apart: From theory to practice. *Journal of New Approaches in Educational Research (NAER Journal), 6*(2), 144–152. Springer. https://doi.org/10.7821/naer.2017.7.237

Coker, J. S., Heiser, E., Taylor, L., & Book, C. (2017). Impacts of experiential learning depth and breadth on student outcomes. *Journal of Experiential Education, 40*(1), 5–23. Sage. https://doi.org/10.1177/1053825916678

Davis, B., & Sumara, D. (2006). *Complexity and education: Inquiries into learning, teaching, and research* (1st ed.). Erlbaum. Routledge. https://doi.org/10.4324/9780203764015

Davis, B., Sumara, D. J., & Luce-Kapler, R. (2008). *Engaging minds: Changing teaching in complex times* (2nd ed.). Routledge.

Davis, B., Francis, K., & Friesen, S. (2019). *STEM education by design: Opening horizons of possibility* (1st ed.). Routledge. https://doi.org/10.4324/9780429025143

Dewey, J. (1938). *Experience and education*. Palgrave Macmillan.

Dollinger, M., Lodge, J., & Coates, H. (2018). Co-creation in higher education: Towards a conceptual model. *Journal of Marketing for Higher Education, 28*(2), 210–231. https://doi.org/10.1080/08841241.2018.1466756

Eisner, E. W. (2002). *The arts and the creation of mind*. Yale University Press.

Engeström, Y., & Cole, M. (2021). Situated cognition in search of an agenda. In *Situated cognition* (pp. 301–309). Routledge.

European Qualifications Framework. (2017). *Official Journal of the European Union*, C 189/15. Retrieved January 9, 2024, from https://eur-lex.europa.eu/legal-content/EN/TXT/HTML/?uri=CELEX:32017H0615(01)&from=EN

Felder, R. M., & Brent, R. (2016). *Teaching and learning STEM: A practical guide*. Wiley.

Friedman, K., & Stolterman, E. (2014). Series foreword. In J. Simonsen, C. Svabo, S. M. Strandvad, K. Samson, M. Hertzum, & O. E. Hansen (Eds.), *Situated design methods* (pp. vii–xii). The MIT Press. https://doi.org/10.7551/mitpress/9936.003.0001

Funtowicz, S. O., & Ravetz, J. R. (2018). Post-normal science. In I. N. Castree, M. Hulme, & J. D. Proctor (Eds.), *Companion to environmental studies* (pp. 443–447). Routledge.

Gallagher, S. E., & Savage, T. (2023). Challenge-based learning in higher education: An exploratory literature review. *Teaching in Higher Education, 28*(6), 1135–1157. https://doi.org/10.1080/13562517.2020.1863354

Gilbert, J. K. (2004). Models and modelling: Routes to more authentic science education. *International Journal of Science and Mathematics Education, 2*(2), 115–130. https://doi.org/10.1023/B:IJMA.0000029402.40497.4f

Goldman, S., & Zielezinski, M. B. (2022). *Design thinking for every classroom: A practical guide for educators*. Routledge.

Gruenewald, D. A., & Smith, G. A. (Eds.). (2014). *Place-based education in the global age: Local diversity* (2nd ed.). Routledge.

Grugulis, I., & Stoyanova, D. (2011). Skill and performance. *British Journal of Industrial Relations, 49*(3), 515–536. https://doi.org/10.1111/j.1467-8543.2010.00779.x

Halverson, E. R., & Sheridan, K. (2014). The maker movement in education. *Harvard Educational Review, 84*(4), 495–504. https://doi.org/10.17763/haer.84.4.34j1g68140382063

Hoffmann, T. (1999). The meanings of competency. *Journal of European Industrial Training, 23*(6), 275–286. https://doi.org/10.1108/03090599910284650

Holmes, K., Gore, J., Smith, M., & Lloyd, A. (2018). An integrated analysis of school students' aspirations for STEM careers: Which student and school factors are most predictive? *International Journal of Science and Mathematics Education, 16*(4), 655–675. Springer. https://doi.org/10.1007/s10763-016-9793-z

Honey, M. (Ed.). (2013). *Design, make, play: Growing the next generation of STEM innovators*. Routledge. https://doi.org/10.4324/9780203108352

Hughes, J. (2000). *Ecology and historical materialism*. Cambridge University Press. https://doi.org/10.1017/CBO9780511490262

James, W. (1975). *The meaning of truth* (Vol. 2). Harvard University Press.

Jantsch, E. (1972). Towards interdisciplinarity and transdisciplinarity in education and innovation. In L. Apostel (Ed.), *Interdisciplinarity: Problems of teaching and research in universities* (pp. 97–120). CERI/OEC.

Johnson, C. C. (2013). Conceptualizing integrated STEM education. *School Science and Mathematics, 113*(8), 367–368. https://doi.org/10.1111/ssm.12043

Kelley, T. R., & Knowles, J. G. (2016). A conceptual framework for integrated STEM education. *International Journal of STEM education, 3*(11) Springer, 1–11. https://doi.org/10.1186/s40594-016-0046-z

Kennedy, T. J., & Odell, M. R. (2014). Engaging students in STEM education. *Science Education International, 25*(3), 246–258.

Kolb, D. A. (1984). *Experiential learning: Experience as the source of learning and development* (1st ed.). Prentice Hall.

Kossybayeva, U., Shaldykova, B., Akhmanova, D., & Kulanina, S. (2022). Improving teaching in different disciplines of natural science and mathematics with innovative technologies. *Education and Information Technologies, 27*(6), 7869–7891. Springer. https://doi.org/10.1007/s10639-022-10955-3

Kristensen, M. L. A., Larsen, D. M., Seidelin, L., & Svabo, C. (2024). The role of mathematics in STEM activities: Syntheses and a framework from a literature review. *International Journal of Education in Mathematics, Science and Technology, 12*(2), 418. https://doi.org/10.46328/ijemst.3357

Larsen, D. M., Kristensen, M., Hjort, M., & Seidelin, L. (2022). STEM- didaktik – et internationalt, systematisk review om STEMundervisningens didaktik [STEM didactics—An international, systematic review of STEM education didactics]. *MONA—Matematik- Og Naturfagsdidak-tik, 22*(1), 6–22. https://tidsskrift.dk/mona/article/view/131923

Larsen, D. M., Kristensen, M. A., & Svabo, C. (2024). Expanding the horizons of mathematics in STEM: Diverse mathematical roles in STEM activities. In J. Anderson & K. Makar (Eds.), *The contribution of mathematics to school STEM education*. Springer.

Lave, J., & Wenger, E. (1991). *Situated learning: Legitimate peripheral participation*. Cambridge University Press. https://doi.org/10.1017/CBO9780511815355

Latour, B. (1999). On recalling ANT. *The Sociological Review, 47*(1_suppl), 15–25. https://doi.org/10.1111/j.1467-954X.1999.tb03480.x

Latour, B. (2007). *Reassembling the social: An introduction to actor-network-theory*. Oxford University Press.

Lee, D., Morrone, A. S., & Siering, G. (2018). From swimming pool to collaborative learning studio: Pedagogy, space, and technology in a large active learning classroom. *Educational*

Technology Research and Development, 66, 95–127. Springer. https://doi.org/10.1007/s11423-017-9550-1

Li, Y., Schoenfeld, A. H., Sessa, A. A., Graesser, A. C., Benson, L. C., English, L. D., & Duschl, R. A. (2019). Design and design thinking in STEM education. *Journal for STEM Education Research, 2*, 93–104. Springer. https://doi.org/10.1007/s41979-019-00020-z

Margot, K. C., & Kettler, T. (2019). Teachers' perception of STEM integration and education: A systematic literature review. *International Journal of STEM Education, 6*(2), 1–16. https://doi.org/10.1186/s40594-018-0151-2

McGill, I., & Beaty, L. (2013). *Action learning: A practitioner's guide*. Routledge. https://doi.org/10.4324/9781315042480

Miner, A., & Nicodemus, B. (2021). What is situated learning?. Situated learning in interpreter education: From the classroom to the community, 6(1), 17–39. Springer. https://doi.org/10.1186/s40594-018-0151-2

Morrison, J. C. (1978). Vico's principle of verum is factum and the problem of historicism. *Journal of the History of Ideas, 39*(4), 579–595. https://doi.org/10.2307/2709443

Nadelson, L. S., & Seifert, A. L. (2017). Integrated STEM defined: Contexts, challenges, and the future. *The Journal of Educational Research, 110*(3), 221–223. Taylor and Francis. https://doi.org/10.1080/00220671.2017.1289775

National Intelligence Council. (2021). *Global trends 2040: A more contested world.* [Report] Office of the Director of National Intelligence. https://www.dni.gov/files/ODNI/documents/assessments/GlobalTrends_2040.pdf

National Research Council. (2013). *Next generation science standards: For states, by states*. The National Academies Press. https://doi.org/10.17226/18290

Nicholls, A. (2018). *Managing educational innovations*. Routledge.

Norman, D. (2010). Why design education must change. Core77.com. Retrieved September 30, 2019, from https://www.core77.com/posts/17993/why-design-education-must-change-17993

OECD. (2019). *Transformative competencies for* 2030. *OECD Future of education and skills 2030 concept note*. Retrieved from https://www.oecd.org/content/dam/oecd/en/about/projects/edu/education-2040/concept-notes/Transformative_Competencies_for_2030_concept_note.pdf

Olsen, B. (2012). *Archaeology: The discipline of things*. University of California Press.

Penprase, B. E. (2020). *STEM education for the 21st century*. Springer. https://doi.org/10.1007/978-3-030-41633-1

Peters, B. G. (2017). What is so wicked about wicked problems? A conceptual analysis and a research program. *Policy and Society, 36*(3), 385–396. Taylor and Francis. https://doi.org/10.1080/14494035.2017.1361633

Pickering, A. (1995). *The mangle of practice: Time, agency, and science*. University of Chicago Press.

Pickering, A. (2010). *The cybernetic brain: Sketches of another future*. University of Chicago Press. https://doi.org/10.7208/chicago/9780226667928.001.0001

Pickering, A. (2025). *Acting with the world: Agency in the Anthropocene*. Duke University Press.

Putnam, H., & Putnam, R. A. (2017). *Pragmatism as a way of life: The lasting legacy of William James and John Dewey*. Harvard University Press.

Rathje, W. L., Shanks, M., & Witmore, C. (Eds.). (2012). *Archaeology in the making. Conversations through a discipline.* Routledge. https://doi.org/10.4324/9780203083475

Revans, R. (2017). *ABC of action learning*. Routledge. https://doi.org/10.4324/9781315263533

Risopoulos-Pichler, F., Daghofer, F., & Steiner, G. (2020). Competences for solving complex problems: A cross-sectional survey on higher education for sustainability learning and transdisciplinarity. *Sustainability, 12*(15), 6016. https://doi.org/10.3390/su12156016

Rorty, R. (1982). *Consequences of pragmatism*. University of Minnesota Press. https://doi.org/10.1057/978-1-137-60083-7_1

Sanders, M. E. (2009). STEM, STEM education, STEMmania. *The Technology Teacher, 68*(4), 20–26. http://hdl.handle.net/10919/51616

Sawyer, R. K. (2005). *Social emergence: Societies as complex systems*. Cambridge University Press. https://doi.org/10.1017/CBO9780511734892
Sayes, E. (2013). Actor–network theory and methodology: Just what does it mean to say that nonhumans have agency? *Social Studies of Science, 44*(1), 134–149. https://doi.org/10.1177/0306312713511867
Schmidt, J. R. (2019). Hvem definerer STEM i skolen og i skoleforskning? [Who defines STEM in schools and in educational research?]. *MONA—Matematik- Og Naturfagsdidaktik*, (2), 70–88. https://tidsskrift.dk/mona/article/view/114698
Shanks, M., & Svabo, C. (2018). Scholartistry: Incorporating scholarship and art. *Journal of Problem Based Learning in Higher Education, 6*(1), 15–38. https://doi.org/10.5278/ojs.jpblhe.v6i1.1957
Shaughnessy, J. M. (2013). Mathematics in a STEM context. *Mathematics Teaching in the Middle School, 18*(6), 324–324. https://doi.org/10.5951/mathteacmiddscho.18.6.0324
Silberman, M. L., & Biech, E. (2015). *Active training: A handbook of techniques, designs, case examples, and tips* (4th ed.). Wiley. https://doi.org/10.1002/9781119154778
Silva, E. C. E., Lino-Neto, T., Ribeiro, E., Rocha, M., & Costa, M. J. (2022). Going virtual and going wide: Comparing team-based learning in-class versus online and across disciplines. *Education and Information Technologies, 27*(2), 2311–2329. https://doi.org/10.1007/s10639-021-10683-0
Sinatra, G. M., Heddy, B. C., & Lombardi, D. (2015). The challenges of defining and measuring student engagement in science. *Educational Psychologist, 50*(1), 1–13. https://doi.org/10.1080/00461520.2014.1002924
Sternberg, R. J. (Ed.). (1998). *Handbook of creativity*. Cambridge University Press. https://doi.org/10.1017/CBO9780511807916
Sutaphan, S., & Yuenyong, C. (2019). STEM education teaching approach: Inquiry from the context based. *Journal of Physics: Conference Series, 1340*(1), 012003. https://doi.org/10.1088/1742-6596/1340/1/012003
Svabo, C. (2021). Living world dynamics: Or what Brian Eno can teach us about knowing in a complex world. In E. Brandt, T. Markussen, E. Berglund, G. Julier, & P. Linde (Eds.), *Proceedings of Nordes 2021: Matters of scale* (pp. 384–389). https://doi.org/10.21606/nordes.2021.42
Svabo, C., & Shanks, M. (2013). Archaeology and photography: A pragmatology. In A. González-Ruibal (Ed.), *Reclaiming archaeology: Beyond the tropes of modernity* (pp. 89–102). Routledge.
Svabo, C., Larsen, D. M., Borch, K. B., Svendsen, M. W. H., & Kristensen, M. A. (2024). *STEM-didaktik med fokus på matematik: til grundskole, gymnasie og dagtilbud* [STEM didactics with a focus on mathematics: For primary schools, secondary schools, and early childhood education]. Syddansk Universitetsforlag.
Tan, O. S. (2003). *Problem-based learning innovation: Using problems to power learning in the 21st century*. Gale Cengage Learning.
Thibaut, L., Ceuppens, S., De Loof, H., De Meester, J., Goovaerts, L., Struyf, A., et al. (2018). Integrated STEM education: A systematic review of instructional practices in secondary education. *European Journal of STEM Education, 3*(1), 02. https://doi.org/10.20897/ejsteme/85525
Tytler, R. (2020). STEM education for the twenty-first century. In J. Anderson & Y. Li (Eds.), *Integrated approaches to STEM education: An international perspective* (pp. 21–43). Springer. https://doi.org/10.1007/978-3-030-52229-2_3
Uhrmacher, P. B. (2009). Toward a theory of aesthetic learning experiences. *Curriculum Inquiry, 39*(5), 613–636. https://doi.org/10.1111/j.1467-873X.2009.00462.x
Van Berkel, K., & Manickam, A. (2020). *Wicked world: Complex challenges and systems innovation*. Routledge. https://doi.org/10.4324/9781003154495
Vaughn, M. (2021). *Student agency in the classroom: Honoring student voice in the curriculum*. Teachers College Press.
Whitehead, A. N. (1978). *Process and reality*. The Free Press.

Wong, S. C. (2020). Competency definitions, development and assessment: A brief review. *International Journal of Academic Research in Progressive Education and Development, 9*(3), 95–114. https://doi.org/10.6007/IJARPED/v9-i3/8223

World Economic Forum. (2023). *Putting skills first: A framework for action.* Retrieved January 9, 2024, from https://www.weforum.org/publications/putting-skills-first-a-framework-for-action/

Wrigley, C., & Straker, K. (2017). Design thinking pedagogy: The educational design ladder. *Innovations in Education and Teaching International, 54*(4), 374–385. Taylor and Francis. https://doi.org/10.1080/14703297.2015.1108214v

Zairul, M. (2020). A thematic review on student-centred learning in the studio education. *Journal of Critical Reviews, 7*(2), 504–511. https://doi.org/10.31838/jcr.07.02.95

Zeidler, D. L. (2016). STEM education: A deficit framework for the twenty-first century? A sociocultural socioscientific response. *Cultural Studies of Science Education, 11*(1), 11–26. https://doi.org/10.1007/s11422-014-9578-z

Zhou, C. (2012). Fostering creative engineers: A key to face the complexity of engineering practice. *European Journal of Engineering Education, 37*(4), 343–353. https://doi.org/10.1080/03043797.2012.691872

Zhou, C. (2016). Fostering creative problem solvers in higher education: A response to complexity of societies. In C. Zhou (Ed.), *Handbook of research on creative problem-solving skill development in higher education* (1st ed., pp. 1–23). IGI Global. https://doi.org/10.4018/978-1-5225-0643-0.ch001

Connie Svabo PhD, Professor of STEM Education and Science Communication, Director of STEM Center for Research in Science Education and Communication (FNUG), Department of Mathematics and Computer Science, University of Southern Denmark.

Connie is a transdisciplinary scholar and creative practitioner with expertise in learning environment design and experience. She is the founding Director of STEM Center for Research in Science Education and Communication (FNUG), Section Leader and member of the management team of the Department of Mathematics and Computer Science, chair of the SDU research programme in Science and Mathematics Education and Communication. She promotes Creative Pragmatics as a model for knowing in a complex world and underpins this stance with a practice that ranges from art over education to design and consultancy. Her portfolio includes geological UNESCO heritage sites Stevns Klint and the South Fyn Archipelago, as well as the Danish National Gallery Open Air Museum and the museum of natural history, Naturama. Previously a Continuing Studies Instructor at Stanford University and Faculty at Roskilde University, where she directed the research centre Experience Lab, chaired the study programmes in Communication Performance Design 2013–2016 and served on the academic board. She has published books on STEM education, sustainability, competence development, design pedagogy and digital life. She holds a PhD in Social Science from Roskilde University's multidisciplinary programme in Society, Space and Technology.

Michael Shanks PhD Doc, Professor of Classics and Archaeology, Stanford University.

Acknowledged as one of the most original and influential of contemporary archaeologists, Michael is a specialist in long-term humanistic views of design and creativity, innovation and social change, while exploring connections across the sciences and arts in a hybrid field of theatre/archaeology. He is senior faculty in Stanford University's programs in Science Technology Society, Urban Studies, Writing and Rhetoric, Archaeology, Classics, and in the Center for Design Research, part of Stanford's d.school in the School of Engineering. He directed Stanford Humanities Lab, pioneering innovative project-based pedagogy in the arts and humanities, and helped found the Revs Program at Stanford, connecting the archaeological automobile with contemporary car design. Founding faculty in Stanford's Archaeology Center, Michael is a director of Stanford's research group Foresight and Innovation, and, for two decades until its retirement in 2023, he worked with mediaX, the affiliates' program that connected Stanford with industry through technology and innovation. He served on the city of Rotterdam's International Advisory Board from 2008 to 2019 and works with many companies and organisations in developing cultures of innovation. In another mobilisation of Creative Pragmatics, Michael continues to pursue archaeological fieldwork in the northern borders of the Roman empire and beyond.

Chunfang Zhou PhD, Associate Professor of STEM Education and Science Communication, Center for Research in Science Education and Communication, Department of Mathematics and Computer Science, University of Southern Denmark.

A researcher and practitioner working on creativity, STEM education, and teacher education. Chunfang has an interdisciplinary and cross-cultural educational background: she finished a Bachelors Degree (Industry Automation and Information Science) and Masters Degree (Philosophy of Science and Technology) in China; during 2008–2012, Chunfang finished her doctoral research on *Group Creativity in Engineering Education in a Problem and Project-Based Learning (PBL) environment* at Aalborg University, Denmark. She is a board member of Danish Development Research Network (DDRN), and works as a senior consultant in the PBL (Problem-Based-Learning) Teaching Innovation and Research Center at Northeastern University (NEU), China. Chunfang locates her research in Science, Technology and Society (STS), with a focus on creativity and STEM education. Since 2013, Chunfang has authored and co-authored over 110 peer-reviewed publications; she is the editor of *Handbook of Research on Creative Problem-Solving Skill Development in Higher Education* (IGI Global, 2016), the author of *Introducing Problem-Based Learning for Creativity and Innovation in Chinese Universities* (IGI Global, 2020), and an editor in the book series *Arts, Creativities and Learning Environments in Global Perspective* (Brill/Sense). Chunfang is an editorial board member of *Discovery Education* and the *European Journal of Education and Pedagogy*, as well as advisory board member for the *International Journal of Engineering Education*.

Tamara Carleton PhD, Universitetslektor in Mechanical Engineering, Blekinge Institute of Technology.

Tamara is an award-winning professor, industry advisor, and expert in radical innovation and strategic foresight. Recognised as one of the Top 50 Women Leaders in Education (in 2023 and 2024) by Women We Admire, she teaches at multiple business schools and technical universities around the world. In addition, she holds the UNESCO Chair in Anticipatory Leadership for Innovative and Better Futures at Tecnológico de Monterrey in Mexico. Tamara is also the founder and CEO of the Innovation Leadership Group, which helps leaders and teams to build innovation capacity and map bold futures. She has written several seminal books in innovation, including *Building Moonshots: 50+ Ways to Turn Radical Ideas into Reality* and the *Playbook for Strategic Foresight and Innovation*. She holds a doctorate in mechanical engineering from Stanford University.

Gabrielė Characiejienė, Research Assistant at the STEM Center for Research in Science Education and Communication (FNUG), Department of Mathematics and Computer Science, University of Southern Denmark.

Gabrielė holds an MA in Philosophy of Education, University College London, UK, and BA in Political Science, Vilnius University, Lithuania. She has worked at the Ministry of Education and Science of Lithuania, as a teacher in a secondary school, and for the most part, at an NGO Centre for School Improvement, supporting teachers, schools and policy on educational change. She was a Teach First Lithuania fellow and later the CEO of the programme. She has contributed to projects on educational leadership, civic education, and digital innovation in schools.

Chapter 2
Creative Pragmatics—Situated Performances of Knowing

Michael Shanks, Connie Svabo, Chunfang Zhou, and Tamara Carleton

2.1 Pragmatism and Situated Knowing

Pragmatism emphasises the practical application of knowledge and its continuous evolution through experience. It offers insights into how individuals interact with and make sense of the world through their actions and experiences and highlights the dynamic nature of human experience. Pragmatist educational philosophy extends these ideas to the field of education, emphasising the importance of active learning, experiential engagement, and the integration of theory and practice (consider Pickering, 2010; Putnam & Putnam, 2017; Rorty, 1982; Dewey, 1938). This perspective advocates for educational practices that are responsive to the complexities of real-world situations, promoting a learning approach that is both adaptive and reflective. The focus of education is shifted from learning programs that are abstract and disconnected to practical and contextualised understanding (Engeström & Cole, 2021; Eisner, 2002; Halverson & Sheridan, 2014; Lave & Wenger, 1991).

Key thinkers like Dewey, Putnam, and Eisner have influenced this perspective by advocating for an educational approach that is deeply intertwined with the learner's environment and experiences. As a natural extension of pragmatist educational philosophy, the concept of Creative Pragmatics favours a disposition that values

M. Shanks (✉)
Archaeology Center, Stanford University, Stanford, CA, USA
e-mail: mshanks@stanford.edu

C. Svabo · C. Zhou
STEM Centre for Research in Science Education and Communication (FNUG), Department of Mathematics and Computer Science, University of Southern Denmark, Odense, Denmark

T. Carleton
Department of Mechanical Engineering, Blekinge Institute of Technology, Karlskrona, Sweden

C. Svabo et al. (eds.), *Creative Pragmatics for Active Learning in STEM Education*, Contributions from Science Education Research 14,
https://doi.org/10.1007/978-3-031-78720-1_2

experiential learning, context-rich exploration, and adaptable problem-orientation (Kolb, 1984; Engeström & Cole, 2021; Lave & Wenger, 1991; Svabo, 2021). This extension embraces pragmatism's core ideals, integrating them with a profound respect for the interplay between creativity, practical utility, and the dynamic complexities of our ever-evolving world.

John Dewey's philosophical approach highlights the transactional relationship between individuals and their environment, emphasising the reciprocal nature of experience and action. Dewey's ideas resonate deeply with Creative Pragmatics, as they underscore the interconnectedness of learning, problem-solving, and adaptation. Dewey's transactionalism provides a theoretical anchor for Creative Pragmatics by emphasising the active, context-sensitive, and dynamic nature of human engagement with the world (Dewey, 1938). Creative Pragmatics, echoing this ethos of pragmatist educational philosophy, embodies a dynamic and adaptive approach to knowledge-building and problem-based inquiry.

This pragmatist perspective encourages a dynamic understanding of knowledge, recognising that learning is not a static acquisition of facts but a continuous, situated process of engagement and reinterpretation (Dewey, 1938; Putnam & Putnam, 2017; Rorty, 1982).

We propose Creative Pragmatics as a rich and multifaceted concept that draws from various philosophical, theoretical, and interdisciplinary fields. One point of departure is in pragmatist philosophy (especially in its manifestation in philosophy of education), which asserts that the value and meaning of knowledge, concepts and theories are best understood and evaluated in relation to their applications and outcomes (Peirce, 1903/1998; James, 1975; Dewey, 1999; Rorty, 1998).

2.2 Transdisciplinary Approach

Building on the pragmatist philosophy's emphasis on practical application and contextual understanding, Creative Pragmatics naturally extends to interdisciplinary and transdisciplinary engagement. This approach integrates knowledge across diverse fields to address complex challenges, promoting a holistic understanding.

Creative Pragmatics is both practical and philosophical, encompassing ontology, epistemology, and methodology in knowledge-making and knowledge-sharing. It involves the intertwining of research and creation, exploring scientific products and formats, and the expositions of knowledge in academia and research-creation (Chapman & Sawchuk, 2015). It also includes schizoid and impure knowledge making (Svabo, 2016), knowledge performances (Svabo & Bønnelycke, 2020) and the aesthetic and multimodal production of research (Svabo et al., 2020).

Creative Pragmatics forms part of a longstanding inquiry into both philosophical and practical aspects of thinking, learning, and knowing. The inquiry encompasses questions such as: How is knowledge developed, created, made, and produced? What roles do materialities, formats, and genres play in knowledge media and performances? How do knowledge media co-constitute knowledge claims? What are the rhetorical and communication design components of research, learning, schooling, and education? How can we conceive of schooling and education in process and relationally sensitive ways? How do universities and other higher education institutions reach out to and engage with their students or non-academic audiences?

These inquiries delve into how media engage in cognition, how materiality influences the formation of the mind, and how technologies shape processes of knowing, thinking, learning, education, and society (Law, 1993; Latour, 1987, 1998, 1999, 2007). They draw on insights from cognitive science, design, archaeology, and related fields, considering, for example, how people and computers collaborate in reasoning, how interactions with musical instruments create distributed pedagogical and caring systems, and how mind has evolved in an evolutionary perspective.

Such questions revolve around the entanglements of mind, media, and milieux, emphasising that learning is not confined to our minds alone; it is influenced by where, with whom, and what we learn. The people and things we learn with, along with the location and site, shape our understanding.

Science and technology studies (STS) have built empirically informed insights about science and technology in the making, highlighting the numerous forces and entities involved in the production of scientific facts, as well as the instrumentality and role of 'non-humans' in knowledge-making processes. From our perspective, these studies point towards a pragmatist philosophical tradition and methodology conceived as abduction, as well as towards process philosophy, mediation ontology (Latour, 2007; Serres & Latour, 1995; Serres, 1995; Stengers, 2023), ontological theatre (Pickering, 2010), and practice and performance (Pickering, 1995).

Incorporating insights from STS, Creative Pragmatics aligns with a pragmatic theory of education, connecting educational practices with scientific inquiry and exploration. It embraces interdisciplinary and transdisciplinary approaches, encouraging collaboration across diverse fields to address complex challenges and promote holistic understanding. Creative Pragmatics draws on multiple, overlapping fields and disciplines as a transdisciplinary approach. It assumes interventionist, creative, practice-based, and practice-oriented, design-based and design-oriented, participatory methodological approaches, drawing inspiration from ethnography, design thinking, systems thinking, and cybernetics. This approach has a special affinity for methodologies that interrogate, open up, and make visible the research process itself; in old Greek, '*meta hodos*' refers to the *path* to research (Shanks & Svabo, 2018).

2.3 Science and Technology Studies

A manifestation of pragmatist philosophy and transdisciplinarity may be seen in the intellectual tradition of science and technology studies, a field that illuminates how knowledge is socially, culturally, and historically constructed. Science and technology studies (STS) provides a valuable lens for understanding how knowledge is both constructed and applied, emphasising the performative nature of knowing.

How might we understand the ways that knowledge is acquired and how it relates to decision making and action? Since the 1970s, studies of science and technology have shown that research-based knowledge is very much something that is fabricated. Rather than a process of discovering and then representing the way the world works, building knowledge is complex productive work, requiring, among other things, institutional and project management, funding for research teams, instruments and facilities, and the rhetoric of publication and persuading others of the validity of claims to knowledge (Gibbons et al., 1994). This has been explored in great detail since Latour and Woolgar's ethnography of laboratory science back in the 1970s (Latour & Woolgar, 1979). It is also amply demonstrated in historical studies of science and technology such as the investigation by Steven Shapin of Robert Boyle's pneumatic experiments in the seventeenth century (Shapin, 1984). STS emphasises that knowledge is a performative act, where human and non-human actors interact to construct what is accepted as scientific fact. Coming to know is a social achievement involving efforts of stabilising and fabrication (Latour, 1987; Law, 1993). Knowing is rooted in local and specific circumstance (Haraway, 1988; Suchman, 1987) and demands competencies in context sensitivity. Processes of knowing are emergent, situated, negotiated and mediated highlighting the dynamic interplay between human and non-human actors (Latour, 1999, 2007).

Science and technology studies have established that knowledge is embedded and part of complex lifeworlds. We use the concept lifeworld here to emphasise the constant ongoing recreation of *all* aspects of experience, including experiences of non-human actors (Sayes, 2013). Reality as a whole is always 'in the making' (Pickering, 2010, with reference to William James, 1975; also Whitehead, 1929). Research and knowledge building, in this field of research, are performative practices (Kołtun, 2015; Bell, 2006).

A pragmatic perspective questions the understanding of knowledge building as one founded on the definition and representation of an object world separate from one who investigates, where the world is something which is fundamentally knowable, where reality is available for inspection and can be broken down into, reduced to, relationships of causality.

Reductive determinism is a specific and limited way of pursuing the building of knowledge (Wrigley, 2019). Understanding the world as a machine is one option among many of how to make sense of the working of the world. STS suggests that other options involve a pragmatic, performative, and non-representational model of knowledge building as practice, engagement, relationship, and intervention. For those familiar with the three modes of "knowledge production", as outlined by Gibbons and his colleagues in the early 1990s, what we are emphasising is similar to Mode 2 describing knowledge that is context-driven, problem-focused and interdisciplinary (Gibbons et al., 1994). In building knowledge we are embedded in, challenged by, unpredictable engagements with the world.

The emphasis on performance in STS aligns with the principles of Creative Pragmatics, which views knowledge not as a static commodity but as an active, ongoing process of engagement and adaptation. If we are to navigate its challenges, a complex lifeworld requires competencies and tactics for acting on the basis of insecurity and incomplete information. Tactics are needed for navigating in unpredictability and in complex processes of change. This kind of competency hinges less on communicating accounts of the world as basically knowable and predictable, and more on the ability to mobilise, infrastructure, assess, reconcile and intervene. By using the metaphor of navigation we invoke more dynamic, polyphonous, processual and relational realities (Serres, 1995). STS supports Creative Pragmatics by illustrating how knowledge creation is not merely about accumulating facts but is fundamentally about the dynamic interaction of ideas and actions within complex systems. The purpose is no longer how to pin down the world, but how to move with it. Some of these forms of movement are captured by verbs such as translating, displacing, smuggling, hacking and performing (Hughes, 2009). Navigation requires sensitivity to context, to environment in all senses, physical, social, cultural, emotional, imaginative. These are what Dewey (1938) called 'environing conditions'. In this respect knowledge is situated (Simonsen et al., 2014).

We might remind ourselves that knowing is not only about knowledge in its propositional and ostensive forms, but is as much about what, pragmatically, you can do with it in particular situations. Technical know-how, social and political savviness, and communication skills incorporate agency, situation and interaction, as recognised long ago in Aristotelian concepts of techne, phronesis, episteme, and sophia, forms of knowing appropriate to life in civic community. Rhetoric, the ancient art and science of persuasion, is part of this performative, pragmatic perspective, supplying the member of a community with the tools to make a case so as to persuade, to effect change, in particular circumstances of dialogue, directed at specific audiences, working alongside other members of the community (Gross, 2006).

The questions raised in this understanding of knowledge building are pragmatic. As well as "What is this and how is it defined?", one asks "What are the options in these particular circumstances?" "What is to be done?" "What might be done?" "For whom?" "On what basis?" "To what end?". These questions belong to transdisciplinary (non-disciplinary, located, transitive) knowledge that transgresses and transports, crosses and smuggles across borders between disciplines and between knowing and acting, between the knower and the known, between places where knowledge is built and applications beyond (Rorty, 1979; Hughes, 2009; Rogoff, 2006).

This focus on the pragmatics and the use value of knowledge building introduces hybridity and intermingling (Hård & Jamison, 2005). The forms in which knowledge is made manifest, such as theories, hypotheses, interpretations, representations, mediation and publications, are mixed up with mundane activities such as managing a lab or research centre, dealing with publishing houses. The scientist is inseparable from their instruments. The chemicals on the lab bench intermingle with the motivations and aspirations of their handlers, as well as the sandwiches they eat for lunch!

For some, and since Plato, this is literally monstrous, because mixing up those who seek to know with the knowledge they produce seems to be a fundamental challenge to the possibility of objective knowledge, free of human values and failings. Separating the knowing subject's mind and everyday life from the object world they seek to know and the forms that knowledge takes, such as representations of causal relations, is necessary to establish pure uncontaminated knowledge, according to such a paradigm (Lottholz, 2016).

Such epistemological and ontological hygiene is but one way of dealing with the messiness of building knowledge. Science and technology studies have introduced many concepts for capturing the messy, monstrous hybridity of knowledge such as actor networks (Latour, 2007), vital matter (Bennett, 2010), assemblages (DeLanda, 2016), cyborg affiliations (Haraway, 1990), the heterogeneous engineering of socio-technical systems (Callon, 1986; Law, 1993; Michael, 2006), the spatial-temporal contingencies of relativity and the entanglements of quantum phenomena (Barad, 2007) as well as the archaeology of things (Shanks, 1992; Olsen et al., 2012; Hodder, 2012).

A point made by Pickering (2010, p. 24) is that theory is not enough to build knowledge. The performance of knowledge is situated, embedded and located, pragmatic and adaptive, and fundamentally creative. Agency plays a crucial role in knowledge-making, transforming theoretical insights into practical actions and solutions. The performance of knowledge is situated, embedded and located, pragmatic and adaptive, and fundamentally creative. Agency is a key feature of knowledge making. Indeed Pickering suggests replacing knowledge with performance.

2.4 Knowing as Performing

Building on the insights from Science and technology studies (STS), the concept of performance is a key component of Creative Pragmatics. We do not limit performance to staged presentations. Performance is an aspect of cultural expression and social interaction, involving rehearsal and planning, iteration, and embodied enactment. Performance is located through formal and informal scenography or scene setting in sites, architectures, environments, and mobilises dramaturgical elements such as scenarios, plots and relationships, incorporating props, artefacts, choreography. This performative nature of knowledge aligns with Creative Pragmatics by encouraging learners to engage with real-world situations, adapt to changing conditions, and co-create solutions with others. Performance can be both reflective, representing or critiquing existing social realities, and constructive, creating new realities or envisioning alternative possibilities.

In its performative character Creative Pragmatics finds foundation in certain standpoints in science and technology studies that can be described as pragmatist and offers a more-than-representational, dynamic, and creative reconfiguration of how research and science work (see Table 2.1).

By adopting a performative understanding of knowledge, Creative Pragmatics encourages learners to become active participants in their educational journey. They are not merely consumers of information but co-creators of knowledge, capable of navigating the complexities of the modern world with creativity and adaptability.

Table 2.1 Understanding science's relationship with the world. This table illustrates contrasting perspectives on the relationship between science and the world, presenting key differences in the representational and performative approaches. The representational view envisions science as an endeavour to depict and mirror nature, emphasising the generation of knowledge aligned with an objective reality. In contrast, the performative stance views science as a dynamic realm of actions and material interactions, suggesting a shift from a passive reflection of reality to an active engagement with the ongoing processes and symmetrical relations inherent in science, technology, and society. (Based on Pickering, The Mangle of Practice, 1995)

Description	Representational	Performative
Nature of science	Science depicts and reflects nature	Science is a realm of actions and performances, grounded in capturing material agency through machines
Purpose of scientific inquiry	Science aims to generate knowledge reflecting reality	Science involves representation and conceptual dimensions, contributing to the production of scientific knowledge
Perception of the world in scientific context	The world is observed through disembodied intellects	The world is a constant realm of activity and forces, where material beings and forces interact
View of science, technology, and society as fields	Science, technology, and society are separate spheres	Science (and technology and society) constitute arenas of both human and nonhuman (material) agency, characterised by symmetrical relations

This approach aligns with the core principles of Creative Pragmatics, which seek to empower individuals to engage with and transform their environments through meaningful and impactful actions.

2.5 Agency

The discussion of agency flows naturally from the focus on performance. In Creative Pragmatics, agency plays a critical role as an extension of performative knowing. We are considering what kind of knowledge, skills and competencies are needed for navigating complexity. Agency, in this context, refers to the capacity of individuals and groups to act purposefully and creatively within complex systems. It highlights the ability to shape and transform one's environment through informed and intentional action.

We follow the field of science and technology studies and the case that knowledge is performance and achievement, and we describe it as creative. In calling it creative, we are not saying that knowledge makes up or invents the world as a kind of fiction relative to those who build knowledge. Instead we are following the line that knowledge and know-how are achievements, creative achievements. Agency, therefore, is the driving force behind these achievements enabling individuals to navigate and negotiate their environments effectively. Knowledge and know-how are made, constructed, built, engineered by people, by social actors or agents, through their embedded and situated engagements and enactments, working with the world they seek to know, and this embeddedness, these engagements include non-human agents and actors (Latour, 1999, 2007; Rathje et al., 2012).

In Creative Pragmatics, agency is closely linked to creativity. This connection highlights the capacity for innovation and adaption in response to changing circumstances. By fostering agency, Creative Pragmatics empowers individuals to engage with complex problems creatively, opening pathways for new ideas and solutions.

In order to explain what we mean, let's explore the question—How is the knower immersed in society, culture, lifeworld? The question might assume a distinction between (social) actors and the "context", such as society, within which they live, work and act. We wish to avoid such a distinction because it makes it difficult to understand where the context, society, comes from, how it is created. Instead we use the concept of agency to connect the individual social actor (or agent) and society (Meyer & Jepperson, 2000).

We draw an important distinction between ostensive and performative concepts of society (Baluch, 2006). The first is where you can point at and define society as something within which people live. An ostensive definition of society involves concepts such as social structure and institutions, norms and values that people have to deal with. Here people and society, knowers and what they seek to know, are kept separate.

A performative understanding holds that society is under constant re-creation. This involves people, as agents, and also all the activity, the energy exchanges with and between heterogeneous assemblages of actors or "actants", a concept that can include both human and non-human actors. Agency (or efficacy) describes the capacity of actants to make a difference to the world they inhabit. This aligns with Creative Pragmatics, which views agency as integral to the dynamic and adaptive processes of knowledge creation. Agency denies the distinction between the micro and the macro, people and society, the psychological and behavioural, and the institutional and the cultural. Society and culture (if we choose to retain the concepts) are emergent, and are not the "context" within which people live (and so neither is "environment") (Latour, 1999, 2007).

The link between agency and creativity in Creative Pragmatics emphasises the importance of innovation and adaptability in navigating the complexities of modern life. Research fields such as ethnomethodology (after Harold Garfinkel, 2022), a key foundation for STS, hold that we are thrown *in medias res*, into what gets to be called society. We are not "in society". We make it, and may change it in our engagements with each other and with things, in our projects and performances, through agency. This is what we mean by creativity—the constant remaking of lifeworld through the agency of social actors or actants. Creative agency implicates all aspects of worldbuilding.

Karen Barad (2007) has made such a case for an ontology and epistemology she calls "agential realism". Agency and power are implicated, because the social lifeworld is constructed, created in, emerges from micro performances. Agency here refers to people's capacities, those of an actor or agent to make things happen, to realise goals and intentions, acknowledged and conscious or not. Implicated also, it is to be noted, is the agency of non-human things.

Agency is intimately associated with making. If complex sociality requires constant probing, exploring, monitoring, much effort is put into maintaining adequate knowledge that one may act appropriately, effectively.

We introduce the distinction between complexity and complication in Chap. 3, Sect. 3.1. Strum and Latour (1987) contrast the complexity of baboon society with human effort directed at reducing complexity to complication in order to bring predictability. Baboons need to constantly monitor and negotiate their lifeworld (knowledge building and agency). Human society has come to find ways of reducing the need to monitor and negotiate the experiences of a complex lifeworld by *delegating* such work to artefacts, material and immaterial. If we follow the likes of Don Norman (2013), we can say that making, material culture, technology, and other systems of artefacts are means of storing and mobilising information and knowledge.

Latour and his actor-network theory (ANT) colleagues give many examples of delegating work and agency to things. Material culture, making and design are key components of what it is to be human because they enable human sociality and are a means of managing complexity. This means that materiality, making, technology, and design are inseparable from the social and cultural. Human being has always been cyborg (Clark, 2003), augmented, prosthetic, distributed through creative experiences of world building, world knowing (Haraway, 1991).

While one may treat it as an individual matter of one's power to get things done, or not, agency is primarily distributed and collaborative co-creation, collectively building lifeworld, with and through others.

Agency may be arrested and curtailed. This is a necessary feature of the distributed, collaborative, pragmatic and located character of worldbuilding. There are always constraints. Insecurity, uncertainty and an incapacity to act effectively to desired ends is certainly a contemporary experience. Of course all sorts of powers may be mobilised to direct change or to maintain continuity, to marginalise some, promote others. Consider, in contrast, the features of creative agency. These will help us see the features of Creative Pragmatics, dealt with in the next section.

2.6 Creativity

Building on the concepts of agency and performance, creativity emerges as an important force in Creative Pragmatics. Creativity is the ability to produce new ideas, approaches and models of thinking and doing (Sternberg, 1998). In Bloom's revised taxonomy (Anderson & Krathwohl, 2001), 'create' is the highest achievable level of cognitive process and educational objective. In the taxonomy, 'create' includes processes 'design, assemble, construct, conjecture, develop, formulate, author and investigate'. According to the taxonomy, being able to act creatively within a subject or discipline reflects the highest attainable level of understanding or knowing. Within Creative Pragmatics, creativity is both a process and an outcome, essential for navigating complex and unpredictable environments. Producing something new or original is an integral part of creativity and it is an important part of Creative Pragmatics.

An implication of 'creative' being at the top of Bloom's revised taxonomy is that Creative Pragmatics also has something to offer to goals of building deep subject- or discipline-specific understanding. This connection underscores the importance of fostering creativity in education, where students learn to integrate knowledge across disciplines and think beyond conventional boundaries.

To create is a capacity which potentially is played out in practically all spheres of life and across the whole life span. Creativity plays a key role in entrepreneurship and innovation. Creativity contributes to physical and psychological health. Creativity is at the centre of societal improvement. Creativity is about ideation, invention, innovation, lust, love and life (Martin & Wilson, 2018). Thus, creativity transcends artistic expression and becomes a driving force in cultural and societal evolution, enabling rapid adaptations across generations. Creativity, hence, is not only about the arts. Creativity is a driving force in cultural and societal evolution which is seen in rapid adaptations taking place from one generation to the next. Creativity is crucial in both reactive, problem solving activities and in proactive, inventive processes (Runco, 2004).

Creative agency tends always to be on the move in constant learning, adapting, exploring in order that one might act. Creativity is an effective and affective response to volatility, unpredictability and change, both at an individual, organisational and societal level, and creativity is in itself also a source for change (Dunne & Dougherty, 2012). In this way, creativity serves as both a catalyst for and a response to change, blurring the lines between continuity and innovation. An insight here is that innovation/change and continuity are not binary opposites, but aspects of the same process of creative agency which contains the potential of both change and continuity.

According to Mark Runco, creativity is a useful and effective response to evolutionary changes, be they classified as cultural, epigenetic or biological (Runco, 2004, p. 658). The most obvious reason for this is that creativity is related to processes of dealing with problems and finding ways to tackle circumstances and situations. A key feature of creative agency here is adaptability, monitoring, being sensitive to circumstance in order that one might (re)act.

Lifeworld pre-exists and is simultaneously reproduced or changed in our creative agency (Elder Jr, 1994). Agency is associated with inherited options and constraints. We act within conditions that are inherited. Though we are path dependent, in debt to the past, there are no set evolutionary pathways. We learn, adapt, and can change (Harris, 2021; Caldwell, 2006).

Creativity is tied to flexibility (Zabelina & Robinson, 2010), both at individual and collective levels, and this flexibility is fundamental for the capacity to cope with advances, opportunities, disruptions and breakdowns, in short, the changes that are part of our contemporary lives (Guilford, 1973).

Rather than focusing solely on the psychology of individual creativity, Creative Pragmatics emphasises the creative capacity to act collectively. We are less interested in the psychology of individual creativity (consider the range of approaches in Kaufman & Sternberg, 2019). We want to foreground the creative capacity to act (see the classic treatment of agency in Giddens, 1986). In the dynamism of environment and social situation, creativity is a crucial capacity, disposition and mode of engagement. How might one act under these circumstances? What can you make out of things? We draw attention to the root semantics of creativity—to bring something into being. The etymology includes the Latin *creare*—to make, produce, bring forth.

Creativity is not only the reactive mental processing that helps people, and other animals, deal with problems or other bundles of circumstances that by design, chance or accident arrive in their lives. Creativity itself is also a source for change. Creativity is a significantly proactive force. It is closely related to play and to other forms of autotelic activity, including aesthetic and erotic expression, as well as games, highlighting its role in fostering exploration and innovation (Pearce, 2006; Sondergard, 1997).

Fundamentally, creativity has to do with generative forces of life (Bergson, 1911; Whitehead, 1929). It is a competence that can be nurtured and developed through intentional practices and educational approaches. Creativity is something you can learn.

2.7 Creativity, Agency, and Learning from Art and Design

Building on the interplay between creativity and agency, the influence of art and design further enrichens the scope of Creative Pragmatics in fostering innovative thinking and problem-exploration. Creativity and agency are essential components of education, equipping students to face the uncertainties of a rapidly changing world (Schnapp & Shanks, 2009). Creative Pragmatics fosters these qualities by encouraging students to take an active role in their learning, empowering them to think critically, explore new ideas, and engage with their environments. This approach helps students develop the confidence and skills needed to navigate uncertainty and complexity, enabling them to become proactive agents who can adapt to new situations and challenges.

Art and design disciplines emphasise the processes of creativity and agency as central to their practice. Art and design thinking play a crucial role in shaping Creative Pragmatics. These disciplines emphasise flexibility, experimentation, and open-ended problem engagement. By integrating these methodologies, students learn to approach problems creatively and collaboratively, developing the ability to adapt to changing conditions and co-create solutions with others.

Drawing on John Dewey's recognition of the educational value of aesthetic experience (Dewey, 2008a, b), late Stanford professor of education Elliott Eisner (2002) emphasised the significance of creative, aesthetic activity in countering mechanised and hyper-rationalised education paradigms. Artistic engagement, encompassing aesthetic production and knowledge practices, encourages a responsive interaction with various forces at play. This heuristic flexibility aligns with creativity's essence and proves essential for effectively navigating complexity (Dewey, 1938; Eisner, 2003; Shanks & Svabo, 2018).

Eisner highlights the concept of "flexible purposing," (Eisner, 2002) inspired by Dewey's idea, as a key aspect of artistry's contribution to learning. This involves learners engaging in artistic activities to develop an understanding of adaptability in their pursuits. Through these activities, learners grasp the fluidity of goals and the emergent nature of work, where materials, tools, and collaborators offer cues for exploration. This process mirrors the unpredictability and self-organisation inherent in dynamic systems. This mindset aligns with the goals of Creative Pragmatics, encouraging students to view challenges as opportunities for sustainable development and innovation.

Flexibility and experimentation are inherent in art and design practice, mirroring the adaptable nature of Creative Pragmatics. Working artistically or in a design-based manner involves 'playing around', coming up with ideas and possibilities, and 'trying things out': "in expert design practice, the design problem is not fixed before the search begins for a satisfactory solution concept. Expert design is more a matter of developing and refining both the formulation of a problem and ideas for a solution in concert, in a process of 'coevolution'" (Dorst, 2015, p. 24). Notably, this process defies a linear approach, as problems and solutions are intricately interconnected. Iterative exploration and flexible engagement with open-ended possibilities are defining characteristics of design practice (Parsons, 2015).

Particularly worth highlighting here is the temporality of this process: it is not sequential, where you first define a problem and then find the solution. On the contrary, the problem-and-solution space involves iterative cycles of exploration, problem definition, and solution trials, where the design problem or challenge remains dynamic and interlinked with the search for solutions (Vermaas et al., 2007). This approach mirrors the iterative and emergent nature of creative problem-solving within Creative Pragmatics.

Importantly, design practice acknowledges that problems are never definitively resolved; rather, they lead to ongoing processes of innovation. There is no absolute security and predictability to the efficiency and ultimate desirability of designs. Solutions often give rise to unforeseen challenges, necessitating strategies like infrastructuring, mitigation, and negotiation that prioritise adaptive and relational interventions (Simon, 1987/1995).

The concept of "infrastructuring" is central to Creative Pragmatics, reflecting a commitment to dynamic, interconnected systems (Bjögvinsson et al., 2012, p. 108; also Karasti, 2014; Star, 1999). Infrastructuring involves interest-driven and relational interventions, rather than absolute, abstract definitions and standpoints, definitive positions taken on problem and solution definition. The competencies in infrastructuring are creating, facilitating and supporting collective processes of sense-making and engagement (Selloni, 2017). In summary, infrastructuring is about building adaptable, sustainable, and participant-centred infrastructures that evolve over time in response to changing needs and contexts, aligning with the flexible and adaptive nature of Creative Pragmatics.

Design is situated, context-sensitive and is about mobilising engagement. In interpersonal issues this, for example, implies working for dialogical processes of exploring and clarifying interests. Dorst talks about design practitioners broadening the "system border": "design contains a process of thinking around the paradox rather than confronting it head-on" (Dorst, 2015, p. 26). This highlights the importance of broadening perspectives and thinking beyond conventional limits. This highlights the system-dimension of design: "over the last 50 years, knowledge of the human capacity for problem-solving has broadened from a narrow concentration on the individual to the study of collective problem-solving processes in organisations. And in the last 20 years, the rise of networked society has sparked interest in collective problem-solving that takes place outside the confines of a single organisation, in networks of stakeholders that can be spread throughout society. The need to address such open, complex, dynamic, and networked problems in society has led to a keen interest in how expert designers approach problems." (Dorst, 2015, p. 24).

Creative Pragmatics fosters a holistic understanding of complex challenges, emphasising interdisciplinary collaboration and innovative approaches. Design practice fosters collective sense-making and engagement, recognising the system-oriented nature of complex contemporary challenges. Expert designers embrace a paradigm that expands the "system border," accommodating the complexities of collective problem-solving in interconnected networks of stakeholders. This approach has gained prominence as society grapples with intricate, dynamic, and networked issues, prompting a reevaluation of design thinking in the context of contextual and systemic considerations (Dorst, 2015). This shift underscores the importance of Creative Pragmatics, supplementing traditional rational decision models with flexibility, adaptability, and an openness to emergent solutions.

Contemporary design practice is oriented towards the contextual and systemic: taking into consideration complex social, technical, and environmental systems. Among other things, this implies a shift: paradigms of problem-solving are increasingly accompanied by open-ended processes of infrastructuring (Bjögvinsson et al., 2012, p. 108). Models of rational decision-making are supplemented by models of flexibility and adaptability—key components of what we are calling Creative Pragmatics.

2.8 Creative Pragmatics and Wicked Scientists

In today's world, challenges are often complex and interconnected, requiring solutions that go beyond traditional disciplinary boundaries. Creative Pragmatics provides a framework for preparing students to become *wicked scientists*—individuals adept at tackling these multifaceted problems. This approach encourages students to draw on knowledge from multiple disciplines, fostering adaptability, creativity, and collaboration. By integrating diverse perspectives, students learn to address problems that lack straightforward solutions, developing skills essential in a rapidly changing world.

Kawa et al. (2021) emphasise the need to train students to handle wicked problems, which are complex, dynamic, and difficult to define. Examples of such problems include climate change, social inequality, and global pandemics. These wicked scientists possess deep disciplinary knowledge and the ability to collaborate across fields and engage with diverse stakeholders, including non-experts. Creative Pragmatics aligns with this vision by emphasising interdisciplinary education and developing relevant competencies, preparing students to contribute to sustainable development in a complex world.

Creative Pragmatics offers a framework for educating students to become wicked scientists by encouraging them to transcend traditional disciplinary boundaries and fostering creativity and agency to address interconnected challenges effectively. It integrates insights from science and technology studies and pragmatism, emphasises the performativity of knowledge and the active role of learners in shaping their educational experiences and societal roles.

This approach encourages students to transcend narrow specialisation and engage with diverse teams, recognising that the world's most pressing problems are not only scientific but also political. By cultivating these competencies, Creative Pragmatics empowers students to become leaders in the pursuit of sustainable and equitable solutions, addressing the grand challenges of our time with creativity, pragmatism, and a commitment to interdisciplinary collaboration.

For a deeper exploration of the competencies required to navigate complexity and uncertainty, and how Creative Pragmatics fosters the development of wicked scientists equipped to address these challenges, see Chap. 3. There, we expand on how educational policies, such as the European Qualifications Framework (EQF), and competence-based approaches in higher education are responding to the growing demands for transversal skills, interdisciplinary collaboration, and holistic problem-solving in an increasingly complex world.

2.9 Summary—Key Principles Underpinning Creative Pragmatics

Creative Pragmatics engages with the design of knowledge, learning and teaching; with the fabrication of facts (Latour, 1987), and with what some call 'knowledge engineering' (Kendal & Creen, 2007); with making sense through making. It is processual, always in negotiation, adapting with lifeworld, always provisional (MacKeracher, 2004).

Creative Pragmatics points to the significance of navigating the dynamics of open-ended systems. Informed by complexity thinking and our Creative Pragmatics stance, drawn from pragmatist philosophy, science and technology studies, and art and design, the key takeaways include a commitment to engaging with ambiguity, learning from diverse sources, and the synthesis of information. The modus operandi of Creative Pragmatics lies in actively shaping and influencing dynamic interplays, emphasising the situated and contingent nature of knowledge building.

After this discussion of situated performance of knowing we might summarise Creative Pragmatics as a set of five rules of thumb, shorthand guides to action.

- Take an adaptive approach. At its core, Creative Pragmatics is dynamic and adaptive. It embraces an ever-evolving, iterative approach to knowledge-building and problem-solving, recognising the fluidity of the learning process and the need for continuous adaptation.
- Be open in embracing contingency and location. Central principles of Creative Pragmatics emphasise openness to diverse perspectives, contingency in response to evolving contexts, and situatedness within specific environments. This acknowledgment of the multifaceted nature of knowledge encourages learners to engage with complex, real-world situations.
- Unpack knowledge as performance. Creative Pragmatics embodies a performative understanding of knowledge generation, rooted in pragmatism. It goes beyond passive representation and prioritises functional utility and practical mobilisation, highlighting the active role of knowledge in shaping and transforming both individual and collective experiences.
- Be interdisciplinary. Drawing inspiration from pragmatist educational philosophy, science and technology studies, and learning from art and design, Creative Pragmatics thrives on interdisciplinary engagement. It integrates insights from various fields to enrich the learning experience and promote a holistic understanding of complex phenomena.
- Be sensitive to transactional relationships. Creative Pragmatics is anchored in Dewey's concept of transactionalism, which emphasises the reciprocal relationship between individuals and their environment. This principle highlights the contextual sensitivity of knowledge-building and problem-solving, emphasising the continuous interplay between learning, action, and sense-making.

Translated into policy and pedagogy these principles offer coordinated and grounded orientations and guidance, as made manifest in the rest of our book. We will return to overviews of Creative Pragmatics in our final Chap. 13.

References

Anderson, L. W., & Krathwohl, D. R. (Eds.). (2001). *A taxonomy for learning, teaching, and assessing: A revision of Bloom's taxonomy of educational objectives*. Longman.

Baluch, W. (2006). *Performing the meaning*. Wydawnictwo UJ. https://doi.org/10.4467/K8742.168/05.06.1564

Barad, K. M. (2007). *Meeting the universe halfway: Quantum physics and the entanglement of matter and meaning*. Duke University Press. https://doi.org/10.5334/opt.081013

Bell, V. (2006). Performative knowledge. *Theory, Culture & Society, 23*(2-3), 214–217. Sage. https://doi.org/10.1177/026327640602300245

Bennett, J. (2010). *A political ecology of things*. Duke University Press. https://doi.org/10.2307/j.ctv111jh6w

Bergson, H. (1911). *Creative evolution* (A. Mitchell, trans.). Henry Holt and Company.

Björgvinsson, E., Ehn, P., & Hillgren, P.-A. (2012). Design things and design thinking: Contemporary participatory design challenges. *Design Issues, 28*(3). https://doi.org/10.1162/DESI_a_00165

Caldwell, R. (2006). *Agency and change: Rethinking change agency in organizations*. Routledge. https://doi.org/10.4324/9780203087565

Callon, M. (1986). Some elements of a sociology of translation: Domestication of the scallops and the fishermen of St Brieuc Bay. In *Power, action and belief: A new sociology of knowledge?* (Vol. 32(1), pp. 196–223). Routledge. https://doi.org/10.1111/j.1467-954X.1984.tb00113.x

Chapman, O., & Sawchuk, K. (2015). Creation-as-research: Critical making in complex environments. *RACAR: Revue d'art Canadienne, 40*(1). https://doi.org/10.7202/1032753ar

Clark, A. (2003). *Natural-born cyborgs: Minds, technologies, and the future of human intelligence*. Oxford University Press. https://doi.org/10.1177/106480460401200312

DeLanda, M. (2016). *Assemblage theory*. Edinburgh University Press. https://doi.org/10.1515/9781474413640

Dewey, J. (1938). *Experience and education*. Palgrave Macmillan.

Dewey, J. (1999). *The essential Dewey* (two volumes edited by Hickman, L. and Alexander, T.). Indiana University Press.

Dewey, J. (2008a). Art as experience. In J. Boydston (Ed.), *The later works of John Dewey, 1925–1953* (Vol. 10, 1934). Southern Illinois University Press.

Dewey, J. (2008b). The philosophy of the arts. In J. Boydston (Ed.), *The later works of John Dewey, 1925–1953* (Vol. 13: 1938–1939, pp. 357–368). Southern Illinois University Press.

Dorst, K. (2015). Frame creation and design in the expanded field. *She Ji: The Journal of Design, Economics, and Innovation, 1*(1), 22–33. Elsevier. https://doi.org/10.1016/j.sheji.2015.07.003

Dunne, D. D., & Dougherty, D. (2012). Organizing for change, innovation, and creativity. In M. M. Mumford (Ed.), *Handbook of organizational creativity* (pp. 569–583). Academic. https://doi.org/10.1016/C2009-0-01842-X

Eisner, E. W. (2002). *The arts and the creation of mind*. Yale University Press.

Eisner, E. W. (2003). Artistry in education. *Scandinavian Journal of Educational Research, 47*(3), 373–384. https://doi.org/10.1080/00313830308603

Elder, G. H., Jr. (1994). Time, human agency, and social change: Perspectives on the life course. *Social Psychology Quarterly, 57*(1), 4–15. https://doi.org/10.2307/2786971

Engeström, Y., & Cole, M. (2021). Situated cognition in search of an agenda. In *Situated cognition* (pp. 301–309). Routledge. https://doi.org/10.4324/9781003064121-12

Garfinkel, H. (2022). *Harold Garfinkel: Studies of work in the sciences* (M. Lynch, Ed.). Routledge. https://doi.org/10.4324/9781003172611.

Gibbons, M., Limoges, C., Nowotny, H., Schwartzman, S., Scott, P., & Trow, M. (1994). The new production of knowledge. In *The dynamics of science and research in contemporary societies*. Sage. https://doi.org/10.4135/9781446221853

Giddens, A. (1986). *The constitution of society outline of the theory of structuration*. Polity Press. https://doi.org/10.1086/228358

Gross, A. G. (2006). *Starring the text: The place of rhetoric in science studies*. Southern Illinois University Press.

Guilford, J. P. (1973). *Characteristics of creativity*. Illinois State Office of the Superintendent of Public Instruction, Gifted Children Section.

Halverson, E. R., & Sheridan, K. (2014). The maker movement in education. *Harvard Educational Review, 84*(4), 495–504. https://doi.org/10.17763/haer.84.4.34j1g68140382063

Haraway, D. (1988). Situated Knowledges: The science question in feminism and the privilege of partial perspective. *Feminist Studies, 14*(3), 575–599. https://doi.org/10.2307/3178066

Haraway, D. (1990). *Simians, cyborgs, and women: The reinvention of nature*. Routledge. https://doi.org/10.4324/9780203873106

Haraway, D. J. (1991). A cyborg manifesto: Science, technology, and socialist-feminism in the late twentieth century. In *Simians, cyborgs, and women: The reinvention of nature* (pp. 149–181). Routledge. https://doi.org/10.1007/978-1-4020-3803-7_4
Hård, M., & Jamison, A. (2005). *Hubris and hybrids: A cultural history of technology and science*. Routledge. https://doi.org/10.4324/9781315024400
Harris, D. (2021). *Creative agency*. Palgrave Macmillan. https://doi.org/10.1007/978-3-030-77434-9
Hodder, I. (2012). *Entangled: An archaeology of the relationships between humans and things*. Wiley-Blackwell. https://doi.org/10.1002/9781118241912
Hughes, R. (2009). The art of displacement: Designing experiential systems and transverse epistemologies as conceptual criticism. In *Footprint, 4, Agency in Architecture: Reframing criticality in theory and practice* (pp. 49–57). https://doi.org/10.7480/footprint.3.1.699
James, W. (1975). *The meaning of truth*. Harvard University Press.
Karasti, H. (2014, October). Infrastructuring in participatory design. *Proceedings of the 13th Participatory Design Conference: Research Papers, 1*, 141–150.
Kaufman, J. C., & Sternberg, R. J. (2019). *The Cambridge handbook of creativity*. Cambridge University Press. https://doi.org/10.1017/9781316979839
Kawa, N. C., Arceño, M. A., Goeckner, R., Hunter, C. E., Rhue, S. J., Scaggs, S. A., Biwer, M. E., Downey, S. S., Field, J. S., Gremillion, K., McCorriston, J., Willow, A., Newton, E., & Moritz, M. (2021). Training wicked scientists for a world of wicked problems. *Humanities and Social Sciences Communications, 8*(1), 189. https://doi.org/10.1057/s41599-021-00871-1
Kendal, S. L., & Creen, M. (2007). *An introduction to knowledge engineering* (pp. 1–25). Springer. https://doi.org/10.1007/978-1-84628-667-4
Kolb, D. A. (1984). *Experiential learning: Experience as the source of learning and development*. Prentice Hall.
Kołtun, A. (2015). *Can knowledge be (a) performative? Performativity in the studies of science*. Wydawnictwo UMCS. https://doi.org/10.13140/RG.2.2.13607.06562
Latour, B. (1987). *Science in action: How to follow scientists and engineers through society*. Harvard University Press. https://doi.org/10.1086/289690
Latour, B. (1998). A few steps toward an anthropology of the iconoclastic gesture. *Science in Context, 10*(1), 63–83.
Latour, B. (1999). On recalling ANT. *The Sociological Review, 47*(1_suppl), 15–25. https://doi.org/10.1111/j.1467-954X.1999.tb03480.x
Latour, B. (2007). *Reassembling the social: An introduction to actor-network-theory*. Oxford University Press.
Latour, B., & Woolgar. (1979). *Laboratory life: The social construction of scientific facts*. Sage. https://doi.org/10.2307/j.ctt32bbxc
Lave, J., & Wenger, E. (1991). *Situated learning: Legitimate peripheral participation*. Cambridge university press.
Law, J. (1993). *Organising modernity: Social ordering and social theory* (1st ed.). Wiley-Blackwell.
Lottholz, P. (2016). Exploring the boundaries of knowledge via hybridity. *Journal of Intervention and Statebuilding, 10*(1), 136–142. https://doi.org/10.1080/17502977.2015.1095204
MacKeracher, D. (2004). *Making sense of adult learning*. University of Toronto Press.
Martin, L., & Wilson, N. (Eds.). (2018). *The Palgrave handbook of creativity at work*. Palgrave Macmillan. https://doi.org/10.1007/978-3-319-77350-6
Meyer, J. W., & Jepperson, R. L. (2000). The 'actors' of modern society: The cultural construction of social agency. *Sociological Theory, 18*(1), 100–120. https://doi.org/10.1111/0735-2751.00090
Michael, M. (2006). *Technoscience and everyday life*. Open University Press.
Norman, D. (2013). *The design of everyday things: Revised and expanded edition*. Basic books.
Olsen, B., Shanks, M., Webmoor, T., & Witmore, C. (2012). *Archaeology: The discipline of things*. University of California Press. https://doi.org/10.1525/9780520954007
Parsons, G. (2015). *The philosophy of design*. Wiley.
Pearce, C. (2006). Games as art: The aesthetics of play. *Visible Language, 40*(1), 66.

Peirce, C. S. (1903/1998). Chaps. 10–16: Harvard lectures on pragmatism. In *The essential Peirce, Vol. 2., The Peirce edition project*. Indiana University Press.
Pickering, A. (1995). *The mangle of practice: Time, agency, and science*. University of Chicago Press.
Pickering, A. (2010). *The cybernetic brain: Sketches of another future*. University of Chicago Press. https://doi.org/10.7208/chicago/9780226667928.001.0001
Putnam, H., & Putnam, R. A. (2017). *Pragmatism as a way of life: The lasting legacy of William James and John Dewey*. Harvard University Press.
Rathje, W. L., Shanks, M., & Witmore, C. (Eds.). (2012). *Archaeology in the making. Conversations through a discipline*. Routledge. https://doi.org/10.4324/9780203083475
Rogoff, I. (2006). 'Smuggling: An embodied criticality'. Retrieved from http://eipcp.net/transversal/0806/rogoff1/en
Rorty, R. (1979). *Philosophy and the mirror of nature*. Princeton University Press.
Rorty, R. (1982). *Consequences of pragmatism*. University of Minnesota Press.
Rorty, R. (1998). *Truth and progress*. Cambridge University Press.
Runco, M. A. (2004). Creativity. *Annual Review of Psychology, 55*(1), 657–687. https://doi.org/10.1146/annurev.psych.55.090902.141502
Sayes, E. (2013). Actor–network theory and methodology: Just what does it mean to say that nonhumans have agency? *Social Studies of Science, 44*(1), 134–149. https://doi.org/10.1177/0306312713511867
Schnapp, J. T., & Shanks, M. (2009). Artereality: Rethinking craft in a knowledge economy. In S. H. Madoff (Ed.), *Art School: (Propositions for the 21st century)* (pp. 141–159). MIT Press.
Selloni, D. (2017). Infrastructuring by design. In D. Selloni (Ed.), *CoDesign for public-interest services* (pp. 141–168). https://doi.org/10.1007/978-3-319-53243-1_9
Serres, M. (1995). *Angels, a modern myth*. Flammarion.
Serres, M., & Latour, B. (1995). *Conversations on science, culture, and time*. University of Michigan Press. https://doi.org/10.3998/mpub.9736
Shanks, M. (1992). *Experiencing the past: On the character of archaeology*. Routledge. https://doi.org/10.4324/9780203973639
Shanks, M., & Svabo, C. (2018). Scholartistry: Incorporating scholarship and art. *Journal of Problem Based Learning in Higher Education, 6*(1), 15–38. https://doi.org/10.5278/ojs.jpblhe.v6i1.1957
Shapin, S. (1984). Pump and circumstance: Robert Boyle's literary technology. *Social Studies of Science, 14*(4). https://doi.org/10.1177/030631284014004001
Simon, H. A. (1987/1995). Problem forming, problem finding, and problem solving in design. In Collen, A., & Gasparski, W. W. (Eds.), *Design and systems: General applications of methodology* (Vol. 3, pp. 245–257). Transaction publishers. (Text of a lecture delivered to the first international congress on planning and design theory, Boston, MA)
Simonsen, J., Svabo, C., Strandvad, S. M., Samson, K., Hertzum, M., & Hansen, O. E. (2014). *Situated design methods*. MIT Press.
Sondergard, S. (1997). Erotic creation and creative eroticism. *Semiotics*, 139–150. https://doi.org/10.5840/CPSEM199726
Star, S. L. (1999). The ethnography of infrastructure. *American Behavioral Scientist, 43*(3). https://doi.org/10.1177/00027649921955326
Stengers, I. (2023). *Making sense in common: A reading of Whitehead in times of collapse* (T. Lamarre, trans.). University of Minnesota Press.
Sternberg, R. J. (Ed.). (1998). *Handbook of creativity*. Cambridge University Press. https://doi.org/10.1017/CBO9780511807916
Strum, S., & Latour, B. (1987). The meanings of social: From baboons to humans. *Information Sur Les Sciences Sociales/Social Science Information, 26*, 783–802. https://doi.org/10.1177/053901887026004004
Suchman, L. A. (1987). *Plans and situated actions: The problem of human-machine communication*. Cambridge University Press. https://doi.org/10.2307/1423221

Svabo, C. (2016). Performative schizoid method: Performance as research. *PARtake: The Journal of Performance as Research, 1*(1), 1–21. Article 7. http://scholar.colorado.edu/partake/vol1/iss1/7/

Svabo, C. (2021). Living world dynamics—or: What Brian Eno can teach us about knowing in a complex world. In E. Brandt, T. Markussen, E. Berglund, G. Julier, & P. Linde (Eds.), *Proceedings of Nordes 2021: Matters of scale* (pp. 384–389). Nordes. https://doi.org/10.21606/nordes.2021.42

Svabo, C., & Bønnelycke, J. (2020). Knowledge catcher: On the performative agency of scholarly forms. *PARtake: The Journal of Performance as Research, 3*(1). https://doi.org/10.33011/partake.v3i1.477

Svabo, C., Pedersen, R. R., Arnfred, A. J., Bønnelycke, J., & Tosca, S. (2020). Hvad er et videnskabeligt produkt? In K. Eskildsen (Ed.), *Humaniora?: En spørgende videnskabsteori* (pp. 105–126). Samfundslitteratur.

Vermaas, P. E., Kroes, P. A., Light, A., & Moore, S. (Eds.). (2007). *Philosophy and design: From engineering to architecture*. Springer. https://doi.org/10.1007/978-1-4020-6591-0

Whitehead, A. N. (1929). *Process and reality: An essay in cosmology*. Macmillan.

Wrigley, T. (2019). The problem of reductionism in educational theory: Complexity, causality, values. *Power and Education, 11*(2), 145–162. https://doi.org/10.1177/1757743819845121

Zabelina, D. L., & Robinson, M. D. (2010). Creativity as flexible cognitive control. *Psychology of Aesthetics, Creativity, and the Arts, 4*(3), 136. https://doi.org/10.1037/a0017379

Michael Shanks PhD Doc, Professor of Classics and Archaeology, Stanford University.

Acknowledged as one of the most original and influential of contemporary archaeologists, Michael is a specialist in long-term humanistic views of design and creativity, innovation and social change, while exploring connections across the sciences and arts in a hybrid field of theatre/archaeology. He is senior faculty in Stanford University's programs in Science Technology Society, Urban Studies, Writing and Rhetoric, Archaeology, Classics, and in the Center for Design Research, part of Stanford's d.school in the School of Engineering. He directed Stanford Humanities Lab, pioneering innovative project-based pedagogy in the arts and humanities, and helped found the Revs Program at Stanford, connecting the archaeological automobile with contemporary car design. Founding faculty in Stanford's Archaeology Center, Michael is a director of Stanford's research group Foresight and Innovation, and, for two decades until its retirement in 2023, he worked with mediaX, the affiliates' program that connected Stanford with industry through technology and innovation. He served on the city of Rotterdam's International Advisory Board from 2008 to 2019 and works with many companies and organisations in developing cultures of innovation. In another mobilisation of Creative Pragmatics, Michael continues to pursue archaeological fieldwork in the northern borders of the Roman empire and beyond.

Connie Svabo PhD, Professor of STEM Education and Science Communication, Director of STEM Center for Research in Science Education and Communication (FNUG), Department of Mathematics and Computer Science, University of Southern Denmark.

Connie is a transdisciplinary scholar and creative practitioner with expertise in learning environment design and experience. She is the founding Director of Center for Research in Science Education and Communication (FNUG), Section Leader and member of the management team of the Department of Mathematics and Computer Science, chair of the SDU research programme in Science and Mathematics Education and Communication. She promotes Creative Pragmatics as a model for knowing in a complex world and underpins this stance with a practice that ranges from art over education to design and consultancy. Her portfolio includes geological UNESCO heritage sites Stevns Klint and the South Fyn Archipelago, as well as the Danish National Gallery Open Air Museum and the museum of natural history, Naturama. Previously a Continuing Studies Instructor at Stanford University and Faculty at Roskilde University, where she directed the research centre Experience Lab, chaired the study programmes in Communication Performance Design 2013–2016 and served on the academic board. She has published books on STEM education, sustainability, competence development, design pedagogy and digital life. She holds a PhD in Social Science from Roskilde University's multidisciplinary programme in Society, Space and Technology.

Chunfang Zhou PhD, Associate Professor of STEM Education and Science Communication, STEM Center for Research in Science Education and Communication (FNUG), Department of Mathematics and Computer Science, University of Southern Denmark.

A researcher and practitioner working on creativity, STEM education, and teacher education. Chunfang has an interdisciplinary and cross-cultural educational background: she finished a Bachelors Degree (Industry Automation and Information Science) and Masters Degree (Philosophy of Science and Technology) in China; during 2008–2012, Chunfang finished her doctoral research on *Group Creativity in Engineering Education in a Problem and Project-Based Learning (PBL) environment* at Aalborg University, Denmark. She is a board member of Danish Development Research Network (DDRN), and works as a senior consultant in the PBL (Problem-Based-Learning) Teaching Innovation and Research Center at Northeastern University (NEU), China. Chunfang locates her research in Science, Technology and Society (STS), with a focus on creativity and STEM education. Since 2013, Chunfang has authored and co-authored over 110 peer-reviewed publications; she is the editor of *Handbook of Research on Creative Problem-Solving Skill Development in Higher Education* (IGI Global, 2016), the author of *Introducing Problem-Based Learning for Creativity and Innovation in Chinese Universities* (IGI Global, 2020), and an editor in the book series *Arts, Creativities and Learning Environments in Global Perspective* (Brill/Sense). Chunfang is an editorial board member of *Discovery Education* and the *European Journal of Education and Pedagogy*, as well as advisory board member for the *International Journal of Engineering Education.*

Tamara Carleton PhD, Universitetslektor in Mechanical Engineering, Blekinge Institute of Technology.

Tamara is an award-winning professor, industry advisor, and expert in radical innovation and strategic foresight. Recognised as one of the Top 50 Women Leaders in Education (in 2023 and 2024) by Women We Admire, she teaches at multiple business schools and technical universities around the world. In addition, she holds the UNESCO Chair in Anticipatory Leadership for Innovative and Better Futures at Tecnológico de Monterrey in Mexico. Tamara is also the founder and CEO of the Innovation Leadership Group, which helps leaders and teams to build innovation capacity and map bold futures. She has written several seminal books in innovation, including *Building Moonshots: 50+ Ways to Turn Radical Ideas into Reality* and the *Playbook for Strategic Foresight and Innovation.* She holds a doctorate in mechanical engineering from Stanford University.

Chapter 3
Creative Pragmatics—Knowing and Competence in a Complex World

Connie Svabo, Michael Shanks, Gabriele Characiejiene, Chunfang Zhou, and Tamara Carleton

3.1 Complexity

Within multiple academic disciplines and throughout the past few decades, there has been a growing orientation towards complexity. Notably, British sociologist John Urry explored various implications of complexity theory in the social sciences, calling this orientation 'the complexity turn' (Urry, 2005). Similar orientations have taken place in education (Davis & Sumara, 2006). Complexity thinking can be traced through social theory, education and organisation studies, as well as physics, biology, mathematics, ecology, chemistry and economics. Across disciplines, the complexity turn features analyses that increasingly highlight complex interactions, systems and dynamics, using terms such as chaos, non-linearity and emergence, as well as complexity. To talk about complexity is to talk about heterogeneous interactions with emergent, dynamic, evolving, self-organising and irreducible features (Capra & Luisi, 2014).

Complexity is related to chaos theory, non-linearity and the study of dynamic adaptive systems, where properties or patterns may emerge spontaneously (Prigogine & Stengers, 1984; Prigogine, 1997). The discussion of complexity, especially in the context of ecosystems and networks, was further popularised in the early 2000s by

C. Svabo (✉) · G. Characiejiene · C. Zhou
STEM Centre for Research in Science Education and Communication (FNUG), Department of Mathematics and Computer Science, University of Southern Denmark, Odense, Denmark
e-mail: svabo@imada.sdu.dk; https://www.sdu.dk/en/forskning/stem

M. Shanks
Archaeology Center, Stanford University, Stanford, CA, USA

T. Carleton
Department of Mechanical Engineering, Blekinge Institute of Technology, Karlskrona, Sweden

C. Svabo et al. (eds.), *Creative Pragmatics for Active Learning in STEM Education*, Contributions from Science Education Research 14,
https://doi.org/10.1007/978-3-031-78720-1_3

American scholar Albert-László Barabási with his mainstream book *Linked: The New Science of Networks* (2002).

In the research literature, complexity typically refers to systems or networks with many movable and interdependent parts (Lu et al., 2015). If we trace the root meanings of the word, *complexity* is the braiding together of different components: *com*-means with or together, and *plek*- (Latin *plectere*) is to plait, braid, fold, weave, enfold, entangle, embrace and entwine. Complexity is the commingling of multiple entities and forces. This means that complexity as a concept is a semantic field that has extraordinarily wide valency.

As British professor of Education Tara Fenwick points out, a definition of complexity can be challenging:

> Complexity science, a highly heterogeneous domain, is impossible to describe accurately in a few sentences. However for purposes of this discussion, complexity can be represented as a radically holistic analysis that does not separate person from context, but shows how all things, (individuals, tools, technologies, ideas and environments) are continually brought forth in dynamic systems or 'assemblages' of 'vital materiality' (Bennett, 2010). These systems emerge in unpredictable ways through non-linear dynamics of mutual interaction and influence, producing a whole that is greater than the sum of its parts. (Fenwick, 2012, p. 14)

Or as education scholars Matthijs Koopmans and Dimitrios Stamovlasis consider:

> When we say complex, we mean that the behavior of a larger systemic constellation cannot be readily reduced to that of its individual members. Consequently, the perspective inspires a holistic view where the behavior of individuals is understood in its larger context. ... When we say dynamical, we mean that current behavior is understood in terms of deviations from past behavior. As a result, the perspective focuses on behavioral change and its determinants, rather than on outcomes frozen in time... When we say systems, we refer to a constellation of individual members who are in a position to interact with each other as a coherent entity. (Koopmans & Stamovlasis, 2016, p. 1–2)

Complexity deals in irreducible dynamics at different scales, in emergence and non-linear consequences. It is not possible to reduce complex interactions or the patterns and movements of complex systems to predictable processes or elementary laws. What happens in complexity is fundamentally unpredictable (Capra & Luisi, 2014; Davis & Sumara, 2006).

In his research, Urry posits that "chaos and order are to be seen as interconnected as large-scale systems move in and through time-space" (Urry, 2005). He believes that "Relationships between variables can be non-linear with abrupt switches occurring, so the same 'cause' can, in specific circumstances, produce different effects" (Urry, 2005).

This does not mean that there aren't patterns. There are patterns in complexity:

> The way that schools of fish move through the water and herds of ungulates roam the savannah provide vivid examples of how individual actions and interactions lead to large-scale patterns. (Yoon et al., 2019, p. 2)

Complexity is not chaos. Moreover, complex is not the same as complicated. A machine with many interacting components is complicated, but this does not mean it is necessarily complex. Complicated machinery and systems are typically designed to be predictable and controlled.

One way of dealing with complexity is to seek to reduce uncertainty by finding ways of building predictability and order into systems, adaptive or not, to find ways of treating what is complex as what is complicated (Strum & Latour, 1987). The distinction between complication and complexity highlights the importance of how actors perceive and act upon complex phenomena. We can understand complicated phenomena like machines by separating them from the complex environments and contexts within which they may be located, but as soon as complicated entities are given context, acting on and changing them can become complex and unpredictable.

A simple example of the distinction and its implications is an observation regarding the Challenger space shuttle disaster (McDonald, 2004). On January 28, 1986, the Space Shuttle Challenger broke apart catastrophically 73 seconds into its flight, killing all seven crew members aboard. The direct cause of the disaster was the failure of the O-ring seals in a joint in the shuttle's right solid rocket booster. While this was what caused the shuttle to break up, the inquiry blamed organisational culture and decision-making processes for failing to safely manage the launch, when it was known that such technical failures were possible in particular circumstances. Technology can deliver complicated machines like the shuttle; Newtonian math can take a machine to the moon. But neither can predict the failings of complex systems of project management.

3.2 The Limits of Reason When Navigating Complexity

It is critical to acknowledge certain features of the concept of complexity: complexity is irreducible to tightly defined and rationalised entities and processes of causation. Complexity can not be made predictable. Complexity cannot be known fully (Davis et al., 2008; Urry, 2005; Sawyer, 2005).

Modes of understanding which hinge on producing deterministic, predictive knowledge through rationalisation have been spectacularly successful as a foundation of (late) modernity, but they only work within tight limits, and there are costs. Focusing on causal relationships between defined variables, attempting to reduce complexity to complication (Urry, 2005) can facilitate control, but at the cost of separating knowledge from the circumstances in which it is fabricated and with which it engages (see in this context Keen, 2018; Morozov, 2014; Gigerenzer, 2022).

In the context of science education, this underscores the need for new models of knowing and competence. While deterministic and predictive knowledge has played a crucial role in scientific advancements, it tends to oversimplify the inherent complexities of natural phenomena. Science education must evolve beyond solely focusing on causal relationships and reductionist approaches, recognising the importance of embracing the features of real-world systems. A pedagogical shift towards holistic understanding, emphasising inquiry-based learning, problem-exploration, and the integration of diverse perspectives, is essential. This approach not only prepares students to navigate the complexities of modern scientific challenges but also ensures that knowledge remains connected to the lifeworld, fostering a more comprehensive and contextually relevant understanding of scientific concepts. Incorporating these principles into science education can contribute with the competence to tackle the dynamic and unpredictable nature of contemporary science and society.

In contrast to ostensive approaches, which treat knowledge as something static and representational, performative understandings emphasise that knowledge is actively constructed through engagement with the world. This distinction becomes crucial when navigating complexity, where rigid, representational models fall short. A performative approach, such as that proposed by Creative Pragmatics, foregrounds action, iteration, and responsiveness to emergent conditions, aligning with the unpredictable nature of complex systems.

3.3 Acknowledging the Uncertainties of Contemporary Scientific Understanding

As we outlined, attention to complexity has grown alongside the development of predictive sciences (Gershenson, 2013; Radford, 2008). Understanding thermodynamic systems and machines (steam engines and after), quantum phenomena, weather and turbulence, for example, has prompted recognition of distinctive features of complex systems. Questions are raised of the relationships between scales of phenomena, tiny to global events and patterns, in fractals, for example, and in some kinds of catastrophic change where small changes lead to massive change. As another example, there is the relationship of the sub-atomic quantum biology of perception and cognition in certain avian species to patterns of global migration (Al-Khalili & McFadden, 2014). Dynamic open systems, such as those involving heat and energy, are constantly adapting through interaction and feedback in specific and changing environments. Characterised by uncertainty around multiple possible trajectories of change and outcome, complexity is intimately associated with multiplicity (Law & Mol, 2002). Hence attention to complex phenomena has generated new kinds of mathematics that deal with multiplicity, nonlinearity, probabilities, chaos, interaction and feedback, changing parameters, information exchange, sensitivity and adaptation to change (Prigogine, 1997).

Strategies that seek to reduce complexity to complication may facilitate action based upon the predictions made possible by simplification and synthesis. It is quite possible to make predictions, but *only* in circumstances of reductive abstraction, when components of a complex system are factored out, and even then such circumstances are very often entangled in more complex and messier matters. If something is complex, there will always be limitations in our understanding of it. It is not possible to generate complete, absolute or final knowledge about complex interactions, only approximations. Again, Newtonian physics can send a spacecraft to the moon, but cannot anticipate the Challenger disaster.

3.4 Challenges and Limitations of Predictive Knowledge in a Complex World

There are four ways that predictive knowledge, built through reductive rationalisation, can fail to deal with complexity:

There are limitations to a mechanistic and rationalist view of our lifeworld. Science and technology can solve problems, yes, but the solutions only address a subset of the whole problem space, which, in a globally connected and complex world, may extend across the entire world in many different ways (Morton, 2013). Complex systems will always actively resist definition and prediction, because they are dynamic and constantly changing.

Again, Newton's laws of mechanics treat the world as a rationally organised and operating machine. They are the basis for designing extraordinarily complicated craft and systems that can take people to the moon. However, Newton's math cannot predict the Challenger disaster, because the machine is embedded in a complex lifeworld. This shows the challenges that arise when applying reductionist approaches to dynamic and complex systems. The contrast between the success of Newton's laws in certain contexts and their failure in predicting the Challenger disaster emphasises the need for a nuanced understanding that considers the intricate dynamics within complex systems.

Unrealistic expectations. Complexity thinking emphasises the limits to our capacity to predict the behaviour and events of complex systems. Complex systems are, by definition, unpredictable. It is unrealistic to expect to control and predict the ecological dynamics of an entire ecosystem, for example. Instead, we should seek ways of navigating the unpredictability and complexity of ecological interactions, recognising that factors such as climate change, species interactions, and environmental fluctuations contribute to a complex and dynamic ecosystem that cannot be fully predetermined.

Of course it is also *unrealistic* to expect Newton's equations and the engineers to predict the failure of a fuel tank seal in the actual project management that was launching the space shuttle. This shows the limitations of predictability even in well-defined engineering contexts.

Awareness. The industrial urban world has shrunk and speeded up in the last couple of centuries with the development of transport, mobility and communication systems. Science and technology are part of extraordinary time-space compression, as well as the capacity to generate and expend energy at a vast scale. Information and communication systems connect us in a global village (Hudson, 2013). The abundance of data may make us believe we can control and manage more aspects of our lives than ever, contributing to the unrealistic expectations just mentioned. But we are, by virtue of the same interconnectedness, aware of the many intersecting complex systems that are the core of the human experience today, aware of failures to manage and predict, institutional and individual, aware that we need to be better at living in complexity and that reductive knowledge is not enough.

This highlights the need to recognise the limitations of reductionist knowledge and improve our competence in navigating complexity.

Institutional inflexibility. National and international institutions of government and administration and corporate structures seem less and less able to handle the scale and complexity of contemporary challenges. Securing policy agreements to deal with climate change seems so difficult to achieve and implement (Gupta, 2010). Calls for more flexible and adaptive policy have been growing over the last few decades, alongside questions raised of the role of the state, market, entrepreneurial innovation and other agencies such as international corporations in addressing complex challenges. Another example is the automotive industry, where design cycles (typically counted in years) and investment in production lines have amounted to inflexibility with respect to changing modes of mobility (ride sharing, for example) and user expectations (that a new automobile will offer the same rapid innovations experienced in consumer electronics). This illustrates the difficulties institutions face in adapting to the scale and complexity of contemporary challenges, and emphasises the need for a reevaluation of the roles played by institutions, including the state, market, and international cooperations, in addressing complex issues.

In a world where complexity resists reductionist and mechanistic approaches, a research paradigm in Creative Pragmatics offers a compelling alternative. Rooted in pragmatist philosophy, and developed with inspiration from science and technology studies, and manifested in practice-based, arts-based, and design-based research, this paradigm emphasises action, iteration, and direct engagement with complex systems. It acknowledges that traditional predictive knowledge may fall short in dynamic environments, where rigid frameworks cannot account for unforeseen variables or emergent behaviours. Instead, work that falls under a creative pragmatic paradigm focuses on practice, engagement, and participation in "learning by doing," where knowledge is generated through active participation, experimentation, and reflection. This not only accommodates the unpredictability inherent in complex systems but also leverages it as a catalyst for innovation and adaptation, embracing the interventionist potential of research.

By engaging directly with the complexities of the world, Creative Pragmatics challenges the limitations of reductive knowledge and fosters a more adaptive, responsive form of inquiry. It encourages researchers and practitioners to embrace the fluidity of complex systems, iteratively refining their understanding through cycles of action and reflection. This approach aligns well with the need for more flexible, adaptive strategies in addressing contemporary challenges, such as those faced by institutions struggling to manage large-scale issues like climate change or technological disruption. In essence, Creative Pragmatics offers a dynamic framework for navigating complexity, one that is attuned to the ever-changing nature of the systems it seeks to understand and influence.

In recognising these challenges, it becomes important to explore alternative models of understanding and developing competence that align with the nature of complex systems. Complexity is increasingly acknowledged in innovation, management and organisation studies (Stacey, 1996; Weick, 1995), where complex, social, technical, mechanical, organisational and environmental issues are creating ambiguous boundaries between design professions and fields (Simon, 1947, 1987/1995) and making it necessary to generate new frameworks and approaches which mix theory and practice in transdisciplinary ways (Buchanan, 1992; Dorst, 2015; Hughes, 2009; Norman, 2013; for the social sciences see Byrne & Callaghan, 2013). Thus, we must ask: what kind of knowledge, skills, and competencies are needed for navigating complexity?

In response to complexity, relying solely on reductive knowledge—which focuses on identifying causal processes to make predictions—is insufficient. Since complexity, by its nature, thwarts exhaustive understanding, the way to handle complexity is pragmatic, through action, iteratively and creatively: in simple terms, learning by doing and by iterative and dynamic sensemaking.

Let's move to consider what kind of knowledge, skills and competencies are needed for navigating complexity.

3.5 Navigating Complexity: The Necessity for New Competencies

As the previous sections have discussed, complexity is an inherent feature of the modern world, whether in ecosystems, social structures, or technological networks. The unpredictable, emergent, and dynamic nature of complex systems challenges the traditional reliance on rational, linear thinking and reductionist approaches. While these methods have been highly effective within certain scientific and technological domains, they fall short when faced with the intricacies and uncertainties of real-world phenomena.

The limits of reason, particularly in attempting to predict or control complex systems, demand a new kind of knower—one who can navigate uncertainty, adapt to unpredictable conditions, and act creatively in the face of ambiguity. Graduates and professionals today must have not only deep disciplinary knowledge but also the ability to think holistically, being able to integrate diverse perspectives, and develop innovative approaches to ill-defined problems. They must become what could be called wicked scientists—individuals who can tackle "wicked problems" that are characterised by incomplete knowledge, conflicting stakeholders, and constantly shifting variables.

In light of complexity thinking and drawing from the European Qualifications Framework (EQF) specification of envisioned learning outcomes, STEM education should be designed to balance the pursuit of highly specialised knowledge, encompassing cutting-edge insights in specific fields, with understanding the interfaces between different fields and promoting a holistic comprehension of complex systems. Cultivating responsibility and autonomy is also crucial, enabling students to manage and transform work or study contexts characterised by complexity and unpredictability.

Key skills and competencies for navigating the complexities of the twenty-first century include problem-solving, leadership, adaptability, teamwork, and systems thinking (Framework for 21st Century Learning, 2019; Risopoulos-Pichler et al., 2020; Latif & Khatun, 2020; Pelly et al., 2020; Planas-Lladó et al., 2021; Brockmüller, 2023), as well as transformative competencies: creating new value, reconciling tensions and dilemmas, and taking responsibility (OECD, 2019b). These competencies contribute to the development of T-shaped graduates and professionals—individuals with both deep disciplinary knowledge and cross-disciplinary competencies—reflecting transversal skills essential for success in a complex world.

3.5.1 Policy Recognition of Complexity

Educational policy frameworks across the globe reflect this reality. The recognition that complexity cannot be fully predicted or controlled has led to an increased emphasis on developing competencies that allow individuals to act in such environments. The European Qualifications Framework (EQF), for example, explicitly highlights the need for students to develop the ability to integrate knowledge, handle complexity, and make informed decisions based on incomplete or uncertain information (European Qualifications Framework, 2017).

At the master's level (EQF level 7), students are expected to manage complex and unpredictable work or study contexts, exercising professional judgement and assuming responsibility for their own contributions. This includes the ability to work across disciplines and communicate with non-experts, a critical skill in addressing the cross-cutting issues that arise in complex systems. At the doctoral level (EQF level 8), the focus shifts to the synthesis of new and complex ideas, as well as the ability to critically evaluate emerging patterns and interactions within dynamic systems.

3.5.2 A Shift Towards Competence-Based Education

The recognition of complexity and unpredictability has sparked a broader shift in higher education towards competence-based learning. As complex systems demand new ways of thinking and acting, graduates must develop transversal competencies—such as problem-solving, critical thinking, collaboration, and adaptability—that enable them to navigate the uncertainties of professional and societal challenges. This shift reflects a growing understanding that education must prepare students not only to possess specific knowledge but also to apply that knowledge creatively and flexibly in dynamic, real-world contexts.

Educational policies have begun to reflect these demands, moving away from purely content-focused curricula and emphasising the development of competencies that support lifelong learning and adaptability in complex environments. For instance, competence-based education prioritises the ability to integrate knowledge across disciplines and adapt to new, unpredictable circumstances—key skills for thriving in a rapidly changing world.

Moreover, this shift towards competence-based learning is not confined to education; it also reflects broader trends in the job market. As employers increasingly seek individuals who can navigate complexity, there has been a noticeable move towards a "skills-first" approach. Initially developed to address skill shortages and promote global mobility, this approach places greater emphasis on competencies over traditional credentials, such as degrees or job titles (Winterton, 2009). The World Economic Forum (2023a) has strongly advocated for this approach, highlighting the importance of holistic skill development in preparing individuals for the evolving demands of the workforce.

This transition towards a competence-based model represents a fundamental shift in how we view education and professional preparation. It underscores the growing recognition that the ability to adapt, collaborate, and engage with diverse perspectives is just as critical as mastering specific technical knowledge. In this way, competence-based education aligns with the broader goal of developing individuals capable of addressing the complex, interdisciplinary challenges that define the twenty-first century.

3.6 Policy Frameworks and Competence-Based Education

The shift towards competence-based education is evident in global educational policies, emphasising higher-order skills and competencies rather than just knowledge. This approach is often encapsulated in the umbrella term "21st Century skills", which names the competencies and skills needed in the changing world (Framework for 21st Century Learning, 2019; Trilling & Fadel, 2009). It helps create a common language for educational aims and responsiveness to societal and job market needs. Another key feature of this approach is its inherently learner-centred nature (Kouwenhoven, 2009). Penprase (2020) describes this paradigm shift as a "Copernican Revolution," moving from a professor-centred to a student-centred model.

In European policy documents and frameworks, competencies are often defined by breaking them down into three categories, showing the broader perspective for what students are expected to learn. For example, in Key Competences for Lifelong Learning (2019), competencies are described as a "combination of knowledge, skills, and attitudes" developed throughout life in various learning environments. This conceptual shift recognises the importance of developing learners' holistic abilities and aligns with current educational policies and frameworks (Table 3.1).

Also, in the European Qualifications Framework (EQF, 2017), learning outcomes are divided into three parts (see Table 3.2): knowledge, skills and the third category is "responsibility and autonomy" (in the earlier version of EQF (2008), the

Table 3.1 Framework for competence in science, technology and engineering. This table shows the components of competence in science, technology, and engineering, covering knowledge, skills, and attitudes. It outlines the foundational understanding of the natural world, emphasising skills like scientific inquiry and technological proficiency. Additionally, it underscores the importance of attitudes such as critical appreciation, ethical considerations, and commitment to safety and sustainability. This framework aligns with the European Commission's Key Competencies for Lifelong Learning. (Source: Adapted from Key competences for lifelong learning, 2019)

Competence in science, technology and engineering refers to the ability and willingness to explain the natural world by making use of the body of knowledge and methodology employed, including observation and experimentation, in order to identify questions and to draw evidence-based conclusions. This competence involves an understanding of the changes caused by human activity and responsibility as an individual citizen.		
Knowledge	**Skills**	**Attitudes**
• Fundamental knowledge of natural world principles, scientific concepts, theories, and methodologies • Understanding of technology, technological products, and processes • Awareness of the impact of science, technology, engineering, and human activity on the natural world • Competence to assess the advances, limitations, and risks of scientific theories and technology in societal contexts	• Understanding science as an investigative process through specific methodologies • Ability to conduct observations and controlled experiments • Use of logical and rational thought to verify hypotheses • Willingness to discard personal convictions when contradicted by new experimental findings • Proficiency in using technological tools, machines, and scientific data to achieve goals • Recognition of essential features of scientific inquiry • Ability to communicate conclusions and reasoning effectively	• Attitude of critical appreciation and curiosity • Concern for ethical issues • Support for safety and environmental sustainability • Consideration of scientific and technological progress in relation to oneself, family, community, and global issues

third category was labelled "competence"). Here skills are understood as "cognitive (involving the use of logical, intuitive and creative thinking) and practical (involving manual dexterity and the use of methods, materials, tools and instruments)" (EQF, 2017), while the third category entails "the ability of the learner to apply knowledge and skills autonomously and with responsibility" (ibid). Thus the emphasis is not only on acquiring the skills, but also on enabling, empowering the learner to take them to action.

Similarly, OECD's (2019a) Learning compass 2030 breaks down competencies as having knowledge, skills, while the third category is attitudes and values. In addition, this model has a set of transformative competencies: creating new value, reconciling tensions and dilemmas, and taking responsibility; and a principle of student

Table 3.2 Learning outcomes, Level 7 (master or equivalent level). European Qualifications Framework (2017)

Knowledge	Skills	Responsibility and autonomy
Highly specialised knowledge, some of which is at the forefront of knowledge in a field of work or study, as the basis for original thinking and/or research Critical awareness of knowledge issues in a field and at the interface between different fields	Specialised problem-solving skills required in research and/or innovation in order to develop new knowledge and procedures and to integrate knowledge from different fields	Manage and transform work or study contexts that are complex, unpredictable and require new strategic approaches; take responsibility for contributing to professional knowledge and practice and/or for reviewing the strategic performance of teams

agency and co-agency (OECD, 2019b). Meanwhile some other models mostly emphasise skills (e.g., Framework for 21st Century Learning, 2019; World Economic Forum, 2023b).

The terms "competence" and "competency", although they have distinct meanings (see, for example, exploration in Holmes et al., 2021), here are used interchangeably. The concepts of "skills" and "competenc(i)es" are related in their approach to moving from more traditional ways of passive learning that focus on imparting information, and in that sense can also be used interchangeably. However, competence is usually understood more broadly, and skills are only a part of a more holistic concept of a competence, which also includes knowledge, attitudes, values, responsibility and action.

There are nuances in the distinction between competence, skills and knowledge:

Historically, competence has shifted from a legalistic notion to dynamic understanding, encompassing an individual's ability to achieve desired outcomes in diverse contexts. As noted by Illeris (2009), competence includes personal attitudes and assessments necessary to adapt and thrive in an ever-changing environment. This broader view integrates personal attributes such as flexibility, creativity, and independence, emphasising a holistic approach that aligns with the challenges of contemporary complex systems. Competence is not limited to mastering a specific area of knowledge; it encompasses the capacity to apply this knowledge in context (Larsen & Svabo, 2002).

According to Illeris (2009), *competence* encompasses a holistic integration of knowledge, skills, attitudes, and values, emphasising the capacity to apply these in complex and unpredictable environments. Competence involves a comprehensive approach. It integrates cognitive, affective, and behavioural dimensions to enable responsibility and effective action across diverse contexts (Illeris, 2012).

In contrast, *skills* refer specifically to the abilities and proficiencies acquired through practice and learning, forming just one component of a broader competence framework. They are essential components, but do not encompass the full scope of competence (Illeris, 2009).

Knowledge encompasses the information and understanding that individuals acquire through education and experience (Illeris, 2018). This concept, however, is challenged by situated, process- and practice-oriented approaches to learning, which critique traditional notions for overlooking the context-dependent and dynamic nature of learning (Brown & Duguid, 1991; Blackler et al., 2003). For this reason we use the term *knowing. Knowing is* a dynamic process that involves situated, embedded, negotiated and contested processes of working with knowledge in practice (Brown & Duguid, 1991). This perspective is reinforced by practice-based approaches to knowing in organisations (Nicolini et al., 2003), which highlight how materiality and practice play crucial roles in constructing and enacting knowledge within organisational learning settings (Svabo, 2009). These approaches emphasise that knowledge is not static, but is actively constructed through interaction with the environment and within specific social and cultural contexts.

Competence and *Knowing* are closely related. They both involve engaging with knowledge in ways that are fluid and responsive to the intricacies of real-world contexts. Competence is about navigating and co-evolving with complex environments, integrating cognitive, affective, and behavioural aspects as part of an ongoing process. Knowing emphasises the active, continuous construction and re-construction of understanding through practice, reflecting a commitment to being in tune with the ever-changing landscape of learning and action.

The understanding of competence as encompassing both knowledge and knowing illustrates their compatibility and interrelation. Both notions highlight the situated nature of human activity, learning, and development (Simonsen et al., 2014).

3.7 Critique of Outcome-Based Educational Policy

A competence-based approach also has some aspects to it that have to be taken into cautionary consideration. Firstly, although addressing the needs of the job market and steering education for economic growth is important, it is sometimes argued that this approach is primarily driven by business and can overlook other, more holistic goals of education (Germaine et al., 2016). Although many of the global taxonomies include personal wellbeing- and society- oriented competence such as (global) citizenship, the language of productivity and efficiency puts the weight of those into question. This debate can also be found in J. Dewey's educational theory (Pring, 2017), him saying that the aims of education being formed from outside are against the idea of democratic education (Obelleiro, 2017; Waks, 2017).

One of the lines of critique for the twenty-first century skills is that most of these, except some of the digital and technological ones, are not actually new. For example, the 4Cs (Critical thinking, Creativity, Communication, and Collaboration) can be traced back centuries (Germaine et al., 2016; Kennedy & Sundberg, 2020). It is notable that even the idea of preparing students for the future that is uncertain can be found in J. Dewey's work (Odell & Pedersen, 2020). The problem with the emphasis on the new and the push for reform without acknowledging the existing practices is that it might devalue the current work of educational professionals, reduce self-confidence of teachers and raise reform fatigue.

There is critique, not so much on the competence-based approach itself, but on the global outcome-based movement that comes mostly from economic goals, if seen in a context of standardisation and marketisation of education. Sahlberg (2006, 2023) coined a term GERM—Global Education Reform Movement, to highlight the shortcomings of standardised, outcome-based and accountability-driven reforms across the globe. Sahlberg's critique of a standardised approach to global reform can find connections to Creative Pragmatics as a suggestion to encompass complexity (Wrigley, 2019). Some simplistic reforms have not shown positive results (for example, standardisation of teaching and learning in schools or increasing competition), while some that were more complex, requiring holistic and context-specific views, were overlooked (collaboration and networking, creative customisation of curriculum, trust-based professionalism). Although the competence-based approach aims for a broader and deeper view of capabilities, standardisation can only take into account what can be easily measured by tests. The standardisation, accountability movement and high stakes testing (Anderson & Li, 2020), although it has an intention to orient education to more desirable outcomes for the job market, has the opposite effect, as it deprives the learning environment from necessary flexibility, time and spontaneity needed to address deeper learning of competencies.

The approach to competencies has a danger of becoming reductionistic. It is important to note that any deeper competence in the process of learning cannot be addressed strictly individually because competencies are deeply intertwined (Hager & Beckett, 2022). More specifically, twenty-first century skills can be criticised for lacking affect and the aesthetic, joy of learning, or for being value-neutral, and even destructive, if not infused with intellectual virtues. (Germaine et al., 2016). The approach to competencies therefore has to take a holistic and integrated view in designing learning and assessment.

In higher education, a competence-based approach can encounter criticism that it devalues complex disciplinary knowledge and might lead to fragmented learning in the form of modules that can get in the way of deep understanding (Kouwenhoven, 2009). On the other hand, Carl Wieman emphasises that the first step to deeper learning in STEM is to "reduce cognitive load" (Penprase, 2020). Wieman bases this statement on meta-analyses of college-level science education that show "... substantially enhanced learning and significantly less failure in courses that encourage 'asking rather than telling' and 'doing rather than sitting.' The most successful practices were those that asked students to apply their knowledge rather than merely to absorb it." (Mintzes & Walter, 2020).

It seems that more research-based dialogue needs to take place among education professionals to find common ground and show that competence-based learning does not undermine, but on the contrary, enhances deeper and more complex knowledge acquisition.

3.8 Creative Pragmatics and Competence-Based Education

We argue that Creative Pragmatics in principle takes a competence-based approach to learning but also addresses its weaker points. It is important for higher education institutions to continue moving towards competence-based learning, while at the same time not to reduce it to just skills, but maintain the holistic view, the complexity of knowledge and agency.

While competence-based learning rightly emphasises practical skills, it tends to overlook the broader dimensions of knowledge and agency: competence-based learning often risks reductionism, reducing complex abilities to isolated skills. Creative Pragmatics addresses this limitation by providing a more holistic perspective, recognising that competencies are interconnected and dynamic. This goes beyond mere skills acquisition, fostering a deeper understanding of knowledge structures and individual agency within complex contexts.

An implication of this turn towards competence—which fundamentally concerns how the individual interacts with the environment—is that education should focus on designing relevant and interesting learning environments, situations, and settings; what Dewey (1938) called "environing conditions". This highlights how individuals interact with their surroundings.

Moving beyond solely focusing on knowledge and skills as individual achievements, competence encompasses the relational and situated character of learning and human development. This includes the ability to engage with and adapt to one's environment, to use knowledge and skills in context-sensitive and relevant ways, and to develop understanding and collaborate with others.

This also implies that designing relevant and engaging learning environments can significantly enhance learning outcomes. These environments can be physical spaces, digital platforms, or social situations that foster exploration, creativity, and interaction. By focusing on the design of learning environments, educators can create experiences that are holistic, including cognitive, emotional, and sensory dimensions, and address the sociomateriality as well as the processual aspects of learning. This aligns with educational theories that emphasise experiential learning and student-centered education.

In this way, Creative Pragmatics enhances the competence-based approach by infusing it with philosophical principles that promote a richer and more dynamic educational experience. By doing so, it mitigates the risk of oversimplification and ensures a more comprehensive view of what constitutes competence. Embracing competence-based learning in science education higher institutions is crucial for empowering students. This approach acknowledges students' right to engage with their education actively, moving away from a traditional professor-centric model in a rigid body of knowledge or skill-centric framework. By focusing on competencies, students gain the tools to explore the world around them, fostering curiosity, critical thinking, and a sense of creative agency. Moreover, competence-based learning with a holistic view aligns with the civic agency of students, emphasising their right to a well-rounded education. It encourages a sense of responsibility towards societal issues, environmental concerns, and ethical considerations within the realm of science.

In summary, competence-based learning in science education should go beyond professional applications, recognising the inherent value of exploring the world and fostering civic agency. Creative Pragmatics values competence-based learning, while emphasising holistic understanding and complex knowledge dynamics.

By combining practical skills with a holistic understanding of scientific concepts, students can embark on a meaningful educational journey that equips them not just for careers but for active and engaged citizenship and agency.

Creative Pragmatics transcends a competence-based approach to education. By grounding Creative Pragmatics in philosophical principles, we can explore its implications for education and competence development, providing a theoretical foundation that goes beyond the instrumental aspects of competence-based learning. This is what we presented in Chap. 2. We dug into the theoretical and philosophical foundations for Creative Pragmatics, beginning with pragmatism and extending to science and technology studies, art, and design. With these lines of orientation, it becomes evident that Creative Pragmatics encompasses a performative understanding of knowing—and gives a special role to creative agency in navigating uncertainty and complexity.

References

Al-Khalili, J., & McFadden, J. (2014). *Life on the edge: The coming of age of quantum biology*. Bantam Press.

Anderson, J., & Li, Y. (2020). Investigating the potential of integrated STEM education from an international perspective. In J. Anderson & Y. Li (Eds.), *Integrated approaches to STEM education: An international perspective* (pp. 1–12). Springer. https://doi.org/10.1007/978-3-030-52229-2_1

Barabási, A.-L. (2002). *Linked: The new science of networks*. Perseus Pub. https://doi.org/10.1119/1.1538577

Bennett, J. (2010). *A political ecology of things*. Duke University Press. https://doi.org/10.2307/j.ctv111jh6w

Blackler, F., Crump, N., & McDonald, S. (2003). Organizing processes in complex activity networks. In D. Nicolini, S. Gherardi, & D. Yanow (Eds.), *Knowing in organizations: A practice-based approach* (pp. 126–150). M.E. Sharpe.

Brockmüller, S. (2023). Structure and measurement of system competence: Promoting systems thinking using analogue and digital models. In G. S. Carvalho, A. D. Afonso, & Z. Anastácio (Eds.), *Fostering scientific citizenship in an uncertain world* (pp. 79–94). Springer. https://doi.org/10.1007/978-3-031-32225-9_6

Brown, J. S., & Duguid, P. (1991). Organizational learning and communities-of-practice: Toward a unified view of working, learning, and innovation. *Organization Science, 2*(1), 40–57. https://doi.org/10.1287/orsc.2.1.40

Buchanan, R. (1992). Wicked problems in design thinking. *Design Issues, 8*(2), 5–21. https://doi.org/10.2307/1511637

Byrne, D., & Callaghan, G. (2013). *Complexity theory and the social sciences*. Routledge. https://doi.org/10.4324/9780203519585

Capra, F., & Luisi, P. L. (2014). *The systems view of life*. Cambridge University Press. https://doi.org/10.1017/CBO9780511895555
Davis, B., & Sumara, D. (2006). *Complexity and education: Inquiries into learning, teaching, and research*. Erlbaum. https://doi.org/10.4324/9780203764015
Davis, B., Sumara, D. J., & Luce-Kapler, R. (2008). *Engaging minds: Changing teaching in complex times* (2nd ed.). Routledge.
Dewey, J. (1938). *Experience and education*. Palgrave Macmillan.
Dorst, K. (2015). Frame creation and design in the expanded field. *She Ji: The Journal of Design, Economics, and Innovation, 1*(1), 22–33. https://doi.org/10.1016/j.sheji.2015.07.003
European Qualifications Framework. (2008). Official Journal of the European Union, C 111/1. Retrieved January 9, 2024, from https://eur-lex.europa.eu/legal-content/EN/TXT/?uri=celex%3A32008H0506%2801%29
European Qualifications Framework. (2017). Official Journal of the European Union, C 189/15. Retrieved January 9, 2024, from https://eur-lex.europa.eu/legal-content/EN/TXT/HTML/?uri=CELEX:32017H0615(01)&from=EN
Fenwick, T. (2012). Complexity science and professional learning for collaboration: A critical reconsideration of possibilities and limitations. *Journal of Education and Work, 25*(1), 1–20. https://doi.org/10.1080/13639080.2012.644911
Framework for 21st Century Learning. (2019). *Partnership for 21st century learning. Battelle for kids*. Retrieved January 9, 2024, from https://www.battelleforkids.org/networks/p21/frameworks-resources
Germaine, R., Richards, J., Koeller, M., & Schubert-Irastorza, C. (2016). Purposeful use of 21st century skills in higher education. *Journal of Research in Innovative Teaching, 9*(1), 19–29.
Gershenson, C. (2013). Facing complexity: Prediction vs. adaptation. In *Complexity perspectives on language, communication and society* (pp. 3–14). Springer. https://doi.org/10.1007/978-3-642-32817-6_2
Gigerenzer, G. (2022). *How to stay smart in a smart world: Why human intelligence still beats algorithms*. MIT Press.
Gupta, J. (2010). A history of international climate change policy. *Wiley Interdisciplinary Reviews: Climate Change, 1*(5), 636–653. Wiley. https://doi.org/10.1002/wcc.67
Hager, P., & Beckett, D. (2022). *The emergence of complexity: Rethinking education as a social science*. Springer. https://doi.org/10.1007/978-3-030-31839-0
Holmes, A. G., Polman Tuin, M., & Turner, S. L. (2021). Competence and competency in higher education, simple terms yet with complex meanings: Theoretical and practical issues for university teachers and assessors implementing competency-based education (CBE). *Educational Process International Journal, 10*(3), 39–52. https://doi.org/10.22521/edupij.2021.103.3
Hudson, H. E. (2013). *From rural village to global village: Telecommunications for development in the information age*. Routledge. https://doi.org/10.4324/9780203933138
Hughes, R. (2009). The art of displacement: Designing experiential systems and transverse epistemologies as conceptual criticism. *Footprint, 4, Agency in Architecture: Reframing Criticality in Theory and Practice*, 49–57. https://doi.org/10.7480/footprint.3.1.699
Illeris, K. (2009). Competence, learning and education: How can competences be learned, and how can they be developed in formal education. In K. Illeris (red.), *International perspectives on competence development: Developing skills and capabilities* (pp. 83–98). Routledge.
Illeris, K. (2012). *Kompetence: Udvikling af menneskers kompetencer i arbejdsliv og uddannelse* [Competence: Development of people's competencies in work life and education]. Samfundslitteratur.
Illeris, K. (2018). *Theories of learning: Learning theories in context*. Routledge.
Keen, A. (2018). *How to fix the future*. Atlantic Monthly Press.
Kennedy, T. J., & Sundberg, C. W. (2020). 21st century skills. In B. Akpan & T. J. Kennedy (Eds.), *Science education in theory and practice: An introductory guide to learning theory* (pp. 479–496). Springer. https://doi.org/10.1007/978-3-030-43620-9_32

Key Competences for Lifelong Learning. (2019). *European Commission, directorate-general for education*. Youth, Sport and Culture, Publications Office of the European Union. Retrieved January 9, 2024, from https://data.europa.eu/doi/10.2766/569540

Koopmans, M., & Stamovlasis, D. (Eds.). (2016). *Complex dynamical systems in education: Concepts, methods and applications*. Springer. https://doi.org/10.1007/978-3-319-27577-2

Kouwenhoven, W. (2009). Competence-based curriculum development in higher education: A globalised concept? In A. Lazinica and C. Calafate (Eds.), *Technology, education and development* (pp. 1-22). InTech. https://doi.org/10.5772/7297.

Larsen, H. H., & Svabo, C. (2002). Fra kursus til kompetenceudvikling på jobbet. In H. H. Larsen & C. Svabo (Eds.), *Fra kursus til kompetenceudvikling på jobbet* (1st ed., pp. 1–14). Djøf Forlag.

Latif, A., & Khatun, J. (2020). Leadership competence for undergraduate nursing students: A concept analysis. *Journal of Nursing Research, Education and Management, 2*(1), 24–39. https://doi.org/10.5281/zenodo.3602544

Law, J., & Mol, A. (2002). *Complexities: Social studies of knowledge practices*. Duke University Press. https://doi.org/10.1215/9780822383550

Lu, Y., Luo, L., Wang, H., Le, Y., & Shi, Q. (2015). Measurement model of project complexity for large-scale projects from task and organization perspective. *International Journal of Project Management, 33*(3). https://doi.org/10.1016/j.ijproman.2014.12.005

McDonald, A. (2004). Lessons learned but forgotten from the space shuttle challenger accident. *Space 2004 Conference and Exhibit*. https://doi.org/10.2514/6.2004-5830

Mintzes, J. J., & Walter, E. M. (2020). *Active learning in college science: The case for evidence-based practice*. Springer. https://doi.org/10.1007/978-3-030-33600-4

Morozov, E. (2014). *To save everything, click here: The folly of technological solutionism*. PublicAffairs.

Morton, T. (2013). *Hyperobjects: Philosophy and ecology after the end of the world*. University of Minnesota Press.

Nicolini, D., Gherardi, S., & Yanow, D. (Eds.). (2003). *Knowing in organizations: A practice-based approach*. ME Sharpe.

Norman, D. (2013). *The design of everyday things*. Basic Books (AZ).

Obelleiro, G. (2017). Democracy without telos: Education for a future uncertain. In L. J. Waks & A. R. English (Eds.), *John Dewey's democracy and education: A centennial handbook* (pp. 46–53). Cambridge University Press. https://doi.org/10.1017/9781316492765.008

Odell, M. R. L., & Pedersen, J. L. (2020). Project and problem-based teaching and learning. In B. Akpan & T. J. Kennedy (Eds.), *Science education in theory and practice: An introductory guide to learning theory*. Springer. https://doi.org/10.1007/978-3-030-43620-9_23

OECD. (2019a). *The OECD Learning compass 2030*. Retrieved January 10, 2024 from https://www.oecd.org/en/data/tools/oecd-learning-compass-2030.html

OECD. (2019b). *Transformative competencies for 2030: Conceptual learning framework*. Retrieved January 10, 2024 from https://www.oecd.org/content/dam/oecd/en/about/projects/edu/education-2040/concept-notes/Transformative_Competencies_for_2030_concept_note.pdf

Pelly, F. E., Wiesmayr-Freeman, T., & Tweedie, J. (2020). Student placement adaptability during COVID-19: Lessons learnt in 2020. Nutrition & Dietetics, 77(4), 481. Wiley. https://doi.org/10.1111/1747-0080.12625.

Penprase, B. E. (2020). *STEM education for the 21st century*. Springer. https://doi.org/10.1007/978-3-030-41633-1

Planas-Lladó, A., Feliu, L., Arbat, G., Pujol, J., Suñol, J. J., Castro, F., & Martí, C. (2021). An analysis of teamwork based on self and peer evaluation in higher education. *Assessment & Evaluation in Higher Education, 46*(2), 191–207. https://doi.org/10.1080/02602938.2020.1763254

Prigogine, I. (1997). *The end of certainty*. Free Press.

Prigogine, I., & Stengers, I. (1984). *Order out of chaos*. Bantam Books.

Pring, R. (2017). John Dewey: Philosopher of education for our time. In L. J. Waks & A. R. English (Eds.), *John Dewey's democracy and education: A centennial handbook* (pp. 340–348). Cambridge University Press. https://doi.org/10.1017/9781316492765.039
Radford, M. (2008). Prediction, control and the challenge to complexity. *Oxford Review of Education, 34*(5), 505–520. Taylor and Francis. https://doi.org/10.1080/03054980701772636
Risopoulos-Pichler, F., Daghofer, F., & Steiner, G. (2020). Competences for solving complex problems: A cross-sectional survey on higher education for sustainability learning and transdisciplinarity. *Sustainability, 12*(15), 6016. MDPI. https://doi.org/10.3390/su12156016
Sahlberg, P. (2006). Education reform for raising economic competitivenessr. *Journal of Educational Change, 7*(4), 259–287. Springer. https://doi.org/10.1007/s10833-005-4884-6
Sahlberg, P. (2023). Trends in global education reform since the 1990's: Looking for the right way. *International Journal of Educational Development, 98*. Elsevier. https://doi.org/10.1016/j.ijedudev.2023.102748
Sawyer, R. K. (2005). *Social emergence: Societies as complex systems*. Cambridge University Press. https://doi.org/10.1017/CBO9780511734892
Simon, H. A. (1947). Administrative behavior; a study of decision-making processes in administrative organization. *The American Political Science Review, 52*(1), 258–259. Cambridge. https://doi.org/10.1017/S0003055400072841
Simon, H. A. (1987/1995). Problem forming, problem finding, and problem solving in design. In Collen, A. & Gasparski, W. W. (Eds.), *Design and systems: General applications of methodology* (Vol. 3, pp. 245–257). Transaction publishers. (Text of a lecture delivered to the first international congress on planning and design theory, Boston, MA, 1987)
Simonsen, J., Svabo, C., Strandvad, S. M., Samson, K., Hertzum, M., & Hansen, O. E. (red.) (2014). *Situated design methods*. MIT Press. Design thinking, design theory. http://mitpress.mit.edu/books/situated-design-methods
Stacey, R. (1996). *Complexity and creativity in organizations*. Berrett-Koehler Publishers.
Strum, S., & Latour, B. (1987). The meanings of social: From baboons to humans. *Information Sur Les Sciences Sociales/Social Science Information, 26*(4), 783–802. https://doi.org/10.1177/053901887026004004
Svabo, C. (2009). Materiality in a practice-based approach. *The Learning Organization, 16*(5), 360–370. https://doi.org/10.1108/09696470910974153
Trilling, B., & Fadel, C. (2009). *21st century skills: Learning for life in our times*. Wiley.
Urry, J. (2005). The complexity turn. *Theory, Culture & Society, 22*(5), 1–14. https://doi.org/10.1177/0263276405057188
Waks, L. J. (2017). A democratic theory of aims. In L. J. Waks & A. R. English (Eds.), *John Dewey's democracy and education: A centennial handbook* (pp. 73–80). Cambridge University Press. https://doi.org/10.1017/9781316492765.011
Weick, K. E. (1995). *Sensemaking in organizations*. Sage.
Winterton, J. (2009). Competence across Europe: Highest common factor or lowest common denominator? *Journal of European Industrial Training, 33*(8/9), 681–700. https://doi.org/10.1108/03090590910993571
World Economic Forum. (2023a). *Future of Jobs Report*. Retrieved January 9, 2024, from https://www3.weforum.org/docs/WEF_Future_of_Jobs_2023.pdf
World Economic Forum. (2023b). *Putting skills first: A framework for action*. Retrieved January 9, 2024, from https://www.weforum.org/publications/putting-skills-first-a-framework-for-action/
Wrigley, T. (2019). The problem of reductionism in educational theory: Complexity, causality, values. *Power and Education, 11*(2), 145–162. Sage. https://doi.org/10.1177/1757743819845121
Yoon, S. A., Goh, S., & Yang, Z. (2019). Toward a learning progression of complex systems understanding. *An International Journal of Complexity and Education, 16*(1), 1–19. https://doi.org/10.29173/cmplct29340

Connie Svabo PhD, Professor of STEM Education and Science Communication, Director of STEM Center for Research in Science Education and Communication (FNUG), Department of Mathematics and Computer Science, University of Southern Denmark.

Connie is a transdisciplinary scholar and creative practitioner with expertise in learning environment design and experience. She is the founding Director of Center for Research in Science Education and Communication (FNUG), Section Leader and member of the management team of the Department of Mathematics and Computer Science, chair of the SDU research programme in Science and Mathematics Education and Communication. She promotes Creative Pragmatics as a model for knowing in a complex world and underpins this stance with a practice that ranges from art over education to design and consultancy. Her portfolio includes geological UNESCO heritage sites Stevns Klint and the South Fyn Archipelago, as well as the Danish National Gallery Open Air Museum and the museum of natural history, Naturama. Previously a Continuing Studies Instructor at Stanford University and Faculty at Roskilde University, where she directed the research centre Experience Lab, chaired the study programmes in Communication Performance Design 2013–2016 and served on the academic board. She has published books on STEM education, sustainability, competence development, design pedagogy and digital life. She holds a PhD in Social Science from Roskilde University's multidisciplinary programme in Society, Space and Technology.

Michael Shanks PhD Doc, Professor of Classics and Archaeology, Stanford University.

Acknowledged as one of the most original and influential of contemporary archaeologists, Michael is a specialist in long-term humanistic views of design and creativity, innovation and social change, while exploring connections across the sciences and arts in a hybrid field of theatre/archaeology. He is senior faculty in Stanford University's programs in Science Technology Society, Urban Studies, Writing and Rhetoric, Archaeology, Classics, and in the Center for Design Research, part of Stanford's d.school in the School of Engineering. He directed Stanford Humanities Lab, pioneering innovative project-based pedagogy in the arts and humanities, and helped found the Revs Program at Stanford, connecting the archaeological automobile with contemporary car design. Founding faculty in Stanford's Archaeology Center, Michael is a director of Stanford's research group Foresight and Innovation, and, for two decades until its retirement in 2023, he worked with mediaX, the affiliates' program that connected Stanford with industry through technology and innovation. He served on the city of Rotterdam's International Advisory Board from 2008 to 2019 and works with many companies and organisations in developing cultures of innovation. In another mobilisation of Creative Pragmatics, Michael continues to pursue archaeological fieldwork in the northern borders of the Roman empire and beyond.

Gabrielė Characiejienė, Research Assistant at the STEM Center for Research in Science Education and Communication (FNUG), Department of Mathematics and Computer Science, University of Southern Denmark.

Gabrielė holds an MA in Philosophy of Education, University College London, UK, and BA in Political Science, Vilnius University, Lithuania. She has worked at the Ministry of Education and Science of Lithuania, as a teacher in a secondary school, and for the most part, at an NGO Centre for School Improvement, supporting teachers, schools and policy on educational change. She was a Teach First Lithuania fellow and later the CEO of the programme. She has contributed to projects on educational leadership, civic education, and digital innovation in schools.

Chunfang Zhou PhD, Associate Professor of STEM Education and Science Communication, STEM Center for Research in Science Education and Communication (FNUG), Department of Mathematics and Computer Science, University of Southern Denmark.

A researcher and practitioner working on creativity, STEM education, and teacher education. Chunfang has an interdisciplinary and cross-cultural educational background: she finished a Bachelors Degree (Industry Automation and Information Science) and Masters Degree (Philosophy of Science and Technology) in China; during 2008–2012, Chunfang finished her doctoral research on *Group Creativity in Engineering Education in a Problem and Project-Based Learning (PBL) environment* at Aalborg University, Denmark. She is a board member of Danish Development Research Network (DDRN), and works as a senior consultant in the PBL (Problem-Based-Learning) Teaching Innovation and Research Center at Northeastern University (NEU), China. Chunfang locates her research in Science, Technology and Society (STS), with a focus on creativity and STEM education. Since 2013, Chunfang has authored and co-authored over 110 peer-reviewed publications; she is the editor of *Handbook of Research on Creative Problem-Solving Skill Development in Higher Education* (IGI Global, 2016), the author of *Introducing Problem-Based Learning for Creativity and Innovation in Chinese Universities* (IGI Global, 2020), and an editor in the book series *Arts, Creativities and Learning Environments in Global Perspective* (Brill/Sense). Chunfang is an editorial board member of *Discovery Education* and the *European Journal of Education and Pedagogy*, as well as advisory board member for the *International Journal of Engineering Education.*

Tamara Carleton PhD, Universitetslektor in Mechanical Engineering, Blekinge Institute of Technology.

Tamara is an award-winning professor, industry advisor, and expert in radical innovation and strategic foresight. Recognised as one of the Top 50 Women Leaders in Education (in 2023 and 2024) by Women We Admire, she teaches at multiple business schools and technical universities around the world. In addition, she holds the UNESCO Chair in Anticipatory Leadership for Innovative and Better Futures at Tecnológico de Monterrey in Mexico. Tamara is also the founder and CEO of the Innovation Leadership Group, which helps leaders and teams to build innovation capacity and map bold futures. She has written several seminal books in innovation, including *Building Moonshots: 50+ Ways to Turn Radical Ideas into Reality* and the *Playbook for Strategic Foresight and Innovation.* She holds a doctorate in mechanical engineering from Stanford University.

Chapter 4
Active and Agentic: Participant-Directed Inquiry in an Introductory Science Programme

Martin Niss

4.1 Introduction

Creative Pragmatics is a rallying cry for educating students to navigate complexity. As the editors write in their introduction, "In such a framework, students are expected to develop competence in handling complex and unpredictable work situations. It is expected that they can assume professional responsibility and with autonomy can initiate and carry out collaboration across disciplines and organisational levels. Students must be able to structure their own work, and the objective is that they develop into self-directed and autonomous individuals. An important component of this is not only to be reactive to externally motivated goals, but also to be able to formulate goals, interests and concerns in self-directed ways and to assess and make informed decisions based on incomplete information and in situations of insecurity, change and risk. This makes it necessary to develop creative competencies in assessing a situation and what is at stake, to draw on experience, communicate with other participants and stakeholders and understand complexity." (Chap. 1, Sect. 1.4, "Policy Highlights the Importance of Handling Complexity").

Martin is an associate professor at Roskilde University, which has a strong tradition of problem-oriented project-based learning. We wanted to invite established scholars like Martin to share some lessons and challenges in the spirit of Creative Pragmatics. His paper and project shows the critical importance of creative agency in Creative Pragmatics—ways that learning can be co-designed, with agency a component of participatory learning in a design process. The case study is transdisciplinary in that the learning program in self-directed open inquiry described by Martin reaches beyond disciplinary fields in working on a real-world complex challenge of shipping. (Editors)

M. Niss (✉)
Department of Science and Environment, Roskilde University, Roskilde, Denmark
e-mail: maniss@ruc.dk

C. Svabo et al. (eds.), *Creative Pragmatics for Active Learning in STEM Education*, Contributions from Science Education Research 14,
https://doi.org/10.1007/978-3-031-78720-1_4

A fundamental aspect of all this is that the graduated students should be agentic so that they have the capacity to take initiative, make decisions and exert control over their actions in an autonomous and purposeful way. However, students are not likely to develop such agency on their own in the traditional teaching models with the latter's focus on teachers delivering knowledge and the educational activities unfolding in particular predetermined ways that lead students to 'mastering' specific scientific 'content'. In such models, students are rarely invited to participate in setting the goals and objects of their activities, choosing the tools, or participating in constructing the rules. So, while the students may be active in their learning process through collaboration and discussions rather than having them passively receive information (Chap. 1, Sect. 1.6 of this book and Lee et al., 2018), they are not afforded agency and thus it is not enough for the development of student agency that the instructional approach supports active learning. OECD (2019) argues that in an educational system that encourages student agency, in contrast, learning involves not only instruction and evaluation but also co-construction where the teachers and students become co-creators in the teaching and learning process. Here, the students acquire a sense of purpose in their education and take ownership of their learning. In other words, such a system allows students to develop competent participation in keeping these activities running and to find allies, design instruments, mobilise the world and so on (van Eijck, 2012). Van Eijck (2012) adds that agency allows students to develop and hence understand how particular elements of knowledge-producing activities in science, such as rules, objects and tools, are used as knots and links in holding together the dynamic flows of these activities.

The purpose of the present chapter is to use the bachelor programme in natural science at Roskilde University as a case of a programme that affords agency to the students to illustrate how this can be done. This programme shares the above ambitions of Creative Pragmatics, in particular, that the programme's students should develop competence in handling complex situations by formulating what the students themselves see as relevant goals and concerns and possess the ability to organise their work towards these goals and concerns in purposeful, self-directed and autonomous ways. In the programme, this is done with a natural science focus so that the complex situations involve a substantially natural science component and so that the students learn to use theories, methods and practices of natural science to handle such situations. While the students employ ready-made science, their autonomy in steering their projects ensures that the program embodies a key aspect of Creative Pragmatics pedagogy according to Pickering (foreword to this book), namely emphasising science in action over finalised, ready-made science.

The programme is characterised by engaging the students in large degrees of inquiry (50% of the study time is devoted to inquiry-based learning), scaffolding by the academic staff and student agency where the students are allowed to make decisions to pursue a goal. The students form groups that do semester-long inquiry-based projects. In the following, we will describe the programme and analyse how it supports the development of the above programme ambitions. More specifically, we will study the following questions:

1. How is student agency in principle afforded in the programme?

2. How does the open inquiry approach of the programme in practice support student agency?
3. How are complex situations handled by the students in the programme?

We employ a qualitative case study method in this chapter as we want to study the case of student agency in a particular programme. Moreover, the questions addressed here are well-suited to a qualitative case study according to Yin (2003), as it fulfils his three specific conditions under which a case study can provide a particularly valuable contribution. Firstly, the present study addresses 'how' questions, namely the above. Secondly, the researcher has no control over the behaviour of the actors; as an exploratory study of an ongoing phenomenon, this study did not attempt to influence participants' behaviour but sought to explore an existing practice in light of current theories. Thirdly, the focus is on a phenomenon in real-time in its natural context, namely how students do project work.

For the first question, we take an analytical approach, where the programme description and documents are analysed to shed light on how agency is theoretically afforded in the programme. For the two remaining questions, we employ other documents and interviews. The documents are written by the project group as the project progresses and they thus allow a characterisation of the development of a project in a naturalistic context as they are written as an integral part of the scaffolding of the students with specific educational purposes. While providing insights into the development of the projects, the written documents do not disclose how the students perceive the development. Thus, a qualitative interview was made with one of the participating students to shed light on such issues as how the student sees such issues as how the goals were set up and on what basis, how decisions were made and on what basis, how control was exerted etc.

The case studied will be the two projects of a single student, anonymised as Anna, that she did as a member of two different groups. For each project, the project groups in the programme must submit three written products where they describe and assess their project so far. In the products, the students share how they see the projects 10 days and 6 weeks within the project as well as when the project is finalised after 4 months.

The interview was conducted with Anna, who participated in this research voluntarily. The interviews were semi-structured, allowing in-depth follow-up of initial responses to questions. The interview dealt with each of the two projects in turn. Many open questions relating to the ideas for the projects, the development of the project and the interaction between the project group and other relevant actors, in particular the supervisor, were asked. The interview lasted approximately 60 min and was recorded. The interview was transcribed and the data generated formed the basis for the analysis.

Methodologically, it is a challenge to describe and analyse a complex form of teaching practice such as the project work in such a bachelor programme. What we try to do, is to present to the reader a sufficiently rich description of the development of project work and provide some illustrative examples of the students' projects to form a concrete basis for our reflections on the three questions presented above.

4.2 Agency

As noted in Chap. 2, Sect. 2.5, agency describes people's capacity to make a difference to the world they inhabit (Latour, 2007), so they make things happen, realise goals and intentions, whether these are recognised and deliberate or not. (Barad, 2007). Several conceptualisations of agency in education exist. Here we will follow Holmes et al. (2020), who start with a definition that an agent is someone who is making decisions to pursue a goal. Furthermore, they argue that students should be afforded goal-driven agency: "Having goal-driven agency can be demonstrated through the decisions one makes: setting goals and planning how to achieve them (planning), carrying out the plan with self-regulation (executing), and reflecting on whether the goals were sufficiently met (reflecting). Moving through these phases with agency is not a strictly linear or straightforward process. One must have the motivation to pursue the goals, and the time and space to enact the plan." (Holmes et al., 2020, p. 2).

Holmes et al. contrast agency and autonomy, where the former refers to having the free will to make choices. In their understanding, an agent may or may not be autonomous. An agent can also be contrasted with an actor, who may be carrying out the intention of someone else with no independent decision-making. In this way, agency may be considered as sitting between acting and behaving autonomously.

Stenalt (2021), drawing on Klemenčič (2015), asks us to analytically differentiate agentic possibilities ("power") from agentic orientation ("will") and account for how agentic possibilities emerge in students and how student orientation influences student actions. So, programmes that aim at supporting agency need to provide agentic possibilities to the students, but the student also needs to have an agentic orientation to use these possibilities; Cavagnetto et al. (2020) argue for the important role played by the educational programme in facilitating that the students have the will to participate in an agentic way. More specifically, Cavagnetto et al. see the will to participate in the light of a cost/benefit analysis on the part of the student, balancing the costs and benefits of participating actively; this analysis is influenced by a multitude of pedagogical decisions and classroom characteristics.

4.3 How Is Agency Afforded in the Programme?

4.3.1 Presentation of the Programme

The bachelor programme in natural science at Roskilde University is based on the so-called Roskilde model (Andersen & Heilesen, 2014; Blomhøj et al., 2014). This model is a version of project- and problem-based learning approach (Tan, 2021; Zhou, 2016) that focuses on student development of ability to tackle real-world problems through authentic, problem-based learning rooted in science and mathematics, thus preparing them for making a meaningful impact on society's most pressing issues (Davis & Sumara, 2006). The programme starts with three semesters-long introductory studies aimed at providing the students with a broad background in natural science before they specialise within specific disciplines in the final three semesters of study. The programme is based on general pedagogical principles such as student-directed project work, problem orientation, and exemplarity (Jensen & Olesen, 1999; Blomhøj & Kjeldsen, 2009). These principles imply that the studies are built around a semester-long project each semester, where the students themselves decide what problem they want to work within their project work. The project work, typically done in groups of four to seven students, is supplemented with more traditional subject-oriented courses. The problem-oriented project work can be framed as authentic inquiry (Buck et al., 2008) where the students have to decide all characteristics, including the problem, theory, procedures, analysis, communication, etc. (see Fig. 4.1).

The projects of the first three semesters are organised around themes that the projects should illustrate: the first-semester theme is the application and function of science in technology and society, the second-semester theme is the interplay between theories, models, experiments, and simulations in the sciences, and the third-semester theme is philosophical reflections on science and mathematics.

This educational model is part of a trend, where the incorporation of diverse project-based approaches is a prevalent feature in contemporary university science education reforms. Frequently, proponents emphasise the potential benefits for students' motivation when engaging with scientific content as their primary argument. Conversely, critics occasionally cite the inefficiency and costs associated with covering the science curriculum through project work as a counterargument against its prominence within science programs. Blomhøj and Kjeldsen (2009) argue that both these arguments miss the point for the programme at Roskilde University. Problem-oriented project work can and should serve other educational purposes than traditional forms of teaching and the educational value of project work goes far beyond

Characteristic	Level 0: Confirmation	Level ½: Structured Inquiry	Level 1: Guided Inquiry	Level 2: Open Inquiry	Level 3: Authentic Inquiry
Problem / Question	Provided	Provided	Provided	Provided	Not Provided
Theory / Background	Provided	Provided	Provided	Provided	Not Provided
Procedures / Design	Provided	Provided	Provided	Not Provided	Not Provided
Results Analysis	Provided	Provided	Not Provided	Not Provided	Not Provided
Results Communication	Provided	Not Provided	Not Provided	Not Provided	Not Provided
Conclusions	Provided	Not Provided	Not Provided	Not Provided	Not Provided

Fig. 4.1 Modes of inquiry (Buck et al., 2008)

motivating students for science studies—even though project work might also have that effect. Instead, project work should be argued for as a way of achieving certain educational goals, including the development of agency.

4.3.2 How Is Student Agency in Principle Afforded in the Programme?

As noted by Blomhøj and Kjeldsen (2009), the students' rights and obligations to exercise influence on their education have been one of the hallmarks at Roskilde University since its establishment. This is fully implemented in the project work, where the students choose and formulate the problems they want to work with, decide on methods, and find and read literature both in the form of textbooks and research papers published in scientific journals. They make models, design experiments, perform empirical and/or theoretical analyses, and structure the entire process.

To facilitate this process, the programme has implemented various measures. First of all, a supervisor is assigned who will guide the project group in their endeavour; the supervisor and project group typically meet once a week during the semester. The role of the supervisor is described in the following way in an official document:

> The project supervisor must naturally be able to provide assistance with specific problems within his or her field of academic competence, or other areas in which the supervisor is interested and qualified. The nature of this assistance may encompass providing answers to spontaneous questions, suggesting relevant literature, and pointing to experts in special areas, and helping in the preparation of presentations. Finally, the supervisor must be able to make academic comments and corrections to the presentations, draft report and final evaluation or examination. This form of guidance will usually be combined with interest and commitment, but it requires, of course, that the group members themselves take an active part in the guidance process – which is to say, they must submit written papers, adhere to agreements and deadlines, prepare for meetings, etc. Experience shows that projects in which guidance, in the form of constructive criticism and written presentations, plays an important role tend to display greater quality than other projects, in terms of both the working process and end product.
>
> Another essential function of the project supervisor is to provide advice on how the project work can be organised, both concerning long-term planning and daily activities. Examples of the latter are: ensuring meeting discipline, the use of logbooks, continuous evaluation of the work, subgroups, handling of group conflicts and the use of the supervisor. (Board of Studies for Bachelor Programmes in Natural Sciences, 2023)

This document illustrates how the programme tries to promote a project-group/supervisor relationship where the supervisor allows the project group to be in charge of the decision-making in the project; the supervisor should supervise the group accordingly.

A second measure implemented in the programme to facilitate the project group's progress is the use of formative evaluations during the semester: the group has their project evaluated by an opponent group and their supervisor at a midterm evaluation taking place about 6 weeks within the semester as well at internal evaluation of their almost completed report 2 weeks before the final deadline. Furthermore, within 10 days of the semester, the project group submit a preliminary project description where they describe their initial thoughts on the project just after they have formed a group; this description includes a preliminary research question and ideas for how to answer that question. Next, the students hand in a midterm evaluation report about 6 weeks into the semester, where they describe their project ideas, including the research question, their methodology and plans. Thirdly, they submit a final report at the end of the semester accounting for the entire project.

Holmes et al. (2020) argue that when the students move through the phases of goal-driven inquiry with agency this is not done in a strictly linear or straightforward process. Hence, the students must have the motivation to pursue the goals and the time and space to enact the plan. In the programme at Roskilde University, the student's motivation to pursue the goal is supposed to arise from the fact that the students are engaged in open inquiry where they determine what questions are worthy of pursuit, the theory to apply and what procedures to follow. It is well-known that both the costs and the benefits for the students of the many degrees of freedom of authentic inquiry are quite high; the experience at Roskilde University is, however, that the students' cost-benefit analysis, whether it is implicit or explicit, leads them to conclude that benefits are larger than the cost. Regarding the required time, the experience is that half of the study time for a semester is ample time to do interesting and relevant open inquiry projects, whereas the space to enact such plans stems from a culture and conscious choice of providing such space at the university campus, where there are rooms designated for group work.

4.4 The Projects of Anna

4.4.1 The First Project

We now turn to the two projects of a single student. For her first project, Anna and four other students chose to gather around a project idea suggested by one of the supervisors. He presented the following idea in terms of some bullet points: "• Several recent incidents involving containers being lost from large freight ships during stormy weather point to a growing problem. • Standard rules have apparently not kept up with increasing ship sizes. • Ships are more frequently loaded to maximum load. • Parametric rolling – a resonance excited by waves – appears to be a key factor. • Can a simple physics analysis point the way to reducing this problem?"

The supervisor, a physicist having glass physics as his research area and by no means particularly invested in freight ships or parametric rolling, deliberately chose to present his idea in this rough, open way as an interesting problem where a simple physics analysis might be relevant, but where he did not suggest how to attack the problem. Hence, the students must make a lot of decisions concerning what it more specifically means to reduce the problem of containers lost at sea and how physics can be brought to bear on the problem in a way that might suggest a solution.

The students quickly appropriated this rough idea as their own. In their preliminary project description written 10 days into the semester, they first elaborated on the societal issue of the problem, describing the scope of the problem, that it is a hazard to seacraft and the environment as well as financial issues for the shipping industry. The project group continued that one cause of containers going overboard is parametric rolling and they named four solutions to this problem from the literature and they posed the question "With these solutions in mind, one wonders why we have recently seen an increase in lost containers. While the answer is almost certainly multifaceted, we wish to understand the situation through a scientific lens, asking the question: **what can a simple physics analysis tell us about parametric rolling and its possible solutions?**"

In addition to their clarification of the societal context of the problem, the project group also started to perform the physics analysis of the situation. They wrote that they seek to create a simple physics model which captures the key variables involved in parametric rolling: for the ship itself these could be the centre of gravity, the centre of buoyancy and angular momentum (roll); for the wave, these could be amplitude, wavelength and frequency.

As the students went more in-depth with the issue, the group changed the direction of the project as they realised that although ship architects have already tried to resolve the problem of parametric excitation for ships, it continued to be an issue for large container ships. In their midterm report, they wrote that it "proved difficult to find concrete reasons why, despite the techniques we have to help prevent it [...], [parametric excitation] continues to be a particular problem for larger, Post-Panamax container ships. In response to this difficulty in finding explanations, our main research question can be formulated as such: **can a basic physics simulation shed light on why PE [parametric excitation] continues to be an issue for container ships?**"

Their final report shows how they answered this question. They made a computer simulation of a large container ship (of so-called C11-class) as this was the kind of ship that was involved in the incident with the largest number of lost containers to date (4293) in 2013. They described the essential physics of the situation when such a ship is in head seas, which they used to make a simulation of the motion of a ship. The simulation involves numerous inputs, some of which come from the literature for a C11-class ship, "but it has been a challenge to determine accurate values for the input parameters which we were unable to locate reference values for." (p. 26) No reference values could be found in the literature for seven parameters, so the students themselves had to go to great lengths to estimate suitable input values based mainly on fitting to available data based on requiring realistic behaviour. In this process, the group made an observation when analysing the simulation output that "surprised" them, namely that parametric excitation was rather sharply confined to certain values of the parameters. When they varied the value for the roll damping parameter, they found that the roll angle dropped off rather sharply over a small interval of damping values. This "suggests that there is a critical roll-damping threshold above which we find virtually no PE and below which we find extreme levels of PE." (p. 42)

They concluded "Our findings reinforce an idea that is present in the literature: that inadequate levels of damping are one of the main reasons why PE occurs in ships […] However, what we have not found reference to is there being a critical damping threshold." (p. 42) This they believed could shed some light on the ongoing issue of parametric excitation in ships.

In the interview, Anna gave her version of the project work. Asked about how the project started, she answered that the initial project idea was given by the supervisor, but it was the project group who specified the question "because it couldn't be too broad […], it's like there was a lot of things involved, just like buoyancy and then damping and like, so we had to, yeah, like, look for a research question, like a more specific one." Hence, her experience agrees with the above conclusion that the supervisor's project suggestion was formulated in general terms. In several places, Anna used expressions like "So that's also how we came up with the research question" as indicators that in her view it was the project group who chose the research question.

Asked about how the project developed and if they followed a clear idea from the beginning, she answered "Yeah, I would say we followed [a clear idea], but we also encountered, like, not issues, but more like along the way [points to consider] […] We kept in view what we had to research about that wasn't so clear at the beginning, but it was clear that we were going to do about what was going on with this, like, motion thing happening, the understanding of that […] and yeah, and it was like, along the way, just really finding out what was necessary to include." Hence, while the group had a direction, namely to understand the motion of the ships, a major task for them was to deal with several points as the project developed, such as what had to be included. Her experience was that the project group was in charge of deciding this direction as indicated by her answer to the interviewer's direct question of who she would say was in charge of making the decisions concerning the directions of the project: "I would say it was mostly two people" namely herself and another group member, Ben. Asked about the supervisor "Did he make the decisions?", she answered, "He suggested a lot of stuff, but at the end, we all had to agree, of course, but it was also Ben giving a lot of ideas." Hence, she placed the agency in the project group.

Prompted to give an example of what needed to be included, she mentioned pitch motion: "So at the beginning, like, we were talking only about roll and roll and roll. But then we found out, like, because it's a coupling between pitch and roll, we had to also get, like, a lot of experience on the pitch motion and then on the coupling equations… That wasn't really the idea at the beginning, but it's also like even [the supervisor] didn't really know that much about it at the beginning, but we all were like, researching together. Basically, we're all working together." So, Anna described how she had a feeling that the supervisor and the group were forming what can best be described as a research unit that dealt with emerging issues in a self-directed and autonomous way at those times. At the same time, Anna acknowledged that she and her group needed reassurance from the supervisor that their ideas were okay: Asked whether the supervisor had to approve the ideas, Anna responded "Well, not like he had to approve, but we felt safer. He thought our ideas were also good, and were good ideas because he had experience and we did not at the moment." Moreover, she stressed several times the encouragement of the supervisor and its great importance for the group: "He [the supervisor] gave us so much encouragement and really made us like the topic". In this sense, the students were not acting in a completely autonomous way but needed the reassurance of the supervisor.

How does the open inquiry approach of the programme in practice support that the students are agentic in this project? Or more specifically, how does the approach support that the students structure their project work by setting up their own goals, planning how to achieve those goals and carrying out the plan in a purposeful, autonomous, and self-directed way in this project? While the initial project idea was not the project group's own, the supervisor, in a move true to the ideal of open inquiry, had formulated the idea in such a way that the idea was not at all researchable: There was not yet a research question let alone a procedure for answering it; as regards theory, he only mentioned that a physical analysis might be relevant. Finally, he did not mention any data. Hence, the open inquiry approach meant it was the task of the students to specify the research question by framing and shaping it according to their interest as well as finding relevant theory, data and procedures. As is clear from the above description of the process and the interview with Anna, the group was also true to the ideal of open inquiry and was quick to appropriate the supervisor's idea as its own by framing and shaping the question to be pursued in their way by setting up their own goal: to find out why the container problem still exists given that the problem is well-known and solutions exist by seeing if a simple physics analysis can shed light on this issue. The goal was set up purposefully so that the project group saw its attainment as relevant to the societal issue of container loss. According to Anna, this societal relevance was important for the group's motivation as "that also like made us feel like we could research on something that could say something valuable about what's going on right now." As an indication of the group's ownership of the project, Anna repeatedly referred to the research question as *their* research question.

With this goal in place, their work during the semester showed that they could pursue it in such a way that they could finalise the project in one semester. More specifically, they planned how to achieve this goal, namely by setting up a physics model based on a literature study of existing research on ships and coding a programme that can simulate the behaviour of the model. Next, and this formed the main part of their work, they carried out this plan in a goal-driven manner: (1) They set up and coded a model that is quite simple yet captures the parametric excitation. (2) To simulate the behaviour of real container ships with their code, they showed their capability of assigning realistic values to the parameters of the model, which is a big task. (3) With their simulation in place, they were capable of obtaining results concerning the behaviour of their simulated ship that reproduces findings in the literature but they also found some results showing the existence of a critical threshold that they believed is not noted in the literature. In the interview, Anna stressed that as the project developed, issues arose that they had to tackle, such as the effect of rolling motion on the pitching motion.

Setting up the goal, planning its achievement and carrying out the plan was done in an autonomous and self-directed way: In Anna's experience, the project group was in charge of the entire process even if the group interacted considerably with the supervisor, who came with ideas and feedback. The group was not completely autonomous as they relied on the encouragement of the supervisor to check that their ideas were valid; they did, however, to a large extent participate in an open inquiry where they were in charge of making the decisions, not the supervisor or somebody else.

How did the project group navigate the complexity of the situation? The students were well aware that they had studied a complicated problem where the theoretical description of the situation that should be captured in a model is very complex: "The damping is determined by many factors such as ship speed, the geometry of the ship's hull, its loading configuration, etc. This inherent complexity coupled with the intricacies of ship hydrodynamics means that the accurate calculation of damping forces is non-trivial. Accordingly, it is possible that at times there is a significant gulf between theoretically approximated values and the real-world damping forces." (p. 42–43).

So, their basic strategy consisted in simplifying the complex and complicated real-world situation to an idealised situation that is simple enough that they can draw valid conclusions about this situation, but where the connection between the idealised situation and the real world is perhaps less clear. However, the students are confident enough to suggest that their conclusions of the existence of a damping threshold might shed light on why container ships lose containers: "If a critical damping threshold exists, one consequence would be that errors in damping calculation could be punished severely with extreme levels of PE." (p. 43)

4.4.2 The Second Project

For her next project, Anna and two of her fellow students from the previous project together with a fourth student decided to continue to work on the topic of roll damping of ships, but from a different angle. In their project description 10 days into the semester, they described their idea in the following way:

"The inspiration for this project comes from a first-semester report, which investigates parametric excitation in the rolling motion of containerships. One observation made from this project is that parametric rolling appears to be more of an issue in larger ("Post-Panamax") ships. One possible explanation is that, while smaller ships presumably have adequate levels of roll damping, this level of damping drops off as the ship gets larger. One possible explanation for this is that roll damping does not scale at the same rate as ship size. We wish to investigate the validity of this explanation and have formulated the following preliminary research question: **how does roll damping scale with the size of a floating body?**"

They also wrote a few words about their plans:

"In order to answer this question, we will build differently sized scaled models of simple floating bodies and conduct a series of experiments where we analyse their roll behaviour. As roll damping is a complex phenomenon, we can simplify the concept by focusing on the basic shape of a cylinder. This allows us to focus only on roll damping brought about by skin friction, avoiding the more theoretically dense topic of eddy damping and added mass. We can build our models in the RUC [Roskilde University] workshop, perhaps utilising their 3D printing facility, and conduct the experiments in a water tank, using a camera to record roll behaviour. We can feed our experiment data into a simple computer model, using this to draw conclusions about how roll-damping scales with cylinder size. If feasible we may move onto more complex shapes – ones which resemble real ship hulls."

Here they formulated a crucial idea that they pursued throughout the project, namely to study floating cylinders rather than shapes more realistic of ships and only look at the latter if the experiments of the simple shapes were successful; their experimental set-up would involve such cylinders in a water tank. To assist them in this endeavour, they would enrol the people in the workshop at the University. The students also stated another line of inquiry in their preliminary description, namely, to make a simple computer model, which they also pursued throughout the project. The project group was assigned a supervisor, who was another physicist working on glass physics.

A few weeks later, in their midterm report, the project group had worked further along the lines indicated above and had landed on the following formulation of the research question: "**What effect does scale have on the roll damping motion of a ship's hull**?" In this formulation, which became their final one, they now referred explicitly to the hull of ships. More importantly, they made an important addition to the cylinders, namely adding weight in the form of a metal rod inside of each cylinder to get the cylinders to act as damped pendulums with a restoring torque so that the cylinder will by itself rotate to return to its initial position, just like a ship would. In short, they wanted their design to be a so-called damped pendulum well-known in physics. Furthermore, they now elaborated on the relationship between their experiment and real ship hulls:

"We have designed our experiment according to the time and resources we have available. That means that we have had to make a number of compromises away from the optimal experimental model. For instance, it would be ideal to conduct an experiment using scaled replicas of actual ship hull forms [...]. A real ship's hull is both non-cylindrical in cross-sectional area but is also differently shaped along the ship's length. Our experiment, on the other hand, uses cylinders which are identically shaped along the cylinder's length. While our cylinders do not completely map to reality, we believe that they are similar enough to be able to glean worthwhile data and insights from."

In the midterm report, they also expressed a clearer idea of how they envisioned experimenting: "We are conducting a series of experiments involving cylinders, submerged in a 2.0 m x 1.0 m x 1.0 m water tank, located on RUC [Roskilde University] premises. These objects are mounted on an axis and allowed to rotate freely while a video camera captures their movement. The intention is that these cylinders come in a satisfactory range of sizes, so that the scale effects can be studied." More specifically, they expected to study five cylinders of diameters of 30 mm to 300 mm. Furthermore, the group explained how they intended to use measurements for empty cylinders to determine the so-called skin friction on the cylinders as well as how the attachment of so-called bilge keels, a sort of fin used to reduce a ship's tendency to roll, would allow them to determine another type of roll friction, namely so-called eddy-making damping.

In the midterm document, they added that "So far we have not made any of the shapes yet, as we have struggled to come up with a design for the support/base for our tank to hold our ships (shapes), so we can study them, but we hope so soon have a design ready." They did resolve this issue, but only after they had dismissed one design of the housing. In their final report, they write "We tried using this for the first experiments, but realised it did not work, as the housings would not be perfectly aligned, so we came up with a new design [...]. With the first housing our idea was to position a rod throughout the centre of the whole cylinder, but with the new housing design, we would not need this anymore. The new housing design was proved to be working, as there was not any noticeable friction from the offset angle between the bearings, unlike the latter housing design, so we began the experiments." (p. 36)

Another challenge encountered was that for various reasons they could only get access to three cylinders with given dimensions, so their initial idea of using five cylinders with specifications given by the project group was somewhat amputated, but they carried on with the available materials.

They did various measurements on these materials, and with those data, they found the amount of damping of each cylinder, with and without a bilge keel. Next, they studied how this amount of damping scales. To do this, it was necessary to calculate the moment of inertia for each of the cylinders so that they could isolate the damping coefficient from the equations which govern the motion of a damped pendulum. Furthermore, they made a Python simulation to compare our results from the different-sized cylinders and to confirm that their physical experiments fit the theory. They modelled the cylinders and their behaviour with the equations found in the literature and ran simulations with the properties derived from physical experiments. From this, they obtained figures that allowed them to compare the computer simulation and the results of their physical experiments. The conclusion was that although a linear relationship between damping and cylinder size was anticipated, the experiment results indicate that this relationship appears to be governed by a power law. Hence, while their results did not serve as an explanation as to why larger ships seem to experience inadequate levels of rolling, they did provide insight into the scale effects of damping.

In the interview, Anna gave her version of the project work, emphasising how this project originated in the students' wish to work on an issue that had arisen in the previous project. This meant that they had thought about what they wanted to do "But yeah, but we actually did have a very clear idea from the beginning because we... It came from [the first project], so we already were thinking, OK, we're going to do this right after [the first project]." Still, Anna had the experience that the project developed considerably during the semester:

"But then it never goes out as planned from the beginning, so there were a lot of changes along the way. Then if the data were bad, we needed to see what was wrong, how could we improve it. Then, I remember, like, we were going to do like cylinders on water and then again like the analogue of like pendulum motion and the ships and we had to figure out how we were going to restoring force to these cylinders. So first we needed to do like more research on the physics behind it to get the idea of how we're going to do the experiment and then... Yeah, it was a lot of changes along the way. Like with friction, like with the bearings, that, what if it's in water? Also what if it gets rusted? We have to be like very fast [in] all [of] the process or get new bearings like be changing them all the time and so yeah."

To effect all these changes, the group had to make many decisions. They had originally envisioned, for instance, that they would make a 3D print of a ship's hull, but they realised that was too complicated, so they would represent the hull by a cylinder instead. Other decisions concern the bearings in the experiment, the size of the cylinders etc.

In this project, the supervisor was less involved in guiding the project compared to the previous supervisor, which according to Anna did have positive effects: "I think it made it easier even not having such an involved supervisor because we were almost all of us coming from [the previous project] and was like... The theoretical [side], we already knew and now it was more about like how to make it experimental [side], right?" Hence, according to Anna, the project group did not need help regarding the theoretical aspects of the project and would be self-directed in that respect.

Concerning the experimental side, the project group interacted with the University's workshop. Asked about the role of the workshop, Anna said "They also gave ideas like how to attach everything to the tank or for example, we hadn't, we hadn't thought about that the friction of the bearings and all this was going to impact [the set up] and they said OK, we need frictionless bearings". Moreover, the people of the workshop also gave directions like "this is what you can do and this you can't do." Hence, according to Anna, the workshop would assist when consulted by the project group but would not direct the course of the project. Another example is when the workshop built the experimental set-up, where the people of the workshop "would come to us with more questions, like, what is it that you need on this and this and this?"

This description demonstrates that Anna had the experience that the project group was in charge of making the decisions regarding the project. A further example of this was when Anna and the interviewer talked about the process that led to the representation of the ship hull by a cylinder and the interviewer asked her "Can you say something about who was involved in that process of thinking about it?" Anna answered "I fairly think the four of us [of the project group] were as involved. It was a great group. We were all completely into it." Shortly after, she added, "So all the decision-making was by the four of us." Anna did not mention anything about the project group's need for assurance from the supervisor or others that their ideas were good.

According to Anna, time was an important constraint for the project group. Early in the interview, she said "We were already warned that an experiment takes time."

This warning informed their decision-making as the project developed. In Anna's words "it's like everything was changed along the way to what was doable for us and not just doable, but like also on time because we needed to be, yeah, realistic with the time."

How did the open inquiry approach of the programme support that the students structured their project work by setting up their own goals, planning how to achieve those goals and carrying out the plan in a purposeful, autonomous, and self-directed way in this project? The initial project idea was their own, i.e., to study how a ship's damp rolling scales with ship size, and this idea determined their goal from the very beginning and they were autonomous and self-directed in setting up this goal. Having established this goal, the project group planned how to achieve the goal, combining a theoretical and an experimental approach. On the experimental side, they decided to study cylinders of various sizes and with different loads in a water tank rather than more realistic shapes with more complicated cross-sections that may change the length direction of the ship. This decision allowed them to study the problem within the constraints afforded by the university, but as the project group notes, it implies that they focus on a single aspect of roll damping on ships while neglecting others. On the theoretical side, they decided they should consider the ship to be analogous to a damped pendulum, so they could make a computer program based on this. The project group took all these decisions on their own before they were assigned a supervisor, as indicated by their preliminary report. According to Anna, what allowed them to work in such a purposeful, self-direct way when setting up the goals and their plan, was their experiences with their previous project.

When it came to executing the plan, they carried out this plan in a goal-driven manner even if "there were a lot of changes along the way," according to Anna, as they had to respond to many challenges. Firstly, they designed and executed the experiment, including making decisions on how to determine the roll damping from their set-up, how to simulate the weight distribution of the ship and how to include the bilge keel. Moreover, they encountered various issues in this step: (1) they could only use three cylinders with fixed dimensions rather than five cylinders of their own choice; (2) The design of the housing of the cylinders caused difficulties and their first design did not live up to their requirements. According to Anna, they received considerable help from the people of the workshop here, but always in a way where the project group was directing the inquiry and the people of the workshop would assist according to the needs of the project group. Secondly, the project group had to analyse the data collected in the experiment and confront their results with the group's theoretical derivations of how the coefficients are expected to scale with size. Hence, they determined the damping coefficients, plotted them as functions of cylinder radius and confronted them with their theoretical derivations to see if the data agree with the theory. The plan was carried out in a purposeful way where the project group was clear about how their decisions had limited their inquiry, but also how time would constrain their possibilities. This was done in a self-directed and autonomous way that seemed to be even more so than the first project as they relied much less on the reassurance of the supervisor, who was less involved in the process.

How did the project group navigate the complexity of the situation? While the complexity of the previous project had to do with the theoretical description of the situation, the complexity of the present project had to do with how to make an experimental set-up that resembles a ship and at the same time allows experiments with clear-cut results to be made. The way the project group tackled this situation is analogous to the previous group as they idealised the situation so that they draw valid conclusions concerning that situation; this meant that also for this project the connection between the idealised situation of cylinders and the complicated shapes of real ships is not completely clear.

4.5 Conclusions

The answer to the question of how agency is afforded in principle in the programme is that it tries to promote three ideals: Firstly, the students are supposedly motivated to formulate, plan and obtain their own goals in a self-directed and autonomous way by the fact that they are engaged in open inquiry where they determine what questions are worthy of pursuit, what theory to apply and what procedures to follow. Secondly, they are assisted in this endeavour by a supervisor who should act to allow the project group to be in charge of the decision-making in the project. Thirdly, they are given an appropriate amount of time and space to support that interesting and relevant open inquiry projects can be done with semester-long projects and campus facilities that allow group work.

The analysis of Anna's two projects demonstrates how the open inquiry approach facilitates that Anna and her group members are put in situations where they have to be agentic. The two projects are quite different as one concerns the application of natural science to a societal issue, whereas the other is a more traditional experimental project. This illustrates the breadth of the open inquiry approach for student agency. Interestingly, the way the students navigate the complexity of the two situations is structurally similar, as they reduced the complexity by idealising the situation which enables them to bring natural science to bear on the situation.

So, the results of this chapter indicate that the program's open inquiry approach not only allows the students to be active but also opens for a large degree of student agency in determining the direction of the inquiry. However, it is also clear that the programme's open inquiry approach is quite demanding. Half of the study time is devoted to open inquiry, the campus facilities are designed for group work, the scientific staff are encultured into a specific project-group/supervisor relationship where the students are in charge of the decision-making etc. Given the great importance of educating students to be agentic and as open inquiry has a definite potential in this regard as argued in this chapter, it would be highly relevant to research whether open inquiry can be implemented in a less demanding way than how it is implemented in the programme studied here.

References

Andersen, A. S., & Heilesen, S. B. (Eds.). (2014). *The Roskilde model: Problem-oriented learning and project work* (Vol. 12). Springer.

Barad, K. M. (2007). *Meeting the universe halfway: Quantum physics and the entanglement of matter and meaning*. Duke University Press. https://doi.org/10.5334/opt.081013

Blomhøj, M., & Kjeldsen, T. H. (2009). Project organised science studies at university level: Exemplarity and interdisciplinarity. *ZDM, 41*, 183–198.

Blomhøj, M., Enevoldsen, T., Haldrup, M., & Nielsen, N. M. (2014). The bachelor programmes and the Roskilde model. In Andersen & Heilesen (Eds.), *The Roskilde model: Problem-oriented learning and project work* (pp. 79–105).

Board of Studies for Bachelor Programmes in Natural Sciences. (2023). *Welcome as project supervisor at Nat Bach*. Unpublished internal university document.

Buck, L. B., Bretz, S. L., & Towns, M. H. (2008). Characterizing the level of inquiry in the undergraduate laboratory. *Journal of College Science Teaching, 38*, 52–58.

Cavagnetto, A. R., Hand, B., & Premo, J. (2020). Supporting student agency in science. *Theory Into Practice, 59*(2), 128–138.

Davis, B., & Sumara, D. (2006). *Complexity and education: Inquiries into learning, teaching, and research*. Routledge.

Holmes, N. G., Keep, B., & Wieman, C. E. (2020). Developing scientific decision making by structuring and supporting student agency. *Physical Review Physics Education Research, 16*(1), 010109.

Jensen, J. H., & Olesen, H. S. (1999). Can "the university" be revived in "late modernity"? In H. S. Olesen & J. H. Jensen (Eds.), *Project studies: A late modern university reform. A late modern university reform?* (pp. 9–22). Roskilde Universitetsforlag.

Klemenčič, M. (2015). What is student agency? An ontological exploration in the context of research on student engagement. In M. Klemenčič, S. Bergan, & R. Primozic (Eds.), *Student engagement in Europe: Society, higher education and student governance* (pp. 11–29). Council of Europe Publishing.

Latour, B. (2007). *Reassembling the social: An introduction to actor-network-theory*. Oxford University Press.

Lee, D., Morrone, A. S., & Siering, G. (2018). From swimming pool to collaborative learning studio: Pedagogy, space, and technology in a large active learning classroom. *Educational Technology Research and Development, 66*, 95–127. https://doi.org/10.1007/s11423-017-9550-1

OECD. (2019). *Student Agency for 2030*. https://www.oecd.org/education/2030-project/teaching-and-learning/learning/student-agency/Student_Agency_for_2030_concept_note.pdf

Stenalt, M. H. (2021). Digital student agency: Approaching agency in digital contexts from a critical perspective. *Frontline Learning Research, 9*(3), 52–68.

Tan, O. S. (2021). *Problem-based learning innovation: Using problems to power learning in the 21st century*. Gale Cengage Learning.

van Eijck, M. (2012). Capturing the dynamics of science in science education. In B. J. Fraser, K. Tobin, & C. McRobbie (Eds.), *Second international handbook of science education* (pp. 1029–1039). Springer.

Yin, R. K. (2003). *Case study research: Design and methods*. Sage.

Zhou, C. (2016). Fostering creative problem solvers in higher education: A response to complexity of societies. In C. Zhou (Ed.), *Handbook of research on creative problem-solving skill development in higher education* (1st ed., pp. 1–23). IGI Global. https://doi.org/10.4018/978-1-5225-0643-0.ch001

Martin Niss, PhD, Associate Professor of Physics, Department of Science and Environment, Roskilde University. Presently, he holds the position of Head of Studies for the Bachelor Programme in Natural Science at Roskilde University. He has extensive experience in supervising student projects in natural science as well as facilitating others' supervision processes. His educational research primarily centers around science education, with a specific emphasis on physics education. Research areas encompass topics such as problem-solving in physics, project work focused on problem-solving in natural science, modelling in the natural sciences, and the Nature of Science.

Chapter 5
Artefacts and Creative Learning in Mathematics Education

Amalie Thorup Eich-Høy and Chunfang Zhou

5.1 Introduction

Creative Pragmatics urges us to rethink our approaches to education practices that help students respond to challenges of living and knowing in a complex world. This chapter contributes to that end by exploring a concrete case on a booklet design process adhering to the principles of Creative Pragmatics mentioned in Chaps. 1, 2 and 3. The chapter also addresses the important roles of artefacts in gaining creative learning experience and mediating creative learning environments in mathematics education.

Based on the case, we explore how an artefact mediates creative learning in mathematics education from the point of view of a postgraduate student who participated in designing an urban escape-esque mathematics mystery booklet and who

We felt it was important to bring younger voices into a discussion of Creative Pragmatics. They will carry the field of STEM education forward, in terms of scholarly inquiry and its practical implementation. Here Amalie, a Danish doctoral researcher who is studying how math is taught, collaborates with book editor Chunfang. They explore the artefact and material components of creative, indeed playful, learning environments, challenging the isolation of learners by activating a user-centered design process through situated experiences and in affinity with arts practice. These are key aspects of Creative Pragmatics. Notable also is the reflective autoethnography at the heart of this case study. This exemplifies the pragmatist premise of Creative Pragmatics that we are part of what we seek to understand. Agency is again a key concept that refers to the active involvement of the artifacts in a material learning environment as well as the learners. (Editors)

A. T. Eich-Høy (✉) · C. Zhou
STEM Centre for Research in Science Education and Communication (FNUG), Department of Mathematics and Computer Science, University of Southern Denmark, Odense, Denmark
e-mail: amtho@imada.sdu.dk; chzh@sdu.dk; https://www.sdu.dk/en/forskning/stem

C. Svabo et al. (eds.), *Creative Pragmatics for Active Learning in STEM Education*, Contributions from Science Education Research 14,
https://doi.org/10.1007/978-3-031-78720-1_5

reflected on her own learning experience through autoethnography (Adams et al., 2017). Accordingly, this chapter exemplifies many points of the concept of Creative Pragmatics; we highlight that learning happens through experience and creative processes, through doing in practice and reflecting on it. The learning process is a result of working playfully, exploratively, and experimentally (Svabo & Borch, 2020). In our concrete case, the knowledge of the postgraduate student is produced through participating in the design process of the booklet, named The Gate Key (TGK). We provide the case of TGK because it serves as an example of project-based learning, design thinking, interdisciplinary learning, informal learning, and mathematics education coming together.

The student-designer is the first author (Amalie) of this chapter, and much of the chapter is thus an autoethnographic account of the iterative, reflective, collaborative, and creative learning process surrounding the design of the booklet. Amalie participated in the design process of TGK while studying mathematics at University of Southern Denmark (SDU), and she has since become a researcher in mathematics education. With these multiple identities, the autoethnography is itself a conversation between student-centred approaches to creative learning experiences, user-centred approaches to artefact design, and mathematics education development broadly. In other words, the case in this chapter is taking the student-centred approach to its most literal extremity, bringing the student experience into the forefront of not only the design of a mediating artefact, but also the research on it, asserting an appropriate "I".

Another aspect of the booklet is its use. We explored three rounds of user tests that gave insight into the further design of the booklet, and more broadly into the design of artefact-mediated creative learning environments. Here we are interested in what mediating artefacts do, and to that end we rely on Connie Svabo's development of the study of interactions, made possible using GoPros as a video-ethnographic tool (Svabo, 2008). We will discuss more about the method in a later section.

We explore both the design and the use of the booklet by writing the chapter according to a chronological structure, weaving the student-designer point of view with interpretations of the user point of view. The structure of this chapter thus reflects the intertwined and reciprocal nature of the design and the use of the booklet, and how the artefact is a mediation between users and designers as well as between both learning processes. We write with a purposeful emphasis on process (Svabo, 2016).

Following the interwoven account of and reflection on the process of TGK, we discuss theoretical points relevant to the case and to Creative Pragmatics. We reflect on the importance of designing active artefacts, on the roles of artefacts in mediating play and learning mathematics, and on the values of art. We conclude by encouraging the development of mathematics education from an arts perspective.

Before we move on to the autoethnography, a few words on practicalities are in order. The booklet was a part of the *InterReg* project *MatKult*[1] at SDU. The objec-

[1] See details of project MatKult at www.matkult.eu/start.dk.html

tive of the project was to use mathematics as the basis for cultural exchange across the Denmark-Germany border. The project regarded mathematics as universal abstractions with the potential to transcend language barriers and bring Danish and German students together in shared experiences. TGK was intended as one of these experiences.

A handful of people with a diversity of specialities collaborated on the booklet, and it consequently became a negotiation between people as well as materials. TGK was developed in a relatively short time frame (from March to June 2022), and the autoethnographic writing is mostly retrospective reflections on the design process. In no way was the process a perfect replica of linear, iterative models from approaches such as design-based research (Bakker, 2018). Although it may be enticing to call the process "inspired by" this approach, preemptively swerving any methodological critique, we will instead content ourselves by saying that we did not commit a priori to any design approach or didactic theoretical framework.

Yet, the vocabulary of design-based research is useful for describing this design-turned-research with its iterations and related reflections. Should any inveterate adherent of this method read this chapter, we ask forgiveness for diverging from its doxa. However, our emphasis on adaptability (rather than predictability) is a point in itself, further underlining the complexity of real-world processes and their inability to rely on orthodox models. To reflect this, as Chap. 3 mentioned, design as well as research must instead work with flexible goals.

5.2 The Beginning

The project was born at a morning meeting in the (at the time) newly established STEM Centre for Research in Science Education and Communication (FNUG)[2] at SDU. Connie Svabo, the director of the Centre, introduced the idea of designing some artefact, game, or activity related to mathematics. She assigned the initial task to me and the multi-faceted artist Christian Eiming, whose many gigs involve designing customised mystery booklets. As mentioned, at that time I was studying mathematics and was also an avid Dungeons and Dragons player—Connie did not have to ask me twice to join the design of a mathematics game. So, before I knew it, I was on a Zoom call with Christian, spitballing ideas left and right: "What if it's a treasure trail shaped like a pentagon?", "what if an evil cult led by Hans Christian Andersen has conspired to taking over the city?", "what if a magic artefact able to stop the cult is hidden inside the pentagon?"

By the end of the meeting, my desk was littered with notes and sketches. Most of those notes and sketches never made it to the final product, and while I will not spend pages diving into each detail, it would be misleading to forget the paths that were untrodden. In hindsight, the trail appears straightforward, but what lies behind

[2] See more details of the Center at www.sdu.dk/en/forskning/stem

are character stats, several ghosts, and some questionable task designs of which I *will* go into heavy detail. But first, let me outline the overall design of the booklet that stayed relatively fixed throughout the design process.

In a nutshell, TGK is a pentagon-shaped treasure trail around the city of Odense in Denmark. Each vertex of the pentagon is the site of a mathematics exercise contained in the booklet. Solving an exercise leads the user of the booklet to the next vertex, and so on, giving rise to a total of five individual exercises. During the pentagon-shaped walk, the user maps their trajectory inside the booklet, drawing the regular pentagon from vertex to vertex. The game concludes when the actual, material gate key is found in the very middle of the pentagon, which in the real world is a heavy pillar just outside a supermarket in downtown Odense.

There were lots of reasons for choosing the pentagon, the biggest of them being that the booklet was funded by the *InterReg* project *MatKult* whose logo is a pentagon (ah, the things you do to appease your funder). Being a mathematician, I could not help myself from thematically tying the regular pentagon to incommensurable magnitudes, the golden ratio, Euclid's algorithm, and Fibonacci's number sequence. And so, the third self-contained exercise took shape. I will focus on the development of the third exercise throughout this autoethnographic writing.

Before judging the initial design of the third exercise too hard, understand that I was immensely inexperienced and (maybe a bit too) eager to maximise the learning outcome of the booklet. I remember writing the task while imagining myself at 13 getting my hands on the booklet. I had no doubt in my mind that this teenager would understand and appreciate every bit of the mathematical delicacies I put onto the pages. "Can you *believe* that the ratio between the side and the diagonal of a pentagon is the *golden* ratio? And have you *heard* how that ties in with the Fibonacci sequence? And *how* fun is Euclid's algorithm?", I asked my younger self through the wordy (more than 5000 keystrokes), quite difficult, but in my eyes very thematically elegant and mathematically rich exercise.

I bring up this memory because it seems obvious that my initial design strategy was basically to project myself onto the user, effectively closing the gap between designer and user. Mediation was a non-issue to me. While typing out the exercise, I was so absorbed in my own excitement that it never occurred to me that my own 'designer-centred' experience might not apply to others. To be honest, my 13-year-old self would have probably taken a glimpse at the crammed pages and shrugged.

5.3 The First User Test

To call the first user test a user test is a bit of an overstatement; the 5000 keystrokes never made it to actual users nor to the booklet format. Instead, the document was e-mailed to Dorte Moeskær Larsen, one of my mathematics education colleagues at the Centre. She agreed to run her experienced eyes over my frantic infodump. Her comments led me to the conclusion that the exercise had to be thoroughly rethought.

At the time, we had decided to cater to lower secondary students. Dorte's comments were mainly user-centred with this target group in mind: the Euclidean algorithm is not part of the curriculum in Danish secondary mathematics education, so the users are unlikely to be able to perform the algorithm. Not least because the vocabulary of the exercise was not level appropriate. She suggested using fewer words and more figures.

5.3.1 My Learning Process

Retrospectively, it is clear to see that Dorte did me a kindness by nudging the booklet-based urban walk in a more user-centred direction. At the time, however, I remember feeling like I would be diluting the exercise of its mathematical richness if I departed from my initial concept. I was quite proud of the way I managed to embed the mathematics as not just the aesthetics of the exercise, but as the actual content and method of it. The vertex of the third exercise is located at a church ground in Odense, and I even managed to fit numbers found around the church into the calculations. This elegance—this coherence between the regular pentagon, incommensurable magnitudes, the golden ratio, Euclid's algorithm, and Fibonacci's number sequence—would be lost.

I realise now that these are quite self-absorbed design motives, or at least that I was convinced that I could somehow convey the mathematical beauty to an idealised user with words alone. I was the only one who was excited about using the Euclidean algorithm to find a code with the digits of the golden ratio. 5000 keystrokes about university-level mathematics do not automatically spark joy in the eyes of a teenager. I realise now that one cannot do mathematics on behalf of the learner.

5.3.2 Artefact Improvement

Having ditched the idea of using the Euclidean algorithm, I set out to salvage as much of the original design as possible. I wanted to preserve the solution (the six-digit code, 161803), and I also wanted to include the numbers that can be found around the church grounds. The code, however, had to be found using simpler calculations than ancient algorithms.

The reason the six-digit code was a key part of the design is twofold. Firstly, we had to come up with a way of making each exercise reveal the location of the next vertex. Secondly, Christian owned a webpage that allowed the user to enter a six-digit access code and then receive some information. It was basically a no-brainer to design exercises around this webpage, making unique access codes for each exercise and revealing the next vertex upon correct entry.

At this stage in the design process, the booklet began to take shape as a digital design for printing. The third exercise was a double page spread. Across the middle of the pages were six juxtaposed pentagons. Above and below those pentagons were four text boxes containing rhetorical arithmetical problems using numbers from the church ground. Two of the textboxes pointed to just one pentagon each, while the other two pointed to two pentagons, distributing the solutions of arithmetical problems across the six pentagons in total. Each pentagon was thus meant to represent a digit in the code.

For example, the textbox in the lower right corner (Fig. 5.1) asks the user to perform the following arithmetic. First, find the Bible quote on Theodor Steenbuch's tombstone. Second, divide the Bible chapter number with the sum of digits of the Bible chapter and the Bible verse. Third, multiply the result with the fifth Fibonacci number. This is the result that fits into the sixth pentagon.

This third exercise is obviously just a snippet of the entire booklet, but hopefully it gives an impression of the overall design. 32 pages in total, the booklet was ready for print and for its first proper user test. Being on a tight schedule and having to use readily available resources, I printed the booklet on regular A4 printer paper.

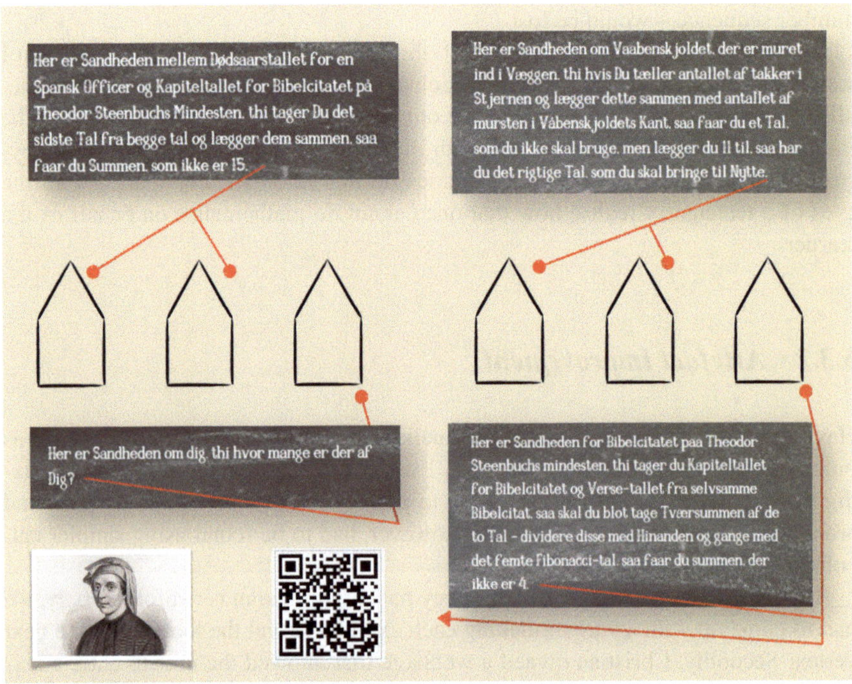

Fig. 5.1 The second iteration of the third exercise in its digital format

5.4 On User Tests

Before getting into the first user test, a brief detour to the method is in order. From this point in the design process and onwards, I documented the user tests in order to analyse the tests in depth. This section explains how the data from the user tests were gathered and later analysed.

The user tests were observed and documented using video-ethnography (Scott & Shrum, 2017). Video-ethnography is a varied method, but for this particular project, the users were wearing GoPros on their foreheads (Svabo, 2008). These recordings give a detail-rich point of view of the users, making it possible to observe their interaction with the booklet, each other, and parts of the environment, while automatically focusing on the part of the environment that was also the focus of the user.

While the GoPros were recording the users' test runs as close as can be, I was tailing them in person, writing field notes, and occasionally conversing with the users. It was sometimes necessary to sacrifice undisturbed user data when they got completely stuck in an exercise, mainly because it was crucial to the design process that the booklet was tested in its entirety in the available time. Thus, as an observer, I was constantly positioning and repositioning myself as both the unobtrusive researcher 1 min, and the intervening mediator the next.

It was my firsthand observations that catalysed our immediate artefact improvements. The video data is useful in writings like this to recall my own memories of the events reliably. This chapter does not just consist of designer-centred analyses, though. We are also interested in user-centred analyses, namely analyses of the users' creative learning process. TGK is a mathematics booklet after all, so obviously it is made with some didactic intent.

To better interpret the users' creative learning processes, it is necessary to break down two prevalent dualisms surrounding thinking: (1) cognition and materiality, and (2) cognition and affect. Creative learning processes are as much a material and an affective matter as it is a cognitive one. Therefore, materiality and affect must be considered in earnest. Video-ethnography accommodates this dualism breakdown especially well. Take for example affect: it is habitual to treat verbal statements as indications of this. Video-ethnography further allows examinations of gestures, tone, facial expressions, and similar bodily clues as manifestations of affect. Likewise, the often-silent materiality of a situation speaks volumes (figuratively speaking) on video.

This sensitivity to the affective contagion of the social and material goes hand in hand with the ecological notion of affordance (Gibson, 1979; Hutchby, 2001). Affordance is a relational term concerning the potential actions that are offered between an artefact and a user in an environment, understood as both opportunities and constraints. The affordance of an environment emerges between actors. From the point of view of the user, affordance constitutes what an environment is, which means that the functions of an artefact manifests through its use.

Noticing the material and affective nature of mathematics agrees with the philosophy of Creative Pragmatics, not reducing mathematical thought to bits in predefined categories. In this line of thinking, mathematics is not an activity happening just in the brain (Larsen et al., 2022; Kristensen et al., 2023). Mathematics and mathematical thinking are embodied and materially distributed. In other words, mathematics is enacted in a way that makes the study of articulations, affordances, and aesthetics not just an epistemological undertaking but an ontological one (Sinclair in Radford, 2014). We analyse creative learning processes by looking for what the subjects do, what the booklet and other artefacts do, what the environment does, and so on.

5.5 The Second User Test

At the time of recording, Helga was an eighth-grader who lived in Odense and agreed to be the first authentic tester of TGK. She met me in the town centre and graciously let me install a GoPro on her forehead. On this spring day, the Danish weather was cold, windy, and rainy, which turned out to be significant for the prototype of TGK. At this point, TGK consisted of 16 pages of double printed paper, so simply turning the pages of the booklet became a matter of defying the elements with half-frozen fingers. These weather conditions combined with the flimsy booklet format meant that it became physically unpleasant and eventually exhausting to interact with the booklet.

By the third exercise, which is the one I have focused on so far, Helga had solved the previous exercises with only minor hiccups. Moreover, she had without fail done the booklet's bidding, including dipping her already cold finger in Odense River and sitting on a wet bench. In other words, her actions reflected a persistent mindset proportional with her success up until now. Her willingness to engage with the whims of the booklet despite the weather conditions at least shows a good degree of motivation.

Another interesting detail about Helga's test run is that she repeatedly positioned herself as a pupil and drew on school experiences as frames of reference for the exercises. For example, the first exercise asked her to use GPS coordinates, and instead of opening the readily available and familiar app Google Maps, she installed Google Earth with the comment, "Oh, we learned how to do that once". The "we" in question were in all probability her class. This indicated to me that she associated TGK with the mathematical practice she knew from school, despite (one might say) the alternative format offered by the booklet.

As mentioned, the idea with the third exercise was to find certain numbers around a churchyard and do simple calculations with them to obtain a code. Inserting the code after the website www.matonline.dk/ would lead her to the next vertex.

When Helga reached the third exercise, her streak of success came to an end. The first thing that went wrong was on account of the way TGK was organised. A long poem about the Pythagorean theorem came between the second and third exercise. It had no connection to the exercises other than tying Hans Christian Andersen to mathematics. Helga spent a quarter of an hour searching for non-existent clues in the poem before I nudged her to turn to the next page.

At this point, Helga's demeanour had turned from persistent to rather despairing, reading aloud from the booklet with immediate outbursts of confusion while ceaselessly sniffling in the cold. The sun was almost setting, and the temperature of the wind had likewise dropped. Because of the hostile architecture, there was no comfortable place to sit, and Helga simply seemed fatigued. It had taken too long. In the end, she found just one out of eight numbers around the churchyard before the GoPro died and I decided to call it quits on grounds of her quivering limbs.

5.5.1 My Learning Process

This tale of woe certainly gave me valuable insights into the continued development of TGK. It was clear to me that the mismatch between the booklet's epistemic affordances and the intended mathematical content completely overshadowed any hope for mathematical learning, given how long Helga spent at the site of the third exercise while not getting to the actual calculations. I considered her frustrated outbursts as testimonies to the didactic design disaster that was the third exercise.

A point that became apparent was that the poor weather seemed to amplify her frustrations about the exercise. During the first couple of exercises which she breezed through, she perseveringly managed the booklet's material properties when turning its pages, and she did not seem too phased by the hostile environment from both the weather and the architecture. However, everything turned sour at once, so the connection between the booklet's affective and epistemic effects seems like a chicken and egg kind of conundrum.

Helga's test run also clearly illustrated to me the discrepancy between the mathematical intention behind the booklet and how mathematics was enacted while Helga used the booklet. I remember thinking that I already stripped the third exercise of so much of its original design, so I was surprised when the new exercise was not as straightforward as I thought it to be. In retrospect, I was clearly still holding on to the core belief that my thoughts would surely imbed itself in the pages. That the pages would contain fixed ideas, zealously scripting actions and pre-defined learning goals. That mathematical ideas could be found behind the words and symbols.

Above all, the reality check that was Helga's run showed how carefully a designer must consider the affordances of the medium if they want to exploit it for mathematics learning. An educational designer cannot completely predict but must at the same time consider the affordances of mathematics learning tools such as a booklet. I learned that a booklet-mediated escape-game cannot rely on text and in the subsequent development, we aimed at relying more on visual representations and concise instructions.

5.5.2 Artefact Improvement

After Helga's test, it was finally starting to dawn on me that the protagonists of any design are the users. Earnestly putting myself in the shoes of the users, I endeavoured to strain out all red herrings and wordings that were mysterious for the sake of being mysterious. The design had to be sleeker: the exercises clearer stated, the booklet better suited for cold fingers lacking fine motor skills, and the pages looking overall inviting.

To this end, Michael Steensen, who is a graphic designer at our institute, was put on the task. He transformed the look of the booklet from gritty typewriter ink on parchment to sweeping pen on geometric shapes, utilising a minimalist colour palette of white, black, and orange yellow. Michael's contribution shows that in design, there is no neutral choice. Colour, typography, and layout don't just impact what one might call the "surface" of the booklet. Or, put differently, the surface of the booklet is an integral part of its very nature.

In this iteration of TGK, we simplified the third exercise by ditching the numbers around the churchyard. Instead, the exercise is contained in the booklet, and the code is made up of certain square- and Fibonacci numbers and a sum of digits. Mechanically, the exercise works in basically the same manner as before: by typing the correct digits after the website, the users gain access to a file that leads them to the next location of the game.

As we see in Fig. 5.2, the exercise spreads across two pages and consists of six empty regular pentagons (one for each digit) and four questions. Two of the questions point to a single pentagon, while the other two questions point to two

Fig. 5.2 The third iteration of the third exercise in its digital format

pentagons. For example, the first question with two pentagons asks for the fourth square number. Here, the users are meant to write "16" using a pentagon for each digit. Another question asks for the fifth Fibonacci number. Here, the users are meant to write "3" in the pentagon.

5.6 The Third User Test

The third iteration of TGK was tested by Helga's classmates (18 in total). They were divided into groups of three of which two were assigned GoPros. Letting several groups loose at once created fruitful social dynamics of both competition and cooperation, which was obviously not the case in Helga's lone test run. We met at their school located relatively close to the first vertex of the booklet. After Helga's test, I was grateful that the weather was mild on this spring morning. I handed out a printed booklet to each group, and they were let loose.

The first group equipped with a GoPro consisted of Anton, Ester, and Birk, and they were the first group to find the first vertex. They behaved highly motivated by the competition element, although this behaviour conflicted with their realisation that the longer they spent on TGK, the less school mathematics they had to do.

The second group equipped with a GoPro consisted of Mille, Jeppe, and Karl, who tackled TGK with a humorous attitude, following the booklet's bidding with a layer of ironic distance while positioning themselves as scouts on an expedition. In that regard, the two groups had quite different approaches.

I will continue to highlight the events around the third exercise. Like Helga, the groups carried out the first two exercises rather smoothly, but once again, the third exercise proved troublesome. The main problem arose from the unfortunate circumstance that the exercise was somewhat visually confusing and phrased ambiguously. For example, is 0 a square number? Is it a Fibonacci number? The exercise assumed that 0 is a Fibonacci number, but not a square number. Not even among mathematicians and certainly not among eighth-graders are these undeniable facts. Depending on your stance on 0, the fifth Fibonacci number could be either 3 or 5. Likewise, the fourth square number could be 9 or 16.

Exactly this ambiguity confused both groups. The first group initially wrote "3" in both pentagons relating to the fourth square number, reasoning that there must be an implied multiplication sign between the pentagons. Why did they do this? I can think of two reasons: firstly, the square number sequence is generated by multiplication. It is more than fair to assume that the juxtaposed pentagons represent the same number multiplied by itself. Secondly, the group's guess, "$3 \cdot 3$", clearly included 0 in the square number sequence—contrary to what the exercise assumed.

The second group mistook square numbers for Fibonacci numbers and initially wrote "2" in both pentagons. The visual clue of the two pentagons, intended to show that the answer had two digits, instead turned out to be an element of confusion. Their mix-up with the different number sequences also indicated that their unfamiliarity with the mathematical content simply was not overcome by the booklet's mediation of it. Instead, both groups relied on their phones, googling the unfamiliar terms, and finding a mixed bag of sources. Especially the question of whether 0 belonged to the different number sequences was inconsistently answered online.

A wrong initial guess might not seem like a big deal. However, by the nature of the technical setup, the six-digit code could either be completely correct, or the website would not give a file at all—"error 404" does not discriminate between slightly or very wrong. This meant that the booklet did not support the method of trial and error. The groups were unable to get any confirmation on their progress by testing their code so far; the first group did not test their code at all.

Unfortunately, the exercise thereby reinforced the notion that mathematics is a matter of getting *the* right answer. This is especially detrimental in an exercise where the inclusion of 0 is so ambiguous (and a worthwhile consideration!). The design, however, left them guessing blindly, and with no sense of progress, their level of confusion rose, inversely related to their level of motivation. Even though the first group was initially happy with TGK as an alternative to school, this feeling shifted after 15 min of being stuck. At this point, the following exchange of words took place[3]:

Anton: Jeg har ikke oplevet en mere forvirrende opgave nogensinde.
I have never encountered an exercise as confusing as this one.
Ester: Det er lige før, jeg hellere vil—du vil hellere lave talrækker?
I would almost rather—you would rather do number sequences?
Anton: Ja. Men det her, det er jo sådan næsten talrækker.
Yes. But this, it's something like number sequences.

While Anton and Ester continued their discussion, Birk had completely disengaged from the exercise. In the interest of time, I chose to intervene by telling them the first three square numbers: 1, 4 and 9. Despite their despair, when both groups eventually got to the right answer with a little nudging from my part, they shouted with joy at the sight of the file and continued to the next vertex of the pentagon, undaunted.

5.6.1 My Learning Process

To be honest, I went into the user test of the third iteration thinking that it would be smooth sailing. Obviously, I was wrong. I thought I was treating the users as protagonists; that I had strained out all red herrings and mysterious wordings. But the users still behaved differently than expected. I remember thinking that my biggest learning outcome was that design will always take me by surprise: "I think I have made one thing, but in the hands of the users, it turns out that I made something quite different!".

[3] The original language of the conversation was in Danish, which we translated into English.

As is probably evident from my autoethnography so far, I was very hesitant and even resistant to changing the direction of my design. I had an initial idea and straying away from it felt like an erosion of that idea. I remember using the metaphor of a veneer to describe the way TGK developed into mathematical aesthetics with no real substance. Without the Fibonacci sequence, the Euclidean algorithm, and their ties to the regular pentagon, was TGK really about mathematics at all?

At some point during the development of TGK, I had a conceptual shift in understanding my role as a designer. Instead of seeing the hiccups in the user tests as failures of my intention, I began seeing them as answers to my questions. I started having a conversation with the booklet and the user tests. "Is the Euclidean algorithm appropriate for this context?", was a question I realised I had gotten the answer to ("NO!") early on. Recognizing the agency of the design itself and how it behaved in user-driven contexts, I let go of my own control over the booklet and instead gave in to the surprise of which direction the design was inevitably going.

In this way, my learning process models the active learning goal of Creative Pragmatics. Through designing a mathematics learning game, I underwent a transformation from trying (and failing) to execute my designer-centred ideas in a complete and controllable manner, to being open-minded enough to create the unintended.

In retrospect, I would also say that my learning process aligns with Eisner's notion of flexible purpose (Eisner, 2002). This process is open-ended and remains opportunistic, having adaptive goals that may evolve with the unpredictable real-world context. In the design of TGK, the user tests modelled the contingent real world, and my role as a designer was simply to be sensitive and attentive to its emergent features.

5.6.2 Artefact Improvement

Based on the test of the third iteration, we realised that words alone offer inadequate mathematical representation. At least it is safe to say that the ambiguous 0 of the square number sequence was not made less ambiguous by representing it with simply the words "square number sequence". In the test run, the booklet initially sparked high motivation, but representational hiccups such as too much text and hard-to-understand mathematics quickly turned the motivation into its inverse sibling, frustration. At the same time, the closed-ended questions in the third exercise reinforced unhelpful attitudes toward mathematics ("either you know, or you don't"). We saw this manifested in how Birk completely disengaged from the booklet while Anton and Ester had no luck finding the six-digit code.

Given that TGK was meant to be a fun and engaging activity in its own right, we knew we had to make some changes to the third exercise. The final iteration can be seen in Fig. 5.3. The most obvious change is probably the number of pentagons, which now matches the number of answers instead of the number of digits (which was also reduced to four). In that way, we tried to prevent users from interpreting two juxtaposed pentagons as multiplication, like both groups did in the test run.

Fig. 5.3 The final iteration of the third exercise in the final format. (Svabo et al., 2023)

We also phrased the question in a way that provided a bit of scaffolding. For example, when asking for the fifth square number, the question now initiates with the words: “The first four square numbers are 1, 4, 9 and 16”. In this way, the users will not have to guess whether 0 is included in the sequence or not. They might also be able to deduce the fifth square number from the sequence even though they never heard its name before. While still being quite closed-ended, the exercise at least provides the opportunity for exploration.

The QR-code on the bottom left is also a new addition. There was a QR-code before, but in the previous iterations, it just led to Fibonacci’s Wikipedia page. This time, the QR-code leads to a document explaining the number sequences in a concise and visual way. In every test, the users relied on their phones when they got stuck, so we wanted to ensure that the users would not get lost in the vast information scape of the internet.

The last improvement is the amount of text, which has been heavily reduced. In this final iteration, we only included information relevant to the exercise in the writing. The entire booklet underwent a similar transformation, becoming less word-heavy, while no step along the urban escape walk was left fickle.

In the meantime, Lisbet Foged, one of my colleagues, made sure the booklet was proof-read, printed, distributed, and advertised. There is no point in developing a booklet like TGK if no one is going to use it, so this part of the process was arguably the most important one. This part of the process also made explicit the flexible purpose of TGK. We had to decide where and to whom the booklet was advertised based on its emergent features. Lisbet arranged for TGK to be advertised in a local café who offered a discount to those who completed the escape game. She also asked several local schools if they were interested in a free mathematics activity.

TGK thereby ended up being somewhat of a hybrid between a family- or tourist-oriented leisure activity and educational material aimed at secondary level mathematics.

Now that the booklet is done, the question arises: what would I change if I were to do it again? If I had the opportunity to improve another iteration of TGK, what would I do differently this time? By the open-ended nature of the iterative, flexible process, a design such as TGK is never completely done. This just happened to be the iteration that was the last one—it is not the one and only "solution" to the task Connie posed in that morning meeting at the Center all these months ago. For one thing, TGK could have been aimed more unified towards schools, matching more clearly with the Danish mathematics curriculum. Others have suggested turning it into an app instead of a booklet, since the exercises rely on websites and QR-codes anyway.

If I were to develop TGK further, I would probably do more to tailor the booklet to schools. Specifically, TGK could benefit from a teacher's guide, so that the teacher can provide sufficient scaffolding before the walk, help when students get stuck, and later elaborate on the mathematical content in the booklet. The user test of the third iteration at least showed me that the social and competitive dynamics that arose from letting a whole class loose at once lifted the experience.

It is surprising to me that at the end of the design process, my ideas for further development go in the school direction. When I was still at the brainstorming phase with Christian, the thought of making the booklet explicitly school-like repelled me. I had this idea that I could make a game where the mathematics would manifest as the actions and strategies of the players. Playing the game would be a sort of mathematical gotcha: "A-*ha*! You were doing mathematics all along. And you were even enjoying it!". But I have come to think that it is rather okay to admit when you are doing mathematics.

5.7 Discussion

The chapter has demonstrated one key point to Creative Pragmatics: the open-ended process of design-based learning cannot be separated into linear, consecutive phases. Illustrative of this, the chapter oscillates between autoethnographic reflections on the design process and interpretations of students' use of the booklet in its different iterations.

In the above, we have seen how Amalie recorded different iterative processes of the design, the use of the artefact, and the improvement of the artefact, TGK. Her conceptual shift in understanding her role as a designer is indicative of how the artefact mediates creative learning as a student project. The following is a discussion of our case's implications for mathematical artefacts as part of Creative Pragmatics and university education. Locating the case of TGK in mathematics education, we also rethink the role of an artefact when facilitating active learning, developing playful learning, and bringing values of arts into mathematics education.

5.7.1 Designing Active Artefacts

In the design process, TGK underwent three iterations in the process to improve the prototype until its last version. The core questions that engaged Amalie in the three iterations were: How do I design TGK as a satisfying artefact from a user perspective? How can TGK engage users in solving mathematics problems in a practical context? How do I best design TGK as a successful mediator between users, environment, and learning? The users or learners hold TGK in their hands in the process of undergoing active learning by solving problems. All their discussions, decisions and activities respond to what they learn from TGK. This means users' creative learning experience is developed out of 'conversations' between users and TGK. The 'conversations' are 'back and forth' processes—TGK 'sets' a scenario of learning activities, 'asks' users to meet task challenges, 'calls for' group discussions, and 'speeds up' or 'delays' decisions. In this sense, we call TGK an 'active artefact'. The artefact 'facilitates' students' active learning by setting scenarios or challenges, motivating participation, and providing hints to solve problems. In this sense, a metaphor for describing a good design of an artefact like TGK could be an 'active facilitator' of students' learning (Zhou, 2014).

Following this metaphor, mediation, as the function of an active artefact, is a more-than-representational process. As Clark (2005) argued, when people collaborate, their communication and coordination acts go beyond linguistic signals and involve the use of material artefacts, locations, and physical spaces, i.e., material signals. An active artefact 'facilitates' students to learn in depth through participation in practice, which was the case for Amalie's participation in designing TGK. She experienced the real design process of an artefact, experimenting with and testing the booklet on authentic users, while collaborating with professionals.

The active artefact should provide students opportunities to change practice and engage in creative learning experience through a process of peer- or mentor-supported experimentation (Fseda et al., 2004). As Beetham (2002) found, people who had actually changed their practice reported that a crucial turning point was often the opportunity to witness the real thing, in the real context, with the real people, in other words, to actually watch a new approach or tool in action. In that way, TGK was an active artefact in Amalie's creative learning process, mediating the knowledge about design and mathematics education.

Usability, functionality, and contextual sensitivity are the main considerations that are involved in Amalie's observation on 'mediation' in pursuit of her improved design solutions. As is evident from Amalie's reflections, users' satisfaction seemed to be a very important consideration in the design process. This consideration might have been overly emphasised, especially since the subjective aspect of evaluation is such a complex aspect of design. According to Jordan (1998), satisfaction reflects the level of comfort users feel when using a product and how well they feel the product manages to achieve their goals. Therefore, a user-centred approach should take both physical and cognitive characteristics into account. The physical aspect is related to users' age, sex, nationality, and occasion, etc. The cognitive aspect involves users' specialist knowledge, attitudes, expectations, and evaluation, etc. These characteristics presented challenges to Amalie's design process by introducing a layer of complexity, but they also provided opportunities or clues that inspired Amalie to improve the 'functionality' of the design. This is also related to 'affordance', the concept that Amalie mentioned much in her learning journey, which enabled her to regard the properties of TGK as they emerged in the user tests.

5.7.2 Mediating Play and Learning Mathematics

As Amalie noted, TGK is a playful artefact that invites users to solve mathematical problems as they journey through the city centre of Odense. In the different stages of user tests, the users experienced different emotions such as motivation and frustration when engaging the challenges. We might say that when creative learning happens, users' emotions are also mediated by TGK in a way of play. To better understand this point, we clarify the notion of play.

As Mainemelis and Ronson (2006) suggested, play as engagement facilitates the cognitive, affective, motivational, and skill dimensions of the creative processes, while play as diversion fosters a psychological and social-relational climate that is conducive to creative learning. This is in line with how Singer (2013) described the characteristics of play: (1) play has value in itself and play gives pleasure; (2) play bonds through rhythms, rules and structure; (3) play gives freedom to change experience; (4) play is circumscribed by fluid boundaries, through rules in time and space; and (5) play is one of the wellsprings of culture.

As Koehler and his colleagues (2011) highlighted, design may facilitate deep-play and learning in-depth. Furthermore, when teachers use 'producing games' as a primary educational method, students' play can become creative learning experiences (Ejsing-Duun & Karoff, 2015). At the same time, when developing play pedagogy, teachers should support students in playing in a safe and challenging environment and adapt to their needs and interests (Singer, 2013). The teacher should pay attention to the quality of playfulness when practising play pedagogy, as it goes hand in hand with the co-construction of meaning and emotion among students. As Chang (2013) argued, 'playfulness' can be interpreted as abilities such as emotional expression and the use of intrinsic motivation, as well as characteristics and behaviours such as naturalness, a sense of freedom, happiness, being childlike, playing or being funny (Chang, 2013). As mentioned, the emotional expression is one of the focuses in Amalie's observations of user tests. Focusing on emotions made her adjust the design direction towards playfulness. The purpose of TGK became pleasure. Instead of adhering to a curriculum, the booklet had to be valuable in itself.

The design process of TGK was also one of playfulness. The design was a kind of problem-oriented learning activity, where the artefact set a situated learning context. By situating the design of the mathematical booklet in a real-world context, Amalie as a student-designer progressively reiterated the conceptual design to meet the demands of the real-world context (Gravemeijer, 2008). TGK bridged her thinking between the real world and the conceptual world, while also facilitating collaboration with others. According to Zhou (2014), students may much better involve themselves in self-directed learning, group coordination, and project management in situated problem-solving contexts. We saw in the autoethnography how Amalie's understanding of mathematics education design improved when dealing with complexity and uncertainty. This is in line with what Zosh and his colleagues (2018) argue, specifically that guided play may result in good academic outcomes, since it harnesses the features of an optimal learning environment as well as joy (positive affect) and iteration (construction of new knowledge through hypothesis testing). We may regard the design of TGK as part of Amalie's mathematical education, and by considering the design as a playful activity, we can see how the design process corresponds to a playful pedagogy striking a balance between 'free play' and 'structured play'. By exploring different playful pedagogies, students are encouraged to try and test their defined hypotheses, revisit understanding and experience, and engage in scripting, modelling, organising, sequencing, representing, creating, extending and socialising, all of which can be meaningful mathematics experiences (Perry & Dockett, 2007).

5.7.3 Rethinking Values of Arts

TGK is the result of Design-Based Research as well as Art-Based Education research. Through a user-centred approach, Amalie identified her design inquiry that accessed educational phenomena through the study and production of visual forms that bridge art and mathematics. As Lähdesmäki and Fenyvesi (2017) suggested, art often offers artistic or emblematic representations of the world; it deals with the everlasting question of the meaning of a thing in the world, while the fundamental educational question is how to enter in dialogue with the world—both disciplines deal with human beings' becoming as subjects through interaction with the material and social world (Lähdesmäki & Fenyvesi, 2017). Broadly, we can understand that the arts are embedded in the practice of educational technology. Through forms of images, multimedia, and video, etc., different kinds of artefacts are thoughtfully and intentionally introduced to and integrated into learning and performance environments in service of greater educational goals (Moore & Tillberg-Webb, 2022).

Mathematics teachers are expected to recognize the importance of arts and pay much attention to art-based pedagogy. The learners of mathematics should develop an appreciation for mathematics, experience the joy of mathematics, and explore the beauty of mathematics. Furthermore, Rezat (2006) noted that the focus of research in mathematics education has extended from the individual student's cognition and knowledge to contextual, socio-cultural, and situated aspects of mathematics learning and knowing. In this sense, arts help to lay the pathway to the extension, while the artefacts may be used to negotiate learning and knowing in between diverse aspects. According to Baird (2015), the integration of arts into mathematics teaching stems from the notion that the arts play a significant role in keeping students engaged in education by encouraging their learning through a physical and embodied way. This means arts invite students to collaborate with peers, by requiring them to respond emotionally and by calling upon their cognitive capacities as they learn in, through and about the arts (Upitis, 2011).

Broadly, education has many similarities with art, such as the fact that its purpose changes according to the process of the act and that the style and its content cannot be separately addressed (Komatsu & Namai, 2022). As Shen et al. (2015) argued, learning of sciences and arts can mutually support each other. On one hand, the students apply their scientific understanding of light to create artistic products, and on the other hand, students use visual aids as means to communicate their scientific ideas and findings both informatively and aesthetically, which also fuel their creative thinking. Eisner (2004) claimed that the aim of education ought to be conceived of as the preparation of artists; the 'artists' are individuals who have developed the ideas, the sensibilities, the skills, and the imagination to create work that is well proportioned, skilfully executed, and imaginative, regardless of the domain in which an individual works. Research from a humanistic view also suggests arts may provide a higher quality of human experience throughout a person's lifetime and helps to cultivate well-rounded individuals that meet challenges of fostering twenty-first century skills (Sousa & Pilecki, 2013). In this sense, STEM education is moving towards STEAM education which calls for the integration of arts into STEM education (Ge et al., 2015).

5.8 Conclusion

Exemplifying the principles of Creative Pragmatics, this chapter explored how an artefact, TGK, mediates creative learning in mathematics education. The student-designer's autoethnographic account of her participation in the design process of TGK showcased the iterative, reflective, and creative learning process surrounding the design of the booklet.

Conducting user tests was a major part of her learning process, in that the tests clearly illustrated the complexities of real-world contexts and thereby the contingency of design. Learning from the user tests, Amalie developed her conceptual framework from a designer-centric perspective to a more user-centric approach to design, recognizing the inability to transfer mathematical ideas in a linguistic and straightforward manner. This is also evident from the development of the booklet itself. The third exercise, which was the focus of this chapter, went from several pages of text to a sleek double page spread relying more on visuals.

In the discussion, we called attention to the complexity in the affordance of an artefact. We introduced the notion of an active artefact, regarding the booklet as a group member of the project, a group member that mediated the creative learning process. Acknowledging the agency of the design corresponds to Amalie's experience of surprise and eventual surrender when the booklet went in directions she did not predict.

In dialogue with theories on design, play, and mathematics education, we contextualised the case in research adjacent to Creative Pragmatics. Designing an artefact fits into active learning pedagogies, by which students learn from creative experiences. In Amalie's authentic participation in practice, she engaged playfully in design, and by reflecting on her experiments, she developed her knowledge on mathematics education. Based on this, we suggest enriching mathematics education with artistic approaches such as this one.

Our chapter also sows the seeds for further research. Firstly, it seems fruitful to take a systematic view to research affordance and artefact design in developing active pedagogies; secondly, the practice of designing or using an active artefact should consider issues of contextual sensitivity, which means we should develop artefact-mediated learning environments that are suitable for particular students and their learning activities; and thirdly, an artistic image should be added to the profile of mathematics education that may enrich future research on artefacts, creative learning, and active mathematics education.

References

Adams, T. E., Ellis, C., & Jones, S. H. (2017). Autoethnography. In J. Matthes, C. S. Davis, & R. F. Potter (Eds.), *The international encyclopedia of communication research methods*. https://doi.org/10.1002/9781118901731.iecrm0011

Baird, D. (2015). *Integrating the arts in mathematics teaching*. University of Toronto.

Bakker, A. (2018). *Design research in education: A practical guide for early career researchers*. Routledge.

Beetham, H. (2002). Developing learning technology networks through shared representations of practice. In C. Rust (Ed.), *Improving student learning through learning technologies*. Oxford Centre for Staff and Learning Development.

Chang, C. P. (2013). Relationships between playfulness and creativity among students gifted in mathematics and science. *Creative Education, 4*(2), 101–109.

Clark, A. (2005). Ways of seeing: Using the Mosaic approach to listen to young children's perspectives. In A. Clark, P. Moss, & A. Kjørholt (Eds.), *Beyond listening: Children's perspectives on early childhood services* (pp. 29–49). Policy Press.

Eisner, E. W. (2002). *The arts and the creation of mind*. Yale University Press.

Eisner, E. W. (2004). What can education learn from the arts about the practice of education? *International Journal of Education and the Arts, 5*(4), 1–13.

Ejsing-Duun, S., & Karoff, H. S. (2015). Creativity and playfulness: Producing games as a pedagogical strategy. *Proceedings of European Conference on Game Based Learning*.

Fseda, R. S., Beetham, H., & Ravenscroft, A. (2004). Active artefacts: Representing our knowledge of learning and teaching. *53 Interesting Ways in Which Colleagues Resist Change, 5*, 16.

Ge, X., Ifenthaler, D., & Spector, J. M. (2015). Moving forward with STEAM education research. In X. Ge, D. Ifenthaler, & J. M. Spector (Eds.), *Emerging technologies for STEAM education: Full Steam ahead* (pp. 383–396). Springer.

Gibson, J. J. (1979). *The ecological approach to visual perception*. Houghton Mifflin Company.

Gravemeijer, K. (2008). RME theory and mathematics teacher education. In D. Tirosh & T. Wood (Eds.), *The international handbook of mathematics teacher education: Tools and processes in mathematics teacher education* (pp. 283–302). Sense Publisher.

Hutchby, I. (2001). Technologies, texts and affordances. *Sociology (Oxford), 35*(2), 441–456. https://doi.org/10.1017/S0038038501000219

Jordan, P. W. (1998). *An introduction to usability*. Taylor & Francis Ltd..
Koehler, M. J., Mishra, P., Bouck, E. C., Deshryver, M., Kereluik, K., Shin, T. S., & Wolf, L. G. (2011). Deep-play: Developing TPACK for 21st century teachers. *International Journal of Learning Technology, 6*(2), 146–162.
Komatsu, K., & Namai, R. (2022). Art=research: Inquiry in creative practice. In K. Komatsu, K. Takagi, H. Ishiguro, & T. Okada (Eds.), *Arts-based methods in education research in Japan* (pp. 1–23). Brill.
Kristensen, M. L. A., Larsen, D. M., Seidelin, L., & Svabo, C. (2023). The role of mathematics in STEM activities: Syntheses and a framework from a literature review. *International Journal of Education in Mathematics, Science and Technology, 12*(2), 418. Article 12(2).
Lähdesmäki, T., & Fenyvesi, K. (2017). Bridging art and mathematics: Introduction. *Aesthetics of interdisciplinarity: Art and mathematics*, 1–25.
Larsen, D. M., Kristensen, M. L., Seidelin, L., & Svabo, C. (2022). Mathematics as the focal point of STEM teaching. In *Twelfth congress of the European Society for Research in mathematics education: CERME 12*. Article 12.
Mainemelis, C., & Ronson, S. (2006). Ideas are born in fields of play: Towards a theory of play and creativity in organisational settings. *Research in Organisational Behaviour, 27*, 81–131.
Moore, S. L., & Tillberg-Webb, H. K. (2022). Mediating artefacts ethical, social, and political dimensions of content and media selection and design. *The Journal of Applied Instruction Design*, 105–119.
Perry, B., & Dockett, S. (2007). *Play and Mathematics. The Australian Association of Mathematics Teachers (AAMT)*. Retrieved on 28 January, 2024 from https://www.semanticscholar.org/paper/Play-and-mathematics-Perry-Dockett/6e574174176a3bfa6f338c40161e9ee8dc1bd849
Radford, L. (2014). Reflections on history of mathematics. In M. Fried & T. Dreyfus (Eds.), *Mathematics & Mathematics Education: Searching for common ground* (Advances in Mathematics education). Springer. https://doi-org.proxy1-bib.sdu.dk/10.1007/978-94-007-7473-5_7
Rezat, S. (2006). The structures of German mathematics textbooks. *ZDM, 38*, 482–487.
Scott, G., & Shrum, W. (2017). Video ethnography in practice: Planning, shooting, and editing for social analysis. *Sage*. https://doi.org/10.4135/9781071802755
Shen, J., Jiang, S., & Liu, O. L. (2015). Reconceptualizing a college science learning experience in the new digital era: A review of literature. In X. Ge, D. Ifenthaler, & J. M. Spector (Eds.), *Emerging technologies for STEAM education: Full Steam ahead* (pp. 61–79). Springer.
Singer, E. (2013). Play and playfulness, basic features of early childhood education. *European Early Childhood Education Research Journal, 21*(2), 172–184.
Sousa, D. A., & Pilecki, T. (2013). *From STEM to STEAM: Using brain-compatible strategies to integrate the arts*. Corwin Press.
Svabo, C. (2008). Artefacts and the performance of an exhibition. *Nordic Journal of Digital Literacy, 3*(2), 113–129. https://doi.org/10.18261/ISSN1891-943X-2008-02-04
Svabo, C. (2016). Performative schizoid method: Performance as research. *PARtake: The Journal of Performance as Research, 1*(1), 1–21.
Svabo, C., & Borch, M. M. (2020). Oplevelseslæring og BioFabLab. *MONA—Matematik- og Naturfagsdidaktik, 1*(2020) https://tidsskrift.dk/mona/article/view/118892/166746
Svabo, C., et al. (2023). *Den Mystiske Portnøgle [Booklet]*. University of Southern Denmark.
Upitis, R. (2011). Engaging students through the arts. *The Literacy and Numeracy Secretariat, 33*, 1–4.
Zhou, C. (2014). A student project as an 'extra group member': A metaphor for the development of creativity in problem-based learning (PBL). *Academic Quarter | Akademisk Kvarter, 9*, 223–235.
Zosh, J. M., Hirsh-Pasek, K., Hopkins, E. J., Jensen, H., Liu, C., Neale, D., Solis, S. L., & Whitebread, D. (2018). Accessing the inaccessible: Redefining play as a spectrum. *Frontiers in Psychology, 9*, 1124.

Amalie Thorup Eich-Høy PhD student at STEM Centre for Research in Science Education and Communication (FNUG), Department of Mathematics and Computer Science, University of Southern Denmark. Amalie has a double degree in Mathematics and Danish language and literature. She is currently doing fieldwork for her PhD project in the didactics of university mathematics.

Chunfang Zhou PhD, Associate Professor of STEM Education and Science Communication, STEM Centre for Research in Science Education and Communication (FNUG), Department of Mathematics and Computer Science, University of Southern Denmark. A researcher and practitioner working on creativity, STEM education, and teacher education. Chunfang has an interdisciplinary and cross-cultural educational background: she finished Bachelors Degree (Industry Automation and Information Science) and Masters Degree (Philosophy of Science and Technology) in China; during 2008–2012, Chunfang finished Ph.D. study on Group Creativity in Engineering Education in a Problem and Project-Based Learning (PBL) environment at Aalborg University, Denmark. She is a board member of Danish Development Research Network (DDRN), Denmark and works as a senior consultant in PBL Teaching Innovation and Research Center at Northeastern University (NEU), China. Chunfang locates her research in Science, Technology and Society (STS), with a focus on creativity and STEM education. Since 2013, Chunfang has authored and co-authored over 120 peer-reviewed publications; she is the editor of Handbook of Research on Creative Problem-Solving Skill Development in Higher Education (published in 2016, by IGI Global), the author of Introducing Problem-Based Learning (PBL) for Creativity and Innovation in Chinese Universities (published in 2020, by IGI Global), and one of editors in book series Arts, Creativities and Learning Environments in Global Perspective, Brill/Sense Publisher. Chunfang is an editorial board member in Discovery Education, European Journal of Education and Pedagogy, and advisory board member in International Journal of Engineering Education.

Chapter 6
Summative STEM Competence Assessment: Affect-Sensitive Learning

Maiken Westen Holm Svendsen and Dorte Moeskær Larsen

6.1 Introduction

In this chapter, we delve into summative assessment in the Danish educational system, particularly focusing on its role in evaluating both cognitive and affective learning dimensions within an integrated science, technology, engineering, and mathematics (STEM) educational framework. Our approach is rooted in Creative Pragmatics to bolster the assessment process. Summative assessment, traditionally known for assessing knowledge and skills, is reimagined here to encompass a broader spectrum of students' STEM competencies. Summative assessment, by definition, is the evaluation of student learning at the conclusion of an instructional period, typically through standardized tests or final projects. This form of assessment often lacks a competency dimension and primarily focuses on knowledge acquisition (Binau et al., 2021).

In the context of Danish mathematics and science education, where STEM has gained prominence (Svabo et al., 2024), the demand for STEM-educated individuals is escalating (Seidelin et al., 2021). This surge is partly due to concerns over a future shortage of qualified STEM professionals (Lykkegaard & Ulriksen, 2019).

A Danish doctoral student and assistant professor team up here to consider the trickier and often overlooked aspect of measuring what matters in STEM education. A consideration of emotions is essential for making Creative Pragmatics work in practice: this paper highlights foundations in theory of affect in social and cultural practice. The approach is on evaluating competence through a rounded and holistic focus on learning experiences in their cognitive, sensory, and emotional aspects, in their materialities and affectivities, in learning environments and contexts. (Editors)

M. W. H. Svendsen (✉) · D. M. Larsen
STEM Centre for Research in Science Education and Communication (FNUG), Department of Mathematics and Computer Science, University of Southern Denmark, Odense, Denmark
e-mail: mwhs@imada.sdu.dk; dmla@imada.sdu.dk; https://www.sdu.dk/en/forskning/stem

C. Svabo et al. (eds.), *Creative Pragmatics for Active Learning in STEM Education*, Contributions from Science Education Research 14,
https://doi.org/10.1007/978-3-031-78720-1_6

STEM education, however, is not just about imparting discipline-specific knowledge; it is about fostering competencies such as critical thinking, collaboration, and creativity McLoughlin et al. (2020). A shift towards competencies as educational outcome has been emphasized for over two decades in Danish education (Dolin, 2002; Niss & Højgaard Jensen, 2002). And from that a pivotal question arises: How can our assessment methods effectively capture not only discipline-based knowledge and skills but also broader, transdisciplinary STEM competencies?

Assessing students for interdisciplinary work and integrating STEM teaching and assessment into a discipline-centric educational system presents challenges as it demands a shift from specialized, siloed approaches to more holistic, interconnected methods of learning and evaluation. Moreover, students often perceive mathematics and science as intimidating, impacting their motivation and interest (Gilbert, 2004; Ruffell et al., 1998; Svabo et al., 2024).

In addressing these challenges, this chapter advocates for summative assessment methods that (a) reflect disciplinary learning goals, (b) reflect interdisciplinary STEM competencies and (c) take into account the affective domain of students. Utilizing data from the LabSTEM+ research and development project, we explore assessments in a seventh-grade mathematics and science class engaged in a full-day STEM course. Our analysis is anchored in a theoretical framework that integrates cognitive and affective dimensions of learning STEM, aiming to assess STEM competencies and contribute to what we term "*Affect-sensitive STEM assessment*." This approach represents a shift toward more holistic evaluation practices in STEM education.

Evaluating cognitive competencies requires more than traditional tests; it calls for engaging students in activities where they can actively demonstrate their skills. The students' affective domain is crucial for student performance and involves the emotional experiences of students during learning. Creating a positive emotional environment is essential for effective evaluation, as it influences students' ability to demonstrate their competencies. A supportive, encouraging learning environment enhances student engagement, motivation, and confidence, fostering active learning for the students.

6.1.1 Assessment in STEM Education

To gain a deeper understanding of how STEM assessment practices can be developed, it is beneficial to examine their current state. We focus on STEM assessment practices within a Danish context, as this is also the context of our case study.

In Denmark, there is a significant emphasis on formative assessment. Formative assessment is described as evaluation or feedback *for* learning, provided throughout the academic lifetime, with the aim of enhancing the student's learning process and continuously improving it (Andresen, 2017; Christiansen et al., 2020; Conradsen, 2021; Dolin et al., 2017). While there is a consensus on the importance of formative assessment, there is equal agreement on the need to refine summative assessment, as it remains a necessary component of our education system (Jess, 2005; Madsen & Levinsen, 2007), and is viewed as important for both teachers, politicians and parents.

Summative assessment has been criticized for not being competency-oriented (Højgaard & Sølberg, 2023; Nielsen & Dolin, 2016) and for not capturing the STEM competencies educational systems aim to develop and test in students. Educational researchers suggest that the route forward lies in integrating formative elements into summative assessments, allowing a portion of the learning process to influence the final grade (Bjørkli & Arnesen, 2015). This integration is often seen in examination formats commonly used in higher education, such as portfolio exams and project exams, where students present and discuss a process or project they have engaged with during the semester (Andersen, 2008). In a Danish primary seventh grade class, integrating formative elements into summative assessments might involve students working on interdisciplinary STEM projects that contribute to their final grade. For example, a project combining science, math, and technology would be assessed not only on the final outcome but also on the learning process, including problem-solving and collaboration.

6.2 Research Questions

Competence orientation that is sensitive towards affect is especially pertinent in an era where education is becoming increasingly competitive, and students' engagement, motivation, and emotional investment in their learning are recognized as pivotal factors in their overall educational experience (Nielsen & Dolin, 2016). If one momentarily forgets about examinations and solely focuses on teaching, the development of teaching has moved from a mental constructivism approach building on Piaget to a more social constructivist and pragmatic approach building on the work of Vygotsky and Dewey as beneficial for learning. Gibson (1966), among others (Reed, 1991; Shaw & Bransford, 1977; Young, 1993), advanced approaches to learning going from isolated from the environment to learning and knowledge being an activity that is always contextualized and a part of the individual–environment interaction (Barab & Dodge, 2008). These thoughts have a positive influence on the way educators teach STEM subjects today, and this research aims to also transfer these thoughts into the summative assessment practice as well.

Within this context, we explore the role of the affective domain in co-constructing learning, with a particular focus on the development of STEM competencies. Thus, our research question is:

1. *How do students experience a situation of STEM assessment?*
2. *Which competencies unfold in a situation of STEM assessment performed in the project LabSTEM+?*

To answer these questions, we must delve deeper into how we can integrate aims and goals of monodisciplinary subjects and interdisciplinary STEM teaching while also integrating both cognitive and affective dimensions within STEM assessment practices, while acknowledging and valuing the dynamic and iterative nature of learning in a modern STEM educational context.

6.3 Theory for Analyzing Summative Assessment

The theoretical framework in this chapter consists of two main sections.

The first section emphasizes knowledge, skills, and competencies, drawing inspiration from McLoughlin et al. (2020) work on the cognitive aspects of learning in STEM education. This perspective underscores the importance of equipping learners with the necessary tools to understand, apply, and innovate within STEM fields. It emphasizes both the foundational knowledge, skills, and competencies that students acquire within discipline-based education, as well as the STEM-specific competencies that emerge when these disciplines are integrated. This dimension is crucial as it pertains to the tangible abilities and understanding that students gain from their educational experiences. It is about the "what" of learning—what students know, what skills they have developed, and what competencies they can demonstrate.

The second section delves into the affective dimension of learning, grounded in the affect theory of Schindler and Bakker (2020). This dimension recognizes the emotional and motivational components of learning, acknowledging that students' feelings, attitudes, and values play a crucial role in their educational journey. It recognizes that students are not just passive recipients of information but are emotionally engaged beings (Schindler & Bakker, 2020). This dimension addresses the "why" of learning—why students feel a certain way about a topic, why they are motivated to delve deeper, or why they might feel apprehensive or enthusiastic about certain subjects.

In essence, by integrating these two theories, the framework provides a holistic view of the student learning experience while being assessed. It ensures that assessments do not just measure rote discipline-based knowledge but capture the depth and breadth of students' cognitive achievements and emotional engagements, offering a more holistic understanding of their STEM competencies.

In the next sections, we will describe the two theories.

6.4 STEM Assessment and the Cognitive Dimension

The first component of the theoretical framework is the cognitive aspect of the assessment. Here, we focus on the goals and aims for education, delving into competencies as this is the missing link in the current assessment, since competencies are often overlooked due to more traditional assessments that tend to emphasize the recall of knowledge rather than the ability to apply that knowledge in practical, real-world contexts. First, it is essential to note that, unlike other perspectives on STEM teaching, our approach to STEM emphasizes an interdisciplinary understanding of STEM education, meaning that STEM teaching involves integrating at least two of the four STEM disciplines (Svabo et al., 2024). However, very few summative assessments in the Danish school system are interdisciplinary. Thus, current examination practices and assessment methods can limit the extent to which we can introduce STEM into the school system (Honey et al., 2014).

Few exam formats in the Danish education system allow a straightforward integration of interdisciplinary STEM. One exam format in primary schools combines physics, chemistry, biology, and geography. At secondary level there are specific high schools that have a science exam combining chemistry, geography and biology. However, none of these exams at primary or secondary level integrate mathematics, as this subject traditionally has its own exams consisting of both an oral and a written examination.

We aim to inspire more of these interdisciplinary exams, arguing that the objectives of individual subjects are not diminished but rather have the opportunity to be more fully realized by students in these interdisciplinary exam formats, while also constructing situations where STEM competencies can arise. Exam formats such as portfolio and project exams are ideal for interdisciplinary STEM work, as they invite the student to choose what is important.

In recent years, research in scientific and mathematics assessment has emphasized the importance of testing for competencies such as mathematical thinking, problem-solving, modeling, reasoning, representation, formalism, communication, and the use of mathematical tools and aids in mathematics (Niss & Højgaard, 2019) (see Table 6.1), and competencies such as inquiry, modeling, perspectivation, and communication in science subjects (Højgaard & Sølberg, 2023).

Table 6.1 Overview of Mathematical, science and STEM competencies. Common competencies and individual competencies

Mathematical competence (M)	Science competence (S)	STEM competence (STEM)
Communication	Communication	Communication
Problem Solving	Problem Solving	
Inquiry	Inquiry	
Symbol and formalism	Perspectivation	Metacognitive
Aids and tools		Self-regulation
Reasoning		Collaboration
Representation		Critical thinking
		Innovation and creativity

It is also widely recognized in the literature that STEM approaches in teaching allow students to develop a wide range of competencies beyond subject-oriented competencies (Bailey et al., 2015; Blackley & Howell, 2019; Bybee, 2013), but what competencies are we referring to when discussing STEM education? Broadly, ENCoRE (2005) describes core STEM competencies as:

> … [T]he set of knowledge, skills, and/or competences an individual has acquired and/or can demonstrate after completing a learning process. Learning outcomes are statements of what a learner is expected to know, understand, and/or be able to do at the end of a period of learning. (ENCoRE, 2005, p. 11)

McLoughlin et al. (2020) describe what the report refers to as 'core STEM competencies', which, in addition to the competencies specific to the discipline, also include:

1. Collaboration—the ability to collaborate with others on STEM topics.
2. Problem-solving—the ability to solve STEM-related problems.
3. Innovation and creativity—the ability to think creatively and innovatively.
4. Critical thinking—the ability to critically assess problem-solving processes.
5. Self-regulation—the ability to take responsibility and show autonomy.
6. Communication—the ability to communicate and present.
7. Metacognitive competencies—the ability to adapt and be flexible.

In Table 6.1, the competencies in STEM, Mathematics, and Science are presented alongside each other, highlighting both shared elements and distinct characteristics across these domains. The comparison reveals an integration between the specialized competencies of Mathematics and Science and the broader, interdisciplinary skills emphasized in STEM. STEM competencies encompass a wide range of skills crucial for addressing complex, real-world problems. They include the ability to collaborate effectively, which is essential in interdisciplinary projects that require diverse expertise. Problem-solving in STEM is not limited to theoretical contexts but extends to practical applications, encouraging innovation and creativity. Critical thinking and metacognitive competencies are emphasized, highlighting the importance of not only solving problems but also reflecting on and adapting problem-solving processes. Additionally, self-regulation is a key competency, as it involves taking responsibility for one's learning and demonstrating autonomy in complex tasks. Communication is another vital competency in STEM, involving the clear and effective presentation of ideas across different disciplines and to varied audiences. Mathematics and Science competencies, while also addressing problem-solving and communication, have a more specialized focus. Mathematics emphasizes rigorous mathematical thinking, reasoning, and the use of symbols and formalism. It also includes modeling, which involves representing real-world situations mathematically. Science competencies similarly include modeling and emphasize inquiry, a process that involves exploring scientific questions through experimentation and observation. Both fields value clear communication, though in Science, there is an added focus on placing scientific knowledge in broader contexts, a competency referred to as perspectivation. The integration between Mathematics and

Science on one hand, and STEM on the other, is evident in the way STEM builds upon and expands these discipline-specific competencies. STEM education integrates the precise and structured approaches of Mathematics and Science with broader, more flexible competencies such as collaboration, creativity, and adaptability. This integration not only equips students with deep knowledge in specific disciplines but also prepares them for the interdisciplinary and dynamic nature of real-world challenges. Thus, while Mathematics and Science provide the foundational knowledge and methods, STEM competencies facilitate the application and extension of this knowledge in innovative and collaborative contexts.

6.5 The Affective Field in Relation to Assessment

To delve deeper into students' affective experiences during assessment, we turn to Maike Schindler and Arthur Bakker's (2020) description of *affective fields*. This model describes various affective factors influenced during learning situations, akin to a magnetic field, where these factors interact and influence each other (Fig. 6.1).

The affective factors include:

1. Emotions: Emotional responses to tasks or interactions.
2. Self-efficacy: A person's trust in their abilities to handle tasks.
3. Beliefs: Deep-seated attitudes about the subject.
4. Interest/motivation: The drive to engage in the subject.
5. Values: Principles guiding attitudes and behavior.
6. Attitude: Overall disposition toward the subject.

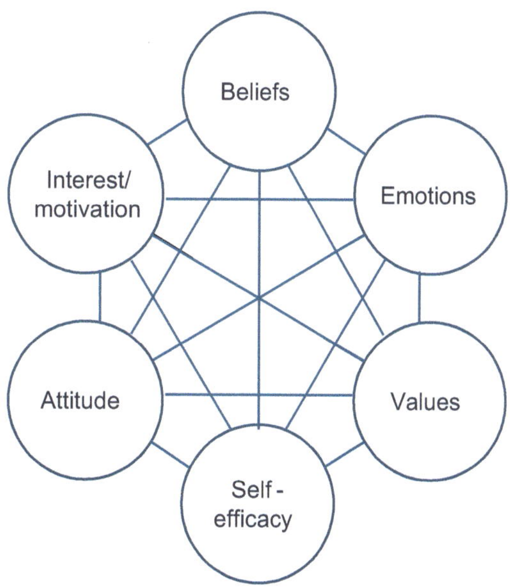

Fig. 6.1 Own representation of the affective field by Schindler and Bakker (2020)

Understanding these factors provides insights into students' experiences during assessments, highlighting the importance of considering both cognitive and affective dimensions for a holistic evaluation.

6.5.1 Affect and Cognition: An Interconnected Relationship

When assessing students' cognitive dimension, it is tempting to focus solely on measurable outcomes, such as test scores or task performance. However, this narrow lens can overlook a crucial aspect of the learning experience: the second component of the framework—the affective dimension.

Affective factors encompass the emotional and attitudinal dimensions of learning, including elements such as motivation, self-efficacy, and feelings (Schindler & Bakker, 2020). In this chapter, we support a comprehensive approach to assessment that integrates both cognitive and affective factors. By braiding the affective dimension into the STEM assessment process, educators aspire to capture a more holistic snapshot of students' learning experiences, while having a focus on following the interests of the students. This perspective acknowledges the intertwined nature of emotions and attitudes with cognitive aspects, recognizing their profound influence on learning outcomes.

Drawing from Wickman et al. (2022), we explore the interconnected dynamics between affect and cognition in the students' learning trajectory. It is emphasized that emotions and attitudes are not isolated from cognitive processes; they reciprocally shape and influence one another. Notably, emotions and attitudes exert a significant impact on students' engagement, interest, and motivation within science education (Schindler & Bakker, 2020). Positive affective experiences, such as curiosity and fascination, have the potential to amplify cognitive processes, leading to a deeper grasp of scientific concepts and enhanced retention of knowledge (Schindler & Bakker, 2020).

Wickman et al. (2022) exploration of aesthetic experiences underscores the significance of affective factors in education. Engaging with visually captivating materials, participating in hands-on experiments, and being immersed in interactive learning activities can elicit positive affective responses. These experiences, as Wickman et al. (2022) suggest: "[O]ffer a richer and more actionable insight into the relation between affect and conceptual learning than mental constructs" (Wickman et al., 2022, p. 718). Consequently, students develop a more integrated and holistic comprehension of scientific phenomena, aligning seamlessly with the principles of Creative Pragmatics.

In essence, the integration of affective factors into the assessment process offers a lens to recognize that emotions and attitudes are not peripheral but central to effective learning (Rosiek, 2003). By evaluating both cognitive and affective dimensions holistically, educators can nurture a richer and more comprehensive educational journey, fostering students' engagement, motivation, and profound understanding of the subject matter.

6.6 The LabSTEM Project

The LabSTEM+ project, initiated in 2023 as a continuation of the LabSTEM project from 2020–2022, focuses on the research-based development of STEM laboratories in Denmark (Svabo et al., 2024). These laboratories serve as collaborative spaces where teachers participate in workshops to design and test integrated STEM activities.

While the technical details of the LabSTEM project are extensive, what is particularly relevant for this study is how the project facilitates the development and examination of STEM assessment practices, specifically in relation to both cognitive and affective dimensions of learning. The project's guiding principles, developed during the original LabSTEM initiative and described by Svabo et al. (2024), emphasize creating meaningful STEM learning experiences that are participant-centered, real-world focused, and integrated across the STEM disciplines. These principles provide a strong foundation for exploring how STEM activities not only meet curriculum objectives but also engage student's affective dimension and cognitively. One of the innovative aspects of the LabSTEM+ project is its focus on incorporating affect-related perspectives into the assessment of STEM activities. This approach recognizes the importance of student agency and the influence of the physical learning environment on the learning process. By integrating these affective dimensions, the project allows for a holistic assessment of STEM activities, considering not just the cognitive outcomes but also how students' emotions and engagement impact their learning experience.

6.7 A Case from the LabSTEM Project with Focus on Assessment

In LabSTEM, 26 STEM activities were developed to primary and lower secondary school. In this chapter we have selected to present one particular case because it clearly prioritized evaluation as part of the activity. This activity took place at a hydroelectric power plant and was specifically aimed at fostering the students' understanding of this location and the functioning of a hydroelectric power plant. The activity was designed for a seventh-grade class by three teachers of the class, and it was designed as a whole day field trip to a nearby hydroelectric plant. This field day was structured with three different types of activities initially, all leading up to the fourth and final evaluating activity, which specifically involved building their own hydropower plant.

The first activity focused on students expressing their opinions on whether more artificial water channels should be dug to increase water flow to the hydropower plant, thus generating more green energy, even though it would impact the biodiversity of the area. Students were required to position themselves under the statement they found most favorable and then proceed to provide the best possible arguments to support their position.

The second activity consisted of a photo safari where students were tasked with capturing images of and explaining the various components of the power plant. This included turbines, generators, water supply systems, eel ladders, etc. The aim of this exercise was to familiarize the students with the different components of the plant. Subsequently, the teacher spent some time explaining how the hydroelectric power plant itself operated.

The third activity was divided into four distinct workshops. These workshops aimed to provide students with a detailed examination of individual parts of the plant and to utilize mathematics to explain and enhance the efficiency of a hydroelectric plant. Each workshop presented an overarching question to be solved:

Workshop 1: The question posed was: “How much water can you transport from the top of the reservoir to the bottom in 10 minutes, and at what speed?” In this workshop, each student was given a bucket to fill with water, and then they were instructed to run downhill with the filled bucket. The mathematical task was to calculate the amount of water they had managed to transport down the hill and the speed of the water (Fig. 6.2).

Workshop 2: The task was to measure the velocity of the water stream going out of the hydroelectric plant. In this activity, the students were required to experiment by placing various objects such as cream, leaves, sticks, etc., in the water and measuring the time it took for them to move a certain distance (Fig. 6.3).

Workshops 3: The task here was to investigate the question: “How much water can the water supply pipe hold?” which means that the students had to measure the circumference and length of the pipe and calculate its volume (Fig. 6.4).

Fig. 6.2 Students collecting water from a large water reservoir in a bucket

Fig. 6.3 Students measuring the velocity of the water stream going out of the power plant

Fig. 6.4 The water supply pipe, transporting water from the reservoir to the turbines

Workshop 4: The task here was to find the volume of the water reservoir on top of the pipe. This reservoir is box-shaped. Here, the students were required to measure the length of the box and the depth using a string with attached stones, as well as the width of the box (Fig. 6.5).

The final and evaluating activity was linked specifically to the task "How can we build the most efficient hydroelectric plant by using different materials (different pipes, jugs of water, small turbines, and voltmeters)?" This part can be considered an engineering project and an evaluation of the focus of the day. The intention was that the students should use all the knowledge they had learned throughout the day to build their own mini-hydroelectric plant. The students began with an ideation phase where they considered how to construct their own hydroelectric power plant based on the materials available to them. Subsequently, they built and tested their ideas in an iterative process. Finally, all groups were required to demonstrate how much power they could generate by running water through their pipes. The task was framed as if they were all engineers tasked with constructing a more environmentally friendly energy source (Fig. 6.6).

Fig. 6.5 The water reservoir

Fig. 6.6 Students building their own hydrogenic power plant, with pipes and turbines, by letting the water run from the top of the hill through the pipe to the turbine

6.7.1 Method

In this study, we employed a case study approach because it focuses on achieving an in-depth and thorough understanding of the teaching situation. The intention is to incorporate the context of the case to better comprehend the various variables influencing the situation. We consider the examined case as an intrinsic case (Stake, 1995), the aim being to understand the specific case in depth.

The design is a classic single-case design, where we try to understand the case as a holistic entity. To conduct an in-depth examination of this case, the entire field day at the hydroelectric power plant was recorded with a handheld video camera. Given that several of the exercises occurred asynchronously in different workshops, a group was selected that was video filmed throughout the different activities. During the final evaluation exercise, recording was done in slightly different groups to observe various approaches to solving the task.

The utilization of video recordings in research provides numerous advantages, including the capture of rich, contextual data that is difficult to quantify through traditional methods. Video recordings afford researchers the opportunity for in-depth analysis of behaviors and interactions, enhancing the understanding of complex processes. The reproducibility of events through repeated viewing supports reliability and the identification of nuanced details. The visual nature of video recordings aids in communicating complex findings, making them more accessible because of the possibility of replaying the situation multiple times. The authenticity preserved in video recordings further enhances the credibility of the research (DuFon, 2002; Nassauer & Legewie, 2021).

Upon reviewing the videos, we discerned the ways in which students participated in the activities—how they performed, how they employed their mathematical competences, and how they described or enacted their affective fields.

In the next section, we present our analysis of the video recordings. The analysis of the evaluation/assessment exercise was conducted employing a deductive approach, with the focus toward the two distinct theoretical dimensions: cognition and affect. We endeavored to discern the data, aligning them with each of these theoretical dimensions. To bolster intercoder reliability, both authors initially conducted independent analysis, deferring collaborative discussions.

6.8 Analysis: Building a Hydrogenic Power Plant

6.8.1 The Competencies in the Case

The STEM subjects involved in the activity are mathematics and physics. To understand the integration of STEM assessment within a monodisciplinary educational framework, we must first consider the competencies specific to these disciplines presented in Sect. 6.4. In this context, mathematics and physics serve as the primary vehicles for STEM integration, allowing us to explore how disciplinary knowledge and skills can be expanded through a STEM-focused lens.

In the following, we briefly examine the competencies related to these subjects. Mathematics, for example, emphasizes problem-solving, communication, and the ability to model and interpret data. Physics, on the other hand, focuses on inquiry and communication, understanding fundamental principles of the natural world, and using experimental methods to investigate the physical phenomena of velocity and power. These subject-specific competencies align with broader STEM goals, yet the challenge lies in how they can be effectively integrated to enhance cognitive development within a traditionally monodisciplinary curriculum.

To deepen this understanding, we look to the STEM competencies defined by McLoughlin et al. (2020) (Sect. 6.4), which provide a framework for evaluating how well these disciplinary competencies can be aligned with STEM's interdisciplinary objectives. This part of the analysis will primarily focus on the cognitive aspects of STEM integration, examining how the integration of mathematics and physics can foster deeper learning and a more cohesive understanding of these subjects. Specifically, we will explore how cognitive skills such as reasoning, problem-solving, and conceptual understanding are enhanced or challenged when STEM principles are applied within a monodisciplinary context.

6.8.1.1 Mathematics Competences in the Assessment Activity

The activity involved the construction of a hydroelectric power plant designed to assess the students' problem-solving abilities and their comprehension of mathematical concepts, particularly angles and pipe volume. Throughout the process, the roles assigned to the various students varied significantly; some students engaged in problem handling, using mathematics to address the challenges of building the power plant such as the importance of the speed of the water and the importance of having a large volume of water to flow through the pipe. Modeling competence was demonstrated as students created small-scale models of the power plant. They did not make mathematical construction drawings, but they talked about different ways of making the models. The students incorporated different mathematical considerations, such as discussing the volume and circumference of the water outflow hole in these models. Throughout the activity, the students provided empirical justifications for their construction choices, showing competencies within mathematical reasoning, such as arguing why they had constructed the diameter of the pipe smaller at the bottom of the pipe compared to the top.

Communication between students and teachers was actively encouraged. Teachers circulated and asked questions to guide students in their thinking. This was very important for making the competences explicit. Ultimately, students were expected to articulate the reasoning behind the design of their hydroelectric power plants using different mathematical arguments, highlighting variations in their proficiency in conveying their arguments.

In summary, we observed a presentation of the key mathematical competencies except for symbolic and representation competences within this STEM assessment scenario.

6.8.1.2 Science Competencies in the Assessment Activity

The assessment activity at its core served as an investigation into the principles of hydroelectric power generation. During our observation, we witnessed students actively exploring and gathering data, focusing on elements such as examining angles and calculating water velocity. Additionally, they demonstrated modeling competencies by constructing small-scale models of hydroelectric power plants. These models acted as representations of real-world systems, enabling students to test their hypotheses effectively.

Notably, students exhibited perspectivation competencies as they discerned the similarities between their constructed model and an actual hydroelectric power plant, providing them with a multifaceted perspective on the subject.

Communication played a pivotal role in this scenario. Students were required to articulate their reasoning and the scientific principles underlying their hydroelectric power plant designs, both among their peers and with the teacher. Teachers actively facilitated this communication by posing questions that encouraged students to explain their choices and reasoning, ensuring a deeper understanding.

In summary, all science competencies were present at various stages of the activity, creating a comprehensive learning experience that allowed students to apply scientific principles to real-world challenges and fostered effective communication and critical thinking (Table 6.2).

Table 6.2 STEM competencies in the assessment activity

Collaboration	Some students engaged in collaborative problem-solving. They worked together in teams, and some of the students shared ideas, pooled their resources, and collectively addressed challenges related to the problem.
Problem-solving	Some students actively sought solutions by examining angles, calculating water velocity, and creating models. Some of the students created the ideas, some helped in the implementation of the plan, and others just helped in the process.
Innovation and creativity	Some of the students showed creativity and innovation in building their power plants. They designed the model themselves, using the materials available for them. Some fairly effective models were developed, but some of the models were rather ineffective.
Critical thinking	Some of the students critically addressed their problem-solving processes and products, especially when guided by the teacher.
Self-regulation	Some of the students took responsibility for their role in the team and showed autonomy in designing their models. Others just followed their group.
Communication	Many of the students effectively communicated their ideas, reasoning, and design choices to their peers and teachers.
Metacognitive competencies	The groups adjusted their strategies while designing their models.

In our observations, we noted the prominent presence of the STEM competencies presented by McLoughlin et al. (2020), each contributing to the students' engagement and learning experience in this interdisciplinary activity. These competencies also played a pivotal role in enhancing their understanding of the subject matter related to hydroelectric power generation. In the observed activity, we noted that students actively engaged in problem-solving, exploration, and collaboration. Teachers played an active and important role in facilitating communication by prompting students to explain their choices and reasoning. The observation confirms that STEM assessment approaches promote the development of a broad range of competencies, since the STEM activity emphasizes not only subject-specific knowledge but also the development of competencies such as collaboration, critical thinking, and creativity and innovation. In conclusion, the observations reinforce the idea that STEM education through active learning not only cultivates subject-specific competencies but also nurtures a diverse set of STEM competencies, making it a powerful and comprehensive approach to learning and assessment.

6.8.2 *The Affective Field in the Case*

The assessment activity is analyzed with respect to Schindler and Bakker's (2020) affective field and is stated in the Table 6.3.

The analysis, using Schindler and Bakker's (2020) affective field framework, revealed several noteworthy findings regarding students' experiences during a STEM-oriented assessment. Some students perceived this assessment as markedly

Table 6.3 The affective field in the assessment activity

Beliefs	One student said that he found the assessment different from what he is used to. One student found that the inquiry-oriented approach asked for creativity and innovation
Attitude	One student needed more breaks. "It was nice to be on an excursion away from the classroom."
Emotions	Enjoyment (smiling, laughing). Autonomy (students taking the lead). Safe (together as a group). "We used the brain a lot." "I felt a bit stressed due to the time pressure."
Self-efficacy	"Usually I'm not good at math, but I managed to do it in this case." One student felt safer being in it with his classmates. "It was nice to be guided at the beginning."
Values	One group of students noticed that mathematics and science are something that happens outside the classroom, and they are entangled. "Mathematics is hard." "There isn't much math in this."
Motivation/ interest	Several students found it motivating that the task was relatable to the real world. "We got to be creative." "It was nice to be physically active."

different from their customary evaluation methods, signaling a departure from conventional practices and emphasizing a shift toward innovative and creative approaches to assessment that needs to be supported by the teacher. In terms of attitude, students expressed a need for more breaks, possibly indicating the mental demands of the assessment process. However, they welcomed the change in the learning environment, finding value in an excursion outside the classroom. Emotionally, students reported mostly positive experiences during the assessment. Their enjoyment was palpable through expressions of smiles and laughter, suggesting that the assessment engaged them and provided satisfaction. Moreover, the STEM approach seemed to grant students a sense of autonomy and a feeling of safety within the group, creating a supportive and collaborative classroom atmosphere. Some students felt stressed due to the time pressures imposed on them.

A surprising discovery was the increase in self-efficacy among students. Despite prior reservations, one student mentioned an improvement in their math skills, while other students mentioned that they did not relate the problem to mathematics. Collaborative work with classmates appeared to boost their confidence and sense of competence.

Furthermore, students recognized the relevance of mathematics and science to the world outside the classroom. They perceived mathematics and physics as interconnected and essential to real-world experiences, underscoring the importance of practical applicability. Finally, the assessment was deemed relatable to the real world, enhancing student motivation. The opportunity for creativity and innovation seemed to captivate their interest, making the assessment a more appealing and engaging learning experience.

In conclusion, the analysis suggests that the affect and competence-oriented assessment had a positive impact on various aspects of students' experiences and emotions. It promoted creativity, autonomy, and a sense of safety within the group, fostering an increase in self-efficacy. Additionally, it highlighted the real-world relevance of mathematics and science, enhancing motivation and engagement among the students. These findings align with the principles of student-centered and experiential learning, emphasizing the significance of innovative STEM teaching methods that foster motivation and engagement in students, see Svabo et al. (2024).

6.8.3 Analysis Leading to the Affect-Sensitive Learning in STEM Assessment

In the context of our case about affect sensitive learning in STEM assessment, we can observe a fascinating interplay between competencies, and the affective field, revealing the complexity of the learning experience.

In this assessment, students were challenged to not only demonstrate various subject-specific competencies in mathematics and physics but also STEM competencies. Problem-solving skills were essential as they examined angles and calculated the volumes of pipes. The act of constructing the model required them to think creatively, critically assess their design choices, and communicate their findings effectively to their peers and teachers.

This focus of STEM competencies within the assessment practice had a significant impact on the students' affective fields. As they actively engaged with materials and demonstrated their competencies, they experienced a positive affective response. The assessment, though different from their usual methods, became an enjoyable and motivating experience. They found satisfaction in their creative and innovative problem-solving endeavors, fostering a sense of autonomy and safety within the group. The real-world relevance of the assessment enhanced their motivation, sparking interest and enthusiasm.

In this intricate framework of STEM competencies and the affective field, we see how they are tightly interwoven. The affective field is positively influenced as students find engagement, motivation, and a sense of accomplishment in this holistic learning experience. It underscores the idea that effective education is not merely about acquiring knowledge but about creating an environment where competencies can flourish and where learners are emotionally and intellectually engaged in the process (Wickman et al., 2022).

6.9 Discussing the Viability of Students' Competences and Their Experiences When Participating in a STEM Assessment

When we view an assessment situation through the lens of the affective field and STEM competencies, we move beyond the idea of it being merely a yardstick for existing knowledge. Instead, it becomes an opportunity to foster further learning and knowledge creation. An affect-related STEM assessment not only helps students consolidate their existing knowledge but also enables them to develop their competencies and understanding within the specific subject matter. Competencies, when focused effectively on in the assessment, encourage students to not merely replicate information but to apply, analyze, and synthesize it. This approach promotes a deeper level of understanding and ensures that assessments align with the broader goals of education.

However, it is crucial for teachers to help students develop their competencies and to make them visible in the assessment situation to be able to assess them. Focusing on one competence enhances the visibility of it. This is not an easy task for teachers at first, and it necessitates their self-awareness regarding what to inquire and ask questions about in the more open-ended process in which students are now engaged. Incorporating the cognitive aspect, inspired by McLoughlin's work on learning in STEM education, focuses on equipping students with the necessary tools to understand, apply, and innovate within STEM fields. It emphasizes the foundational knowledge, skills, and competencies that students acquire within discipline-based education. Simultaneously, the affective dimension of learning recognizes the emotional and motivational components of learning, acknowledging that students' feelings, attitudes, and values play a crucial role in their educational journey.

The question for our research was how the students experienced a STEM assessment compared to a standard discipline-based assessment. The affective field, as explored through Schindler and Bakker's (2020) framework, provides a dimension to our discussion. Assessments that focus solely on testing existing knowledge often led to negative emotions such as anxiety, stress, and fear of failure (Katznelson et al., 2022). However, when assessments are designed with an emphasis on learning and knowledge creation, the affective field takes on a positive hue. Struyven et al. (2005) showed how students accustomed to traditional assessments formats may find the STEM assessment unfamiliar, potentially leading to discomfort. This was, however, not the case for the students in our case.

The journey into the realm of affect-sensitive STEM assessments brings both promise and challenges. While recognizing that assessments are not mere measurements of past learning but catalysts for future growth, we must navigate the complexities of curriculum alignment, resource allocation, and student experiences. The path forward is clear: summative assessments must continue to focus on learning, embracing the transformative power of interdisciplinary STEM competencies, and the students' affective field while remaining adaptable to meet the evolving needs of education.

6.10 Implications for Education

Introducing affect-related learning in assessment in STEM emphasizes the need to recognize that teachers are not just facilitators of learning but also crucial participants and facilitators in the assessment process. When considering affect in STEM assessments, several challenges and limitations arise, primarily from the perspective of educators. One of the foremost challenges is ensuring that the STEM assessments align seamlessly with established curriculum standards and predefined learning objectives but also the current structures in the school context.

The nature of considering affect in assessment demands that teachers offer timely and constructive feedback to support students' growth. Developing the skills to provide such feedback can be challenging for some educators and may require additional training and professional development efforts. To address these challenges effectively, it is crucial to integrate assessment capability development into teacher education programs. This means not only imparting theoretical knowledge but also providing practical experience in evaluation methods.

Finally, we also believe that there is a need for conducting more experiments in schools to test various types of STEM assessments. This will enable further research in this area, allowing us to gain more insight into how to introduce and implement more alternative forms of summative assessment. Our aim is that it is not only teaching methods in schools that change into new ways of teaching; we also wish to see assessment methods change and follow new educational approaches.

Acknowledgements The data analyzed in this chapter is a part of the LabSTEM+ project (research grant from the Novo Nordisk Foundation, grant number: NNF22SA0081124).

References

Andersen, H. L. (2008). *Eksamen og eksamensformer* [Examinations and forms of assessment]. Samfundslitteratur.

Andresen, B. B. (2017). Feedbackstrategier i matematik [Feedback strategies in mathematics]. *MONA-Matematik-og Naturfagsdidaktik, 3*.

Bailey, A., Kauffman, B., & Subotić, S. (2015, March 16). *Education technology and the 21st-century skills gap*. Retrieved from https://www.bcg.com/publications/2015/public-sector-education-technology-21st-century-skill-gap.aspx

Barab, S. A., & Dodge, T. (2008). Strategies for designing embodied curriculum. In *Handbook of research on educational communications and technology* (pp. 97–110). Routledge.

Binau, C., Dolin, J., Elmose, S., Nielsen, J., & Schmidt, J. (2021). Kompetenceorienteret undervisning i naturfagene [Competence-oriented teaching in the natural sciences]. *MONA—Matematik- Og Naturfagsdidaktik, 21*(3) Hentet fra https://tidsskrift.dk/mona/article/view/128341

Bjørkli, K., & Arnesen, K. (2015). Kombineret summativ og formativ vurdering i matematik [Combined summative and formative assessment in mathematics]. *Uniped, 38*(4), 337–344.

Blackley, S., & Howell, J. (2019). The next chapter in the STEM education narrative: Using robotics to support programming and coding. *Australian Journal of Teacher Education, 44*(4), 51–64. https://doi.org/10.14221/ajte.2018v44n4.4

Bybee, R. W. (2013). *The case for STEM education: Challenges and opportunities.* NSTA-National Science Teachers Association Press.
Christiansen, J. L., Lilius, K., Thynebjerg, K. A., Jensen, M. A. S., Andersson, J., & Kinnerup, L. B. (2020). Evaluering af modelleringsprocessen i naturfagsundervisningen [Evaluation of the modelling process in science education]. *MONA-Matematik-og Naturfagsdidaktik*, 20-20.
Conradsen, K. (2021). Formativ evaluering i læreruddannelsen [Formative assessment in teacher education]. *MONA-Matematik-og Naturfagsdidaktik.*
Dolin, J. (2002). Fysikfaget i forandring [The Physics Subject in Transition]. *PhD thesis. Roskilde University*, Denmark.
Dolin, J., Nielsen, J. A., & Tidemand, S. (2017). Evaluering af naturfaglige kompetencer [Evaluation of scientific competencies]. *Acta Didactica Norge, 11*(3), 2-sider.
DuFon, M. A. (2002). Video recording in ethnographic SLA research: Some issues of validity and data collection. *Language Learning & Technology, 6*(1), 40–59.
ENCoRE. (2005). *Commission staff working document towards a European qualifications framework for lifelong learning.* http://www.encore-edu.org/ENCoRE-documents/consultation_eqf_en.pdf
Gibson, J. J. (1966). *The senses considered as perceptual systems.* Houghton Mifflin.
Gilbert, J. K. (2004). Models and modelling: Routes to more authentic science education. *International Journal of Science and Mathematics Education., 2*, 115–130.
Højgaard, T., & Sølberg, J. (2023). Fostering competence: A narrative case study of developing a two-dimensional curriculum in Denmark. *Journal of Curriculum Studies, 55*(2), 223–250.
Honey, M., Pearson, G., & Schweingruber, H. (Eds.). (2014). *STEM integration in K-12 education: Status, prospects, and an agenda for research.* National Academies Press.
Jess, K. (2005). Konsekvenser af evaluering i matematikundervisning [Consequences of assessment in mathematics education]. *MONA-Matematik-og Naturfagsdidaktik*, (2).
Katznelson, N., Pless, M., & Görlich, A. (2022). Mistrivsel i lyset af tempo, præstation og psykologisering. *Om ny udsathed i ungdomslivet* [Struggling in light of pace, performance, and Psychologization: On new vulnerabilities in youth life], 1. Center for Ungdomsforskning.
Lykkegaard, E., & Ulriksen, L. (2019). In and out of the STEM pipeline–a longitudinal study of a misleading metaphor. *International Journal of Science Education, 41*(12), 1600–1625.
Madsen, S., & Levinsen, K. T. (2007). *Barriere for studerendes læring af holistisk systemudvikling [Barriers to Students' learning of holistic system development].* Department of Informatics INF, Copenhagen Business School. Working paper/Institut for Informatik. Handelshøjskolen i København no. 2-2007.
McLoughlin, E., Butler, D., Kaya, S., & Costello, E. (2020). STEM education in schools: What can we learn from the research?
Nassauer, A., & Legewie, N. M. (2021). Video data analysis: A methodological frame for a novel research trend. *Sociological Methods & Research, 50*(1), 135–174.
Nielsen, J. A., & Dolin, J. (2016). Evaluering mellem mestring og præstation [evaluation between mastery and performance]. *MONA - Matematik- Og Naturfagsdidaktik, 1.* https://tidsskrift.dk/mona/article/view/36369
Niss, M., & Højgaard, T. (2019). Mathematical competencies revisited. *Educational Studies in Mathematics, 102*, 9–28.
Niss, M., & Højgaard Jensen, T. (2002) *Kompetencer og matematiklæring: Ideer og inspiration til udvikling af matematikundervisning i Danmark* [Competencies and mathematics learning: Ideas and inspiration for the development of mathematics education in Denmark]. Undervisningsministeriets forlag.
Reed, E. S. (1991). Cognition as the cooperative appropriation of affordances. *Ecological Psychology, 3*(2), 135–158.
Rosiek, J. (2003). Emotional scaffolding: An exploration of the teacher knowledge at the intersection of student emotion and the subject matter. *Journal of Teacher Education, 54*(5), 399–412.
Ruffell, M., Mason, J., & Allen, B. (1998). Studying attitude to mathematics. *Educational Studies in Mathematics, 35*, 1–18.

Schindler, M., & Bakker, A. (2020). Affective field during collaborative problem posing and problem solving: A case study. *Educational Studies in Mathematics, 105*(3), 303–324.
Seidelin, L., Petersen, C., & Normann, E. (2021). *Anbefalinger til at få flere piger og kvinder i STEM: En analyse af Teknologipagtens signaturindsats 2021* [Recommendations for increasing participation of girls and women in STEM: An analysis of the technology Pact's signature initiative 2021]. Teknologipagten.
Shaw, R. E., & Bransford, J. (1977). Introduction: Psychological approaches to the problem of knowledge. In R. E. Shaw & J. Bransford (Eds.), *Perceiving, acting and knowing* (pp. 1–39). Lawrence Erlbaum Associates.
Stake, R. E. (1995). *The art of case study research*. Sage.
Struyven, K., Dochy, F., & Janssens, S. (2005). Students' perceptions about evaluation and assessment in higher education: A review. *Assessment & Evaluation in Higher Education, 30*(4), 325–341.
Svabo, C., Larsen, M. D., Bergkvist, K., Svendsen, M. W. H., & Kristensen, M. A. (2024). *STEM-didaktik med fokus på matematik—Til grundskole, gymnasie og dagtilbud* [STEM didactics with a focus on mathematics—For primary schools, secondary schools, and early childhood education]. Syddansk Universitetsforlag.
Wickman, P. O., Prain, V., & Tytler, R. (2022). Aesthetics, affect, and making meaning in science education: An introduction. *International Journal of Science Education, 44*(5), 717–734.
Young, M. F. (1993). Instructional design for situated learning. *Educational Technology Research Development, 41*(1), 43–58.

Maiken Westen Holm Svendsen is a PhD student at the STEM Centre for Science Education and Communication (FNUG) at the Department of Mathematics and Computer Science at the University of Southern Denmark. Maiken has a double degree in mathematics and physics, and is doing research within the STEM education field, with a focus on assessment.

Dorte Moeskær Larsen PhD, Associate Professor of STEM Education and Science Communication, Department of Mathematics and Computer Science, University of Southern Denmark. Currently she serves as the project leader for an interdisciplinary development and research project about STEM education (LabSTEM+). Additionally, she is a member of a national expert group dedicated to the development of mathematics education in primary schools (NCUM). Her primary research focus lies in STEM and mathematics education. She has previously worked as both a primary school teacher and a mathematics teacher educator.

Chapter 7
How to Start Strong: Different Approaches for Crafting Project Briefs with Corporate Sponsors for Global Innovation and STEM-Based Team Projects

Tamara Carleton and Falk Uebernickel

7.1 Introduction

How you start matters—in a new job, a love relationship, and even a team project. Both professional teams and student teams have told us numerous times that a strong start sets the tone for the project execution, builds team momentum and confidence, and increases the likelihood of successful project outcomes. In contrast, starting off badly often adds unnecessary obstacles, which delay and demotivate teams who must then clarify project scope, regain team trust, and steer a project back on track.

For many student team projects, the project brief serves as the primary starting point for a project, which gives this document extra importance in setting a project and the team up for success. What makes a good project brief? How are project briefs typically framed for students, and are there better ways to craft a project brief with a corporate sponsor that can be then presented to student teams to own? We consider these questions based on an international data set of project briefs drawn from project-based courses at multiple universities across a decade, aiming to distill

Teaching tools are another important element for putting Creative Pragmatics in action, so book editor Tamara partnered with her Swiss-German colleague Falk in this chapter to consider the classic tool of a project brief in the context of project-based learning. As a starting point, the prompt in a project brief serves as a significant framing device in the student learning process. (Editors)

T. Carleton (✉)
Department of Mechanical Engineering, Blekinge Institute of Technology, Karlskrona, Sweden
e-mail: tamara.carleton@bth.se

F. Uebernickel
Hasso Plattner Institute, Potsdam, Germany

C. Svabo et al. (eds.), *Creative Pragmatics for Active Learning in STEM Education*, Contributions from Science Education Research 14,
https://doi.org/10.1007/978-3-031-78720-1_7

our analysis into practical recommendations for corporate sponsors and teaching teams. While our recommendations primarily relate to project-based courses taught in a university setting, others using project briefs with student or professional teams may find these recommendations useful to adapt to their unique situation.

Moreover, this study topic connects deeply to taking a Creative Pragmatics approach in STEM (science, technology, education, and mathematics) education, which is the core theme in this book. A project brief is an essential tool that frames a team's shared learning experience, setting the tone and scope of what is expected and possible from the project start. By better understanding what defines an effective project brief, STEM educators can offer students more creative and pragmatic agency, empowering students to shape their own learning and growth.

7.2 Background

This study builds on prior research about project-based learning and project briefs, focusing primarily on studies about Stanford's ME310 course and the SUGAR Network, which conduct different yet complementary project courses, which we describe below.

7.2.1 About Project-Based Learning (PBL)

Project-based learning (PBL) is a student-centric pedagogy where students work on a time-bounded project, often in teams, based on real-world problems. The premise is that students learn best when they are actively engaged in the learning process, often directing their own learning through ongoing inquiry and reflection, and when they can apply what they have learned to real-world situations. Various studies (e.g., Kokotsaki et al., 2016; Duschl et al., 2016; Laur, 2017) have shown that when implemented well, PBL can increase student engagement, improve content retention, and produce better outcomes and learning competencies than traditional learning formats.

7.2.2 The Role of Project Briefs in PBL

In project work, a project brief is a well-known document that outlines the key elements expected for project execution, typically describing the goals, objectives, scope, timeline, various roles and responsibilities, and deliverables for that project. In industry, project briefs are used regularly by project managers, freelancers, designers, and others to set up and communicate project requirements with colleagues, vendors, and stakeholders.

A project brief is used similarly in PBL courses to guide students through a project and ensure that they understand and meet the objectives. In corporate-sponsored projects, a project brief is generally drafted first by a corporate sponsor, often with input from the corresponding professor or instructor. Sometimes the brief is defined in collaboration with students; from our personal experience though, we have rarely seen this joint activity. A project brief may also be used to assess student performance. By comparing the final deliverable to the project brief, course faculty and instructors can evaluate whether students have met the project's objectives and assess their understanding of the subject matter.

Research specific to design education have examined the role of a project brief—also called a design brief—in terms of student engagement (Sadowska & Laffy, 2017) or as a digital learning tool (Demirbas and Timur Ogut (2020), among other functions. Kelley (2020) posits that key elements of the standard design brief are not well understood in STEM or design engineering education, plus a major void exists in primary and secondary education, often termed as K-12 levels in the United States. His view is that, at its essence, a design brief is a "basic description of a design problem" before listing problem constraints and criteria for students (ibid., pg. 9). Kelley has found five common elements in a design brief as follows: (a) a picture or graphic to set the context, (b) a short problem statement or scenario, (c) a list of identified constraints and criteria, (d) a list of prototyping materials, and (e) a final statement telling students how to begin the design process. The problem statement or design scenario is especially important because it provides students with a basic description of the problem they should address.

7.2.3 *Stanford's ME310 Course and the SUGAR Network*

At Stanford University, the ME310 engineering course has been taught repeatedly since 1967, so it offers a long history as an award-winning PBL course (Carleton & Leifer, 2009; Carleton, 2019). The full course is approximately nine months long, or three consecutive academic quarters that are each 12 weeks. During the course, student teams work on one corporate-sponsored project the entire time, so that they may take a concept through testing to initial implementation. A student team starts with a project brief typically received in the late autumn term and presents a working prototype back to their sponsor by the following late spring. Teams have averaged 3–4 students in size, and each team works on a different sponsored project.

Student teams are guided by a mix of senior faculty, teaching assistants, and volunteer coaches on various assignments throughout the course. Not only do students learn from their experience within their respective projects, they also learn from observing and sharing feedback with classmates working on the other projects. Across the course history, several companies have returned as sponsors, though not necessarily the next academic year, so these sponsors have their own learning and evolution with the course.

In the mid-2000s, Stanford's ME310 faculty changed the course format by inviting new partners from academic institutions outside the United States—in places such as Australia, Brazil, China, India, Italy, Germany, Mexico, and Switzerland—to jointly support student projects. While some faculty at these institutions were based in engineering departments, others came from different academic areas, such as business and design. That year and in following years, Stanford student teams were partnered with student teams from other academic institutions, so that a corporate sponsor benefited from a larger and internationally mixed team of students. In practice for example, a team of three students from Stanford in the US worked with a team of four students from the University of St. Gallen in Switzerland on the same project sponsored by pharmaceuticals company Merck based in Germany.

Stanford's ME310 course has been often capped annually at 10 projects total, which limited participation from possible academic partners. The growing partner network became keen to engage more students at their home institutions and accommodate different financing needs from local sponsors, so these partners soon joined together on additional new projects. Faculty at these partner institutions adopted the ME310 curriculum in order to better synchronize student milestones and grading evaluations across all projects. These partnerships ultimately led to an informal consortium dubbed the SUGAR Network, which ran in parallel to Stanford ME310 events (Carleton, 2019; Uebernickel & Thong, 2022). The SUGAR Network gradually formalized and has added more academic partners that collaborate regularly with one another (Wiesche et al., 2018).

Various studies of ME310 and SUGAR courses across the decades have examined student learning outcomes (e.g., Berglund et al., 2007; Sjöman, 2014), plus demonstrated how the PBL format has prepared individuals for career success past the course (Sheppard et al., 2023). Notably, Juan Felipe Ruiz—who has firsthand knowledge from his experience participating at two SUGAR schools—examined 118 project briefs from Stanford's ME310 course in terms of their problem space and solution space (Ruiz, 2020). He posits that creating a good design brief is challenging due to the diversity of design processes, problems, and other variables involved, thus thwarting any attempt at standardization. He introduces a typology dependent on problem context, presented visually as a 2 × 2 matrix comparing problem space and solution space along axes from defined to undefined, which then could function as a "thinking/talking tool" with sponsors.

In addition to existing literature, both authors bring personal knowledge of these project courses. One author (Carleton) has documented the history of Stanford's ME310 course and preserved its final report documentation in Stanford Engineering library archives, plus oversaw SUGAR Network operations for two years. The other author (Uebernickel) served as an early academic partner with Stanford ME310 and then as a SUGAR Network cofounder and has been a SUGAR co-chair since 2016. Together, we draw on a deep knowledge base shared across Stanford and SUGAR, including a decade-long research publication series, jointly spearheaded by Stanford University and Hasso Plattner Institute.

While the ME310 and SUGAR pedagogy share multiple tools that structure student project work, we focus on one tool in this study that is directly relevant to project briefs. The "how might we" tool—often abbreviated as HMW—starts an open-ended inquiry with these three words. As many core SUGAR partners joined during the rise of the design thinking movement from the late 2000s through the mid-2010s, they added several design thinking tools to their curricula, especially the HMW tool. (As an aside, while often attributed to design firm IDEO, the HMW technique actually originates with creativity expert Sid Parnes from his 1967 book *Creative Behavior Guidebook.*) The value of using HMW in project briefs is that students start a project by exploring possible options more freely (the "how" part in HMW), defer judgment on the final solution (signified by the word "might"), and use more collaborative language ("we").

7.3 Research Approach

As data from Ruiz was inaccessible, our study examines a complementary data set of 68 project briefs from Stanford's ME310 and SUGAR Network courses from the years 2011 through 2022 in light of two research questions.

7.3.1 Two Research Questions

Our first research question was: how prevalent is the "how might we" (HMW) phrase as a structural prompt and what other types of prompts were most used by sponsors to frame a project brief? We expected that most sponsors relied on HMW given its recent popularity in the PBL and engineering design communities. To assess HMW and other prompts, we used textual analysis based on absence/presence in the project briefing documents.

Our second research question was: how do project briefs compare in terms of their future horizon? In other words, how far out in the future did sponsors aim for project results when challenging students? This question was analyzed using the Four Horizons model, which originates from innovation work by Carleton and Cockayne (2023). The Four Horizons model lays out a spectrum of four horizons mapped to innovation stages, shown visually in Fig. 7.1:

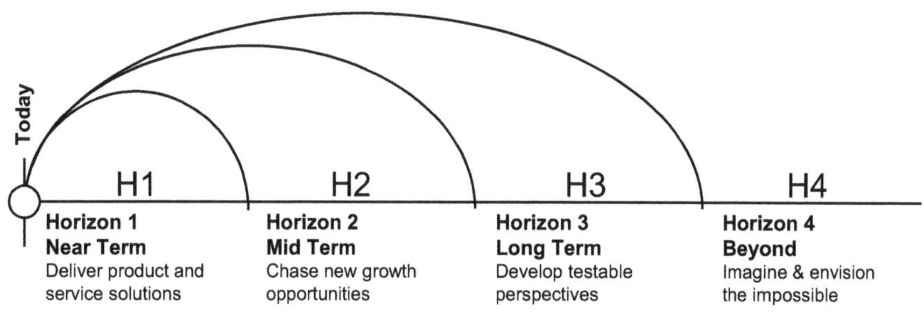

Fig. 7.1 Four Horizons model. (Carleton & Cockayne, 2023)

- Horizon 1 (H1) corresponds to near-term outcomes, such as service improvement or a product extension. Knowledge for a possible solution is readily available and known, though its implementation may or may not be easy.
- Horizon 2 (H2) corresponds to mid-term outcomes, such as different market segments or entirely new product lines.
- Horizon 3 (H3) corresponds to long-term outcomes, such as lab experiments or innovation pilots that require new knowledge to be invented or discovered, thus taking longer to come to fruition.
- Horizon 4 (H4) corresponds to more speculative outcomes, such as far-out visions and what-if ideas, that offer future perspective yet require substantially more understanding before becoming H3 experiments.

By overlaying these horizons on an ambiguity curve, H4 has the highest level of ambiguity going down to H1 that has the lowest level of ambiguity. We analyzed the wording of each project brief, judging its most likely horizon level for the outcomes desired based on existing available knowledge—in engineering, technology, medical science, and other relevant fields—at the time that the brief was introduced to students. (As a related point, Cockayne taught the Four Horizons model as a guest lecturer to Stanford's ME310 teams in 2021, so some students are familiar with it, while others might know of this model through other channels.) While project briefs are intended to provoke more imaginative thinking in student teams, we expected to see project briefs cluster between Horizons 1 and 2. Our assumption was that sponsors desired rough concepts that could be readily adopted for internal implementation. In addition, we felt many sponsors faced internal pressures to show fast and visible value of a university sponsored project to senior management, so they would likely avoid long-term outcomes, focusing more on near-term results.

7.3.2 Combined Data Set

A subset of project briefs was randomly drawn from the ME310 course collections owned by Stanford Engineering library archives across the years of 2011 through 2022, all of which is publicly available. One challenge was that this Stanford collection primarily features final project documentation from student teams, which occasionally includes the original project brief from the course start. We then supplemented Stanford data by contacting academic partners in the SUGAR Network for their personal archival data of project briefs. Over half of SUGAR partners lacked easily accessible records or had not saved them beyond the immediate course year.

Available data of project briefs was collected digitally in various file formats (e.g., PDF, Word, PowerPoint). A total of 19 academic institutions are represented in the data set, ranging from 25 projects maximum to one project minimum per

Table 7.1 Academic institutions in the study listed by project frequency

Academic institution	Country	Total projects
Trinity College Dublin	Ireland	25
Stanford University	United States	23
Kyoto Institute of Technology	Japan	16
Pontificia Universidad Javeriana Cali	Colombia	11
University of São Paulo	Brazil	10
Blekinge Institute of Technology	Sweden	9
Karlsruhe Institute of Technology	Germany	6
Aalto University	Finland	5
Swinburne University of Technology	Australia	5
University of St. Gallen	Switzerland	4
University of Modena and Reggio Emilia	Italy	3
dschool Paris	France	2
Indian Institute of Technology Kanpur	India	2
Indian School of Design & Innovation (Mumbai)	India	2
Norwegian University of Science and Technology	Norway	2
Polytechnic University of Porto	Portugal	2
Linköping University	Sweden	1
National Autonomous University of Mexico	Mexico	1
University of Bologna	Italy	1
University of Science and Technology of China	China	1

institution with a general average of nearly seven projects. As student projects typically span two academic institutions, one institution may provide data on behalf of the partner institution. In addition, five sponsored projects lack a second academic partner. See Table 7.1 for the list of academic institutions, using their respective names in English, listed in terms of project involvement. Note that the project frequency in this study does not represent the institution's actual number of historical projects.

In the data set, project briefs ranged across a decade timespan between 2011 and 2022. See Table 7.2 for the total number of projects listed by academic year (AY), which start in the fall semester (trimester) for projects. The highest number of project briefs was from AY2018-19, occurring before the global COVID19 pandemic.

The corporate sponsors for the projects ranged across industries. The greatest representation came from the manufacturing industry, which sponsored 29% of briefs studied. The next tier of industry representation came from Technology at 16% and then Education, Food, and Retail split across 6–7%. See Table 7.3 for the industry breakdown of corporate sponsors; industries sharing the same distribution percentage are then listed alphabetically within that percentage bracket.

Table 7.2 Student projects listed by academic year in the study

Academic year	Total projects
2012–2013	1
2013–2014	4
2014–2015	4
2015–2016	8
2016–2017	8
2017–2018	8
2018–2019	14
2019–2020	5
2020–2021	5
2021–2022	7
2022–2023	4

Table 7.3 Industry representation by percentage across the project briefs

Industry	Corporate sponsor(s)	Sponsor distribution
Manufacturing	Volvo Construction Equipment, Toppan Printing, Yanmar	29%
Technology	Panasonic, Philips, SAP	16%
Education	Trinity College Dublin E3 Institute, Pontificia Universidad Javeriana Cali	7%
Food	Z-Tech / InBev	6%
Retail	Media-Saturn, Totto, Loblaw Co.	6%
Automotive	Daimler, Renault, Volkswagen	4%
Health	CPHI	4%
Services	OrangeWorks, Grupo Coomeva	4%
Transportation	ANA Airlines, SUMMIT	4%
Public Sector	Enable Ireland	3%
Construction	Spie Batignolles	1%
Defense	Saab	1%
Energy	Osaka Gas Chemicals	1%
Fashion	TriWool	1%
Infrastructure	Xylem	1%
Insurance	Grupo SURA	1%
Nonprofit	Plan International	1%
Pharmaceuticals	Takeda	1%
Telecommunications	SwissCom	1%

7.4 Results

Our analysis showed that corporate sponsors followed no set standard or consistent template across any project briefs, even from the same sponsor in different academic years. Overall, project briefs were kept concise, many starting with a short description about the sponsor's business and/or respective industry to provide initial context to student teams. More than half (57%) of project briefs were described in fewer than 50 words, while the remaining briefs (43%) were longer, some up to several pages in length.

7.4.1 Uneven Inclusion of Common Brief Elements

Although Kelley (2020) identifies five common elements of a design brief as mentioned in our literature review, we found that none of these same elements were consistently used by either ME310 or SUGAR project sponsors. We discuss the average frequency of each element in order of Kelley's list, which to recap are: (a) a picture or graphic to set the context, (b) a short problem statement or scenario, (c) a list of identified constraints and criteria, (d) a list of prototyping materials, and (e) a final statement telling students how to begin the design process. In terms of visuals, slightly over three-quarters (76%) of the project briefs in our study featured no visual of any type—such as a photo, illustration, or diagram—which meant slightly under a quarter (24%) of briefs featured any visual as another way to visually show the opportunity or challenge. No project brief specified prototyping material requirements or instructions on how students should begin their design process. Moreover, a large percentage (82%) of briefs lacked any statistics or data points to help quantify the future opportunity in some way. The few briefs with statistics used them to help students understand the potential market size, forecast industry growth, or show the problem magnitude. A small percentage (9%) of briefs included definitions of some type, such as for a specialized industry term or emerging technology. Nearly a third (30%) of briefs highlighted examples to offer more ideas or variations related to the opportunity or challenge presented, such as describing additional obstacles, possible use cases, or alternative technology functions. Almost a quarter (22%) of briefs presented a direct link between the student challenge and the organizational vision or sponsor's business goals. See Table 7.4 for a summary of the common elements included (or not included) in the briefs.

Table 7.4 Summary of basic elements included in the project briefs

Project brief element	Included	Not included
Visual (diagram, photo, etc.)	24%	76%
Data points or statistics	18%	82%
Definitions	9%	91%
Examples	30%	70%
Direct link to sponsor's vision or organizational goals	22%	78%

7.4.2 Seven Types of Project Outcomes

We found that corporate sponsors described seven different types of project outcomes across the briefs. These outcome categories emerged from the data analysis and are loosely interpreted, as some briefs did not neatly fit into a single category. Process outcomes were the most frequently mentioned in 29% of the project briefs related to creating, reimagining, or improving business operations, product or solution delivery, production optimization, or other ways of doing things. System outcomes were the next most mentioned at 24%, and these outcomes generally relate more to solution modularity, fit with the user setting or environmental context, or infrastructure interdependency. See Table 7.5 for the seven types of outcomes sought by corporate sponsors that were initially identified in this study.

7.4.3 Prompt Wording

Nearly two-thirds (65%) of briefs overall used "how might we" phrasing to present the project opportunity or challenge to student teams. A variant of HMW used occasionally was "How can we…" Most sponsors of SUGAR projects relied on HMW to start their project prompts. In contrast, a much smaller percentage—just over a quarter (26%) of briefs—from Stanford's ME310 course used HMW language.

The remaining briefs (35%) in the study used other phrases to provoke possibility. One group of phrases deliberately started with words that evoked imagination, such as "We dream…", "What if...", or "Imagine that…" A second group of phrases addressed the team project as a challenge, using words such as "Your challenge is..."

Table 7.5 Seven types of project prompts used by corporate sponsors

Outcome type	Percentage of use	Prompt example
Process	29%	How might we affect the supply chain and at-home experience to eliminate food waste?
System	24%	What will the future changeable, reconfigurable living space entail and how will it fit the emerging nomadic lifestyles?
Product	17%	Make an innovative modular product or a system by utilizing modern technologies in order to solve current issues or improve productivity with aquaculture businesses.
Service	17%	Your team's challenge is to create a solution to maintain communications for 7 days of power outages.
Market	11%	How might we reach 1 billion customers by 2030?
Integration	3%	How will future robotic, on-demand vehicles gracefully and safely accommodate wheelchairs and similar devices?
Model	2%	What if the user could use the digital thread of material, formerly known as dirt, to repurpose it and mine it once again from urban landscape and thus save the bedrock?

or "The challenge is..." A third group of phrases was heavily directed and solution focused, using specific words such as "The project goal is...", "Help users to...", or "Study and create...".

7.4.4 *Prompts by Innovation Horizon*

We found that project briefs were largely split across mid- and long-term innovation horizons. In fact, slightly over a third (35%) of briefs focused on H2 outcomes. In contrast, a quarter (25%) of briefs aimed at more near-term outcomes at H1. A small percentage (6%) were identified as wildly open-ended H4 prompts. See Fig. 7.2 for the distribution of project briefs by innovation horizon.

We inferred how much new knowledge was needed and the level of ambiguity inherent in each project brief in order to determine where the brief fit along the horizon spectrum. More explanation within a brief did not necessarily reduce the amount of ambiguity for a student team if there were evident "known unknowns" or even "unknown unknowns" in the solution space. See Table 7.6 for some prompt examples by innovation horizon.

Some curious patterns emerge when considering project briefs by academic year. H1 prompts occurred in waves roughly every two academic years. Most H2 prompts occurred during the mid- and late 2010s, specifically the academic years between 2015–16 and 2018–19, and then drop unexpectedly after the COVID19 pandemic. H3 prompts ranged between 13% to 60% of briefs in a year for most academic years, excluding two years when no H3 prompts were identified. H4 prompts start after the 2016–17 academic year, averaging one H4 prompt roughly every two years. See Fig. 7.3 for a breakdown of briefs by innovation horizon and academic year.

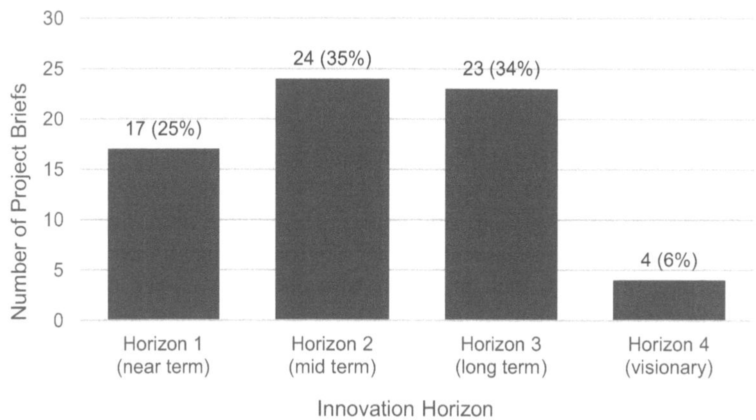

Fig. 7.2 Distribution of project prompts by innovation horizon

Table 7.6 Distribution of project briefs by innovation horizon

Horizon level	Percentage of use	Prompt example
Horizon 1	25%	How might we better integrate people with intellectual disabilities into society?
Horizon 2	35%	Which services for the elderly and for care helpers can we develop around the connected flooring for better aging (in the retirement house or at home)?
Horizon 3	34%	How might we sense and prevent mental disorders through networks of machines and humans?
Horizon 4	6%	How can nature inspire the next generation of water treatment, transport, or analysis technologies? What can we learn from the passive treatment processes of plants and soil? Can we replicate nature's low-energy fluid transportation methods? Where can we mimic nature's self-detection and protection mechanisms for adaptive architectures?

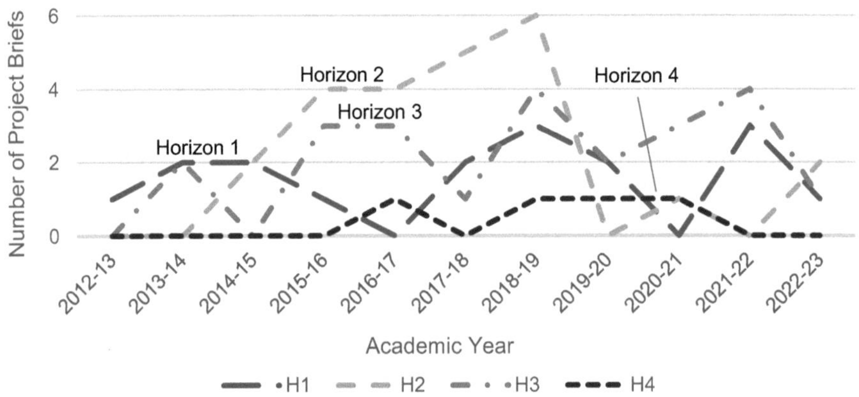

Fig. 7.3 Breakdown of project briefs by innovation horizon and academic year

7.5 Discussion

This study investigated what types of prompts were used by sponsors to frame a project brief project, notably how prevalent the HMW phrase was, and how project briefs compare in terms of their future horizon. Although all project briefs in this study featured a problem statement of sorts, the briefs ranged widely in which elements that they included, such as the use of a photo. Moreover, certain common elements that were identified in the literature review, such as design criteria, were consistently omitted by sponsors. A likely explanation is that the teaching teams for Stanford's ME310 and SUGAR projects addressed these other common elements—namely prototyping materials, deliverable constraints, and design criteria—in the various course assignments throughout an academic year, so sponsors did not need to include these specific elements in their project briefs.

7.5.1 *Prevalence of "How Might We" Prompt*

Across the briefs, the problem statement was presented typically more as a question or thought prompt than a declarative statement per se. Our first research question aimed to identify the prevalence of HMW phrasing in these prompts. Our initial assumption was correct in that most briefs relied on HMW, which we perceive as the influence of the design thinking movement when many core SUGAR academic partners become involved. Stanford's ME310 course has a much longer history and subsequently a larger archive of project briefs, so we expect its teaching team has tested more ways to guide corporate sponsors on how to effectively frame a project brief.

7.5.2 *Range of Innovation Horizons Presented in the Project Briefs*

Our second research question aimed to examine how project briefs compared in terms of their future horizon; however, our expectation was only partially correct. Most project briefs, instead of clustering on Horizons 1 and 2 as hypothesized, clustered across Horizons 2 and 3—and H4 prompts began largely from the COVID19 pandemic on. One possible explanation is that corporate sponsors view student teams as external foils for research and development activities within their organization. While managers at each sponsoring organization must maintain daily job tasks, perhaps they see youth outside their organization able to consider more outlandish and more open-ended questions that fall under Horizons 3 and 4, which can go beyond immediate business objectives. The COVID19 pandemic as a global crisis may also have spurred some corporate sponsors to realize the importance of considering bolder visions or farfetched outcomes. A related explanation is that the Stanford ME310 course has evolved its pedagogy, responding to changing market demands and other factors, thus expanding from narrow engineering problems in the 1970s to more open-ended student challenges through the 2000s and beyond (Carleton & Leifer, 2009).

7.5.3 *Some Implications for PBL Courses and STEM Teaching*

Project briefs are an important tool to start a project. Since many STEM projects and design challenges traditionally focus on existing user needs, our research findings raise key questions on how prepared faculty and student teams are to address more future-oriented PBL challenges with mid or far-term horizons. How much should faculty and coaches help students interpret and question a project brief at the project start? Moreover, how much should faculty push and induce broader thinking in a student team, and do project briefs allow for more forward-looking thinking that extend beyond the innovation horizon as defined within its problem statement, even sponsor's comfort zone?

In a study of Stanford's ME310 projects, Lande (2012) posits that student teams are designing the future while designing *for* the future (emphasis added). He observed that student teams achieved stronger project outcomes when navigating regularly between multiple mindsets as represented by the disciplines of future/foresight, design, engineering, and production—versus operating too tightly within a singular mindset. As such, the project brief offers the first navigation point in a PBL environment, and it is our view that teaching teams can take up the responsibility to encourage student teams to widen their initial mindset for more inventive problem framing (scoping) and subsequent problem solving.

Moreover, we see a connection between the desired innovation horizon and the language used in a project brief. Faculty can help corporate sponsors consider the desired impact of their projects in terms of the innovation horizon and then match the initial prompt phrase accordingly. For example, if a project sponsor wants a more imaginative and ambitious proposal in the H4 realm, then he or she should consider more expansive prompt phrases, such as "We dream…". In contrast, a sponsor seeking more near-term outcomes suitable for H1 should use a more directed phrase in the project brief, such as "Help users to…". HMW could serve as the most flexible phrase targeting mid-term outcomes. Based on this thinking, we outline a two-question guide for faculty and corporate sponsors to use when developing project briefs as part of a PBL course or program. See Table 7.7.

Table 7.7 Two-question guide to PBL project brief development with corporate sponsors

Two-question planning guide for project briefs
1. Which innovation horizon best fits our desired outcome(s), and why?
(a) Horizon 1: Near term
(b) Horizon 2: Mid term
(c) Horizon 3: Far term
(d) Horizon 4: Visionary
2. Which phrase do we feel would provoke students most as a project prompt related to this innovation horizon?
(a) Help users to…
(b) Your challenge is…
(c) How might we…
(d) Imagine that…
(e) We dream…
(f) What if...

7.5.4 Study Limitations

Our study's findings are subject to several limitations. One limit is sample size. Our data set is a partial sample collected from a significantly larger data set—especially Stanford's ME310 course archives that go back to the 1970s—so we may have missed broader historical trends or a greater data range for a specific variable. Another data limitation was unavailable historical SUGAR project briefs, restricting a more comprehensive analysis. Enterprising scholars could remedy this issue for future studies by requesting that SUGAR faculty preserve these briefs. A third limit relates to the research method chosen. While textual analysis is a valuable research tool, a known limitation is researcher subjectivity. Regardless of our best intentions to be systematic, our data analysis is subject to personal biases. There may have been nuanced context when a particular brief text was produced that, if we had known, could have shaped the analysis differently.

7.5.5 Further Research

Project briefs for student teams and professional teams offer a rich area to investigate as this document is used repeatedly to guide new work across universities and industries. By building on our initial study, one future line of inquiry could examine which elements of a project brief are most critical and why. For example, do more visual project briefs—those showing photos, diagrams, and/or illustrations—help teams more effectively visualize the solution space or produce more imaginative results? Another line of inquiry could more thoroughly define and classify the range of project outcomes desired by corporate sponsors in PBL settings beyond our rough list. For example, how do better project outcomes, which are typically measured by course grades, correlate with certain outcome prompts in the project briefs? A third area of research could delve further into the Stanford ME310 archives. Our study provides one slice of project briefs across time, and it would be informative to collect and compare all ME310 briefs across the course history for noteworthy historical trends. A fourth area could examine the role and value of a project brief for problem formulation in PBL courses. These are just some possibilities to expand the research about the essential role of a project brief used in PBL courses, a Creative Pragmatics approach in STEM education, and beyond.

References

Berglund, F., Johannesson, H., & Gustafsson, G. (2007, June). Multidisciplinary project-based product development learning in collaboration with industry. *Proceedings of the 3rd International CDIO Conference*, Cambridge, MA.

Carleton, T. (Ed.). (2019). *ME310 at Stanford University: 50 years of redesign (1967–2017)*. Innovation Leadership Publishing.

Carleton, T., & Cockayne, W. (2023). *Building moonshots: 50+ ways to turn radical ideas into reality*. Wiley.
Carleton, T., & Leifer, L. (2009). Stanford's ME310 course as an evolution of engineering design. In *Proceedings of CIRP design conference 2009*. Cranfield University Press.
Demirbas, D., & Timur Ogut, S. (2020). Re-designing the design brief as a digital learning tool with participatory design approach. *Turkish Online Journal of Distance Education, 21*(1), 83–100.
Duschl, R. A., Arminio, A. A., Jr., & Barton, A. R. C. (2016). *Project-based learning: An integrated science, technology, engineering, and mathematics (STEM) approach*. Routledge.
Kelley, T. R. (2020). The anatomy of a design brief. *Technology and Engineering Teacher, 79*(7), 8–12.
Kokotsaki, D., Menzies, V., & Wiggins, A. (2016). Project-based learning: A review of the literature. *Improving Schools, 19*(3), 267–277. https://doi.org/10.1177/1365480216659733
Lande, M. J. (2012). *Designing and engineering: Ambidextrous mindsets for innovation*. Doctoral dissertation, Stanford University.
Laur, D. (2017). *Project-based learning: Real questions. Real answers. How to unpack PBL and inquiry*. Corwin.
Parnes, S. J. (1967). *Creative behavior guidebook*. Scribner.
Ruiz, J. F. (2020). *Problem of formulating design problems: A typology of design briefs* (publication No. 1895] [Licentiate thesis, Linköping University].
Sadowska, N., & Laffy, D. (2017). The design brief: Inquiry into the starting point in a learning journey. *The Design Journal, 20*(Suppl 1), S1380–S1389. https://doi.org/10.1080/14606925.2017.1352664
Sheppard, S. D., Chen, H. L., Toye, G., Mouallem, A., Lande, M., Shluzas, L., Bunk, T., Elfiki, N., Lamprecht, J. J. L., & Prantl, K. (2023). Decades of alumni: Perspectives on the impact of project-based learning on career pathways and implications for design education. In C. Meinel & L. Leifer (Eds.), *Design thinking research: Innovation–insight–then and now* (Understanding innovation book series) (pp. 25–43). Springer. https://doi.org/10.1007/978-3-031-36103-6_2
Sjöman, H. (2014). *Learning outcomes through global product innovation course in Aalto University* [Master's thesis, Aalto University].
Uebernickel, F., & Thong, C. (2022). Contextualizing design thinking with multiple intelligences: The global SUGAR program as a case. In C. Meinel & T. Krohn (Eds.), *Design thinking in education*. Springer. https://doi.org/10.1007/978-3-030-89113-8_17
Wiesche, M., Uebernickel, F., Byler, E., Garcia-Cifuentes, J. P., Kelly, K.; Suzuki, S., Vignoli, M., Leifer, L , Lang, M., Feldmann, N., Hölttää-Otto, K., Satzger, G., Thong, C., & Krcmar, H. (2018). *Teaching innovation in interdisciplinary environments: Toward a design thinking syllabus*. Proceedings of the 2018 AIS SIGED International Conference on Information Systems Education and Research. https://aisel.aisnet.org/siged2018/13

Tamara Carleton PhD, global expert in radical innovation. Tamara is an award-winning professor, industry advisor, and author of radical innovation and strategic foresight. As a global professor of practice, Tamara is on the faculty of multiple business schools and technical universities around the world. Tamara holds the UNESCO Chair in Anticipatory Leadership for Innovative and Better Futures at Tecnológico de Monterrey in Mexico and was recognized as one of the Top 50 Women Leaders in Education of 2023 and 2024 by Women We Admire. As the founder and CEO of Innovation Leadership Group, Tamara helps leaders and teams to build innovation capacity and map bold futures. She has written several seminal books on innovation, including *Building Moonshots: 50+ Ways to Turn Radical Ideas into Reality* and the *Playbook for Strategic Foresight and Innovation*. She holds a doctorate in mechanical engineering from Stanford University.

Falk Uebernickel has been Head of d-school and Professional Development and Professor of Design Thinking and Innovation Research since April 2023 at the Hasso Plattner Institute (HPI) in Germany. Since 2019, his research has focused on nudging for sustainability and the psychology of digital transformation and innovation in companies. In addition to his position at HPI and a professorship at the School of Management at the University of St. Gallen in Switzerland, Falk is the author and editor of several books and has contributed to over 200 scientific publications. In his professional career, he has accompanied projects for a variety of companies with a focus on product and service development using design thinking and the establishment of digital innovation units.

Chapter 8
Fablabs and the Maker Movement: Learning Through Making

Mads Hobye, Michael Haldrup, Sara Daugbjerg, Sofie Pedersen, Martin Malthe Borch, Nicolas Padfield, Schack Lindemann, Nikolaj Møbius, and Bo Thorning

8.1 Introduction

While prevalent twentieth-century definitions of design tended to equate it with the utilisation of highly specialised knowledge in industrial design (in the US tradition) or the arts (in Europe), a new 'Maker mindset' gained pace at the threshold of the twenty-first. century, with an explicit intention to democratise design and fabrication and tie it into broader social and political spheres of society and everyday life (Ehn et al., 2014). The establishment of MIT's Fablab in 2001 marked the birth of this new model as significant for learning and exploring within an 'established' university context.

Two book editors, Michael and Tamara, are from Stanford University, where a foundational principle in its design community is "build to learn"—meaning that learning best occurs through physical construction, iterative processes of making that connect needs, desires, problems, ideas, materials, tools, outcomes, evaluations. Nine Danish colleagues team up in this chapter to address this build-to-learn power in light of the FabLabs and Maker Movement phenomena and their complementarity to Creative Pragmatics. They establish clearly the programmatic coherence of the maker movement and as expressed in our concept of Creative Pragmatics. Explored are more associations of design process in "epistemologies of the hand", more aspects of transdisciplinary reach in design-based methodologies. This chapter gives more shape and substance to the much-touted principle of embodiment, that active learning is not just cognitive, but also affective and located in corporeality. (Editors)

M. Hobye (✉) · M. Haldrup · S. Daugbjerg · S. Pedersen · N. Padfield · S. Lindemann
N. Møbius · B. Thorning
Roskilde University, Roskilde, Denmark
e-mail: mads@hobye.dk

M. M. Borch
Technical University of Denmark, Kongens Lyngby, Denmark

C. Svabo et al. (eds.), *Creative Pragmatics for Active Learning in STEM Education*, Contributions from Science Education Research 14,
https://doi.org/10.1007/978-3-031-78720-1_8

The number of Fablabs all over the globe (by March 2020: 1750 approved on www.fablabs.io) shows how 'Fablabbing' has become a global social movement (Walter-Herrmann, 2013) in convergence with the 'Maker movement' (Hatch, 2013) rather than 'just' another university outreach project—creating new spaces for learning, innovation, research and self-actualization. In Denmark, the user-maintained map of open Maker spaces (MakerSpace & Fablabs mm, n.d.) and Fablabs counts 47 active publicly accessible spaces, while only ten are on the fablabs.io list. It seems likely there are generally approximately 4–5 times more Maker spaces than "approved" Fablabs (for a brief history of the Maker Movement, see Box 8.1). The Fablab

Box 8.1: The Maker Movement—A Brief History

The Maker movement (Anderson, 2012; Haldrup et al., 2018) has a growing mindset of utilising exploratory approaches to technology for emancipatory purposes. It is an a-disciplinary mindset seeking to create new skills spanning the silos of traditional craftsmanship, artistic genres and academic disciplines.

It goes back to around the early 2000s, emerging as an amalgamation of the twentieth-century do-it-yourself (DIY) culture and increasing access to digital fabrication tools (Anderson, 2012; Davies, 2017). The movement would gain momentum with the growing availability of affordable and accessible technologies such as 3D printers, laser cutters, and microcontrollers like Arduino. The open source software movement and the growth of online collaboration platforms have also contributed to the expansion of the Maker movement worldwide (Anderson, 2012).

In 2001, a collaboration between the Grassroots Invention Group and the Center for Bits and Atoms at MIT Media Lab began to explore how the content of information relates to its physical representation and how an underserved community can be empowered by technology at the grassroots level (Mikhak et al., n.d.). This later resulted in the formation of the Fab Foundation, a formal network of Maker spaces that foster innovation through fabrication technologies and that espouses the idea that in a Fablab one can "make 'almost anything" (The Fab Foundation, n.d.).

Parallel to these events, community-driven hackerspaces and Maker spaces emerged as physical spaces where individuals could come together to share tools, knowledge, and ideas. These spaces provided access to equipment and resources that were otherwise out of reach for individuals, creating hubs for collaboration and innovation through the democratisation of technology (Smith et al., 2016). Many of these workshops engaged with local communities to foster local innovation, questioned the consumerist culture by promoting repairing and personalised fabrication, and discussed new ways of thinking about technology (Davies, 2017; Smith, 2017; Smith et al., 2016). Although these spaces promoted openness and accessibility, studies have shown the majority of their

(continued)

Box 8.1 (continued)

members to be white, male and often with a formal education (Smith, 2017). Technology has often been associated with white masculinity. This legacy hasn't been considered much within the movement until recently, and has given challenges regarding inclusion and diversity (Dunbar-Hester, 2014).

A key catalyst for the Maker movement was the establishment of MAKE magazine in 2005 (Maker Media, n.d.), which brought together a broad community of DIY enthusiasts, artists, engineers and inventors. The magazine provided a platform for sharing projects, tutorials and ideas, fostering a collaborative and open source mindset, inspired by the open source software culture before it. MAKE magazine, owned by Maker Media, has nevertheless been criticised for being a corporation that has capitalised on the Maker movement and has had a strong influence on defining what the Maker movement is and how it should evolve. This contradicts the grassroots and community-driven ethos (Davies, 2017; Smith, 2017; Smith et al., 2016).

The Maker culture that started as a grassroots movement has been adopted by broader society and has inspired public and private institutions, scholars and politicians worldwide to think in new ways of developing local and sustainable means of production to empower communities and citizens technically (Smith, 2017). However, the Maker movement has also been a central reference point for making hyperbolic claims regarding the demise of traditional design, industry and fabrication, as demonstrated by one of the more popular books on the movement by the editor of Wired magazine (Anderson, 2012); for a more nuanced view on the various '3D printing scenarios' of the future see Birtchnell and Urry (2016).

concept, with its focus on hands-on design practices as a vehicle for exploration and experimentation, offers a radical new possibility for solving 'the most practical of all questions we can ask of our civilization' (Dewey quoted in Schelhowe, 2013, p. 95): the integration of 'the mind-body in action' as a model for learning. The Fablab offers practical, playful and explorative perspectives on design that enable dialogue across social spheres and disciplinary boundaries. It is driven by, and a driver of, the availability and access to new and evolving digital production techniques.

The Maker community is, through Fablabs, hackerspaces, Maker spaces, DIY biolabs, biofablabs etc., growing throughout the world. From user-run clubhouses to formal institutions such as libraries, schools and universities, interest in some form of material prototyping and engagement is increasing. For the last decade, people have embraced tools like the laser cutter, 3D printer and Arduino boards as new ways to explore creating 'X' with technology. A fundamental part of this movement is the democratic mindset of knowledge sharing and open usage around the globe. Besides the commonality of a shared interest in using modern tools to explore materials and technology, the field is as eclectic as the number of spaces within it. Each space has its own understanding of what they find valuable to engage with and how. Multiple suggestions of what "making" is, in the form of manifestos and dogmas (see Sect. 8.2), can be found. Although they have overlapping motifs, their intentions differ and can thus be seen as different perspectives and reflections on the same movement. The Maker movement does not have a unified methodological approach, and one can justifiably argue that designerly methodological discussions are not always at the forefront of the everyday conversations in Fablabs and Maker spaces. Instead, focus seems often to be on just trying things out, experimenting and exploring what is possible. This may be an inclusive counter-reaction to a previous somewhat expert-driven exclusiveness: a model with a master and an apprentice, high standards of professional handwork, and expectations of doing things "properly", the "right way". It seems like the Maker movement is more a mindset in the form of a shared set of values and technologies, more than a fixed multi-step methodology. E.g. quick experiments (Haldrup et al., 2018; Hobye et al., 2016; Padfield et al., 2014) are valued over long rigid development processes, and proofs of concept seem more interesting than finished production-ready solutions.

With this premise, it may seem naive to summarise what thinking through making means (or, as we want to term it, Maker thinking). However, it seems safe to say that the widespread movement has something to offer when it comes to creative, designerly and technological engagement. Certainly, when it comes to the local manifestation of the here and now (for the sake of one's own joy of engagement)—but perhaps also on a higher abstraction level, as a more fundamental understanding of how to engage with and enlarge the possibility space of technological advancements—through design, hacking, making, sharing, open sourcing etc. If this is the case, perhaps we should think of the mindset (Maker thinking) as a method in and of itself. This ultimately becomes a question of how we can reframe a mindset as a methodology, and how this can contribute to and expand existing understandings of design methodologies.

In this chapter, we seek to theoretically extrapolate and condense six propositions for "Maker thinking," which we contend capture the principles and procedures that comprise the "methodology" of the Maker mindset. We will do this by simultaneously mapping the numerous existing manifestos and suggestions of the Maker movement/culture and exploring our own experiences with working through Maker methods in connection with the problem-oriented project learning (PPL) approach of Roskilde University (RUC) (see Box 8.2. about the Fablab RUC perspective and Box 8.4 about the PPL perspective). The different manifestos are cultivated in different subsets of the Maker communities (repairing, hacking, fabrication etc.) and thus represent different voices attempting to pinpoint facets of overlapping agendas. By mapping out common themes within the voices, we synthesise overlapping elements that can be considered a shared mindset.

Therefore, we set out first to explore what Maker thinking can bring to the table concerning current methodological approaches to design, and then construct a series of propositions in dialogue with short descriptions of our experience. Attempting to stay true to the principles of the Maker movement, this chapter has been written as a collaborative effort with contributors from multiple Maker communities around the Copenhagen-Roskilde area in Denmark.

The mindsets embedded in Maker spaces/communities share a close affinity with Creative Pragmatics, as outlined in Chap. 1. Both emphasise the complexity and the

Box 8.2: The Fablab RUC perspective

Fablab RUC was established to provide a facility for bringing art, technology and hands-on 'Maker' approaches into academia, not as an optional element in the academic program, but as a vital part of the skills and competencies of future academics, whether training for a degree in technical science, natural science, social science or the arts. Fablab RUC was formally established in the autumn of 2013 but draws on earlier experiences integrating digital fabrication in research and learning (in 2018, it expanded with a biofablab and kitchen/food lab). In 2008, Roskilde University established a general bachelor program in Humanities and Technology (HumTek) in addition to the three existing bachelor programs in Natural sciences, Social sciences and Humanities that, since the early 1970s, have been a core element in Roskilde University's 'Roskilde Model' or PPL learning model.

This new bachelor program was established with the intention of integrating technical science, social science, humanities and design (Nielsen et al., 2014 for an overview of this) into one transdisciplinary program. As with all curricula at RUC, the main form of study relates to student-directed project work—the pedagogical philosophy of Dewey and critical Marxism was introduced in Denmark with the establishment of Roskilde University in 1972 (The "Roskilde Model", Andersen & Heilesen, 2014). At the HumTek bachelor program, this pedagogical model was even closer connected to the original inspiration from European architecture schools and later engineering colleges (Knoll, 1997) that influenced PPL in its early years, drawing inspiration also from the Scandinavian tradition of participatory design (Ehn, 1998).

Hence, establishing workshops at HumTek in 2008 as part of the curriculum reflected a return of craft and hands-on methods to academia. During the last 7 years, Fablab RUC has functioned as an experimental incubator using design and digital fabrication as tools for learning and exploration within a diverse range of academic fields and disciplines. The lab has functioned as a test case for using digital technologies and rapid prototyping as a toolbox with general usability, also in cultural and social sciences. Focusing on problem solving rather than analysis has added new dimensions to the problem-oriented learning approach traditionally used at RUC. Rather than providing suggestions for unequivocal solutions based on critical analysis, projects and coursework developed around Fablab RUC have centred on generating multiple solutions and functional prototypes as well as experimentation with these in real-life user contexts. This has resulted in 'Making' methods being a core element in multiple student projects and courses, from coding workshops to experimental museum installations and large-scale projects such as the Roskilde Light Festival. It also created a heterogeneous and unpredictable learning environment in which teaching, projects external to the university (commercial or artistic), and DIY projects merge into one 'bizarre bazaar', creating unpredictable synergies for technological learning and exploration (Haldrup et al., 2018).

(continued)

Box 8.2 (continued)

From its initial connection to the Humtek Bachelor Program, Fablab RUC has become a central element for multiple educational masters and bachelor programs around the intersection between STEM and the humanities. This is especially true as a part of the Computer Science Science program, Informatics, the Digital Transformation masters program and the Performance Design program. For all those programs, multiple technology and engineering subjects have been applied to real-world experiments and design solutions. The subjects include (but are not limited to) Biohacking, Robotics, the Internet of Things (IoT), Renewable Energy, Computer Vision, Machine Learning, Tangible/Physical Computing, Design Engineering in general, and support for physical experimental setups in the natural sciences.

interwoven nature of contemporary challenges and how they require a holistic perspective on the multifaceted interactions between issues, stakeholders and possibilities. Both depart from models of learning as passive appropriation of theories and methods. In doing this, both Creative Pragmatics and 'maker thinking' draw on the legacy of John Dewey and the inspiration drawn from his pragmatist philosophy in design based and hands-on based forms of learning, for instance in different forms of reform pedagogies (see Andersen & Heilesen, 2014; Dixon, 2020; see in particular the contribution of Nielsen et al., 2014). Acknowledging doing and learning through taking action in complex systems implies thinking through transdisciplinary settings for multifaceted reflection and creativity. Such an approach focuses on developing competencies for a complex world that can understand science as investigative complex processes, rather than ready made theories and methods (outlined in Chap. 3). In this sense Maker spaces and the communities of practice around them can be viewed as already existing 'hybrid hubs' (Haldrup et al., 2018) for experimenting with such complex processes. These are spaces where fuzzy synergies between doing, learning and exploring complex problems are already happening. It is tempting to see this messiness as a flaw that needs to be structured or systematised, instead of a core quality on its own. With our propositions, we seek to highlight the messiness and fuzziness as central to what should be acknowledged and nurtured in them—on the terms of the makers themselves.

8.2 Mapping Maker Manifestos

Most designers recognise that no given methodological model is guaranteed success. A failure could lead to a superficial assumption that we did not deploy the methods correctly, or used the wrong method for the given problem. However, as more experienced designers will know, design thinking and practice is not so much a matter of rigid procedures and skills, as it is about an experimental exploration of possibilities and the practical knowledge emerging from that; what Cross (1982) early coined intuitively "designerly ways of knowing". Rather than fixed and rigid procedures, methodologies can be seen as collections of tools in one's own repertoire. There might be a user experience design method consisting of writing personas, creating wireframes/mockups, rapid digital prototypes, accelerated physical model experiences or using a speak-out-loud test. We consider each element an individual tool and the overarching method a suggested recipe that one can follow. This approach leaves room for methods to be tailored to the given situation even as the project evolves.

This insight also runs through other approaches to learning and education, e.g. different variants of hands-on pedagogics, project-based learning, "agile" methods and the Maker methodology we attempt to describe in this chapter. The experience that emerges from these approaches can be seen as a radical alternative to more rigid ways of thinking about conceptualisation and methods in design and learning. As a replacement for a coherent set of textbooks and methodologies, the practices and procedures of Maker culture are largely to be found in the fast-growing open-source manifestos that flourish on the web, and inspirational talks that can be encountered at various conventions, fairs, festivals, online and at hackathons around the world. In this, the Maker movement resembles more of a grassroots movement than an approach to design thinking and practice. As one of the Maker movement's leading proponents (and propagationists), Dale Dougherty (2016), observes, making is more about fostering a particular mindset than developing rigorous methodologies. Drawing on Stanford psychology professor Carol Dweck (2016)'s book 'Mindset: The new psychology of Success', distinguishing between fixed and growth mindsets, she argues of the fixed mindset "believing that your qualities are carved in stone—the fixed mindset—creates an urgency to prove yourself over and over. If you have only a certain amount of intelligence, a certain personality, and a certain moral character" (Dweck, 2016, pp. 4–6). People who believe in fixed traits are in danger of being measured by failures since the mindset never taught them ways to cope (Dweck, 2016, pp. 38–40). On the other hand, Dweck (2016) writes: "This growth mindset is based on the belief that your basic qualities are things you can cultivate through your efforts, your strategies, and help from others (...)—everyone can change and grow through application and experience" (Dweck, 2016, pp. 6–8). Failures still hurt people with the growth mindset, but are not what defines them, as their basic qualities can be developed (Dweck, 2016, pp. 38–40). "Making" as an educational approach shares significant commonalities with the "hands on" learning and "project based" pedagogy propagated by Dewey and Piaget. "Making" is like these approaches in that it manifests a desire to liberate and emancipate the mind through engaging with the material and social realities at hand.

The idea of thinking in terms of a mindset instead of a fixed method is not unique, but shares similarities with several other approaches, from agile methods to project-based learning. Nor is there a fixed set of parameters defining a "Maker mindset". However, we posit that we can distil a coherent set of defining features from the various manifestos of the Maker movement that can be constructively used as prescriptions for what a "Maker approach" to design and education might look like.

As diverse as the visions, manifestos etc., are, a few recurring formulations emerge (see Box 8.3 for our reflections on including Seminal, Local and Positional manifestos). In Table 8.1, the different Maker manifestos have been thematically

Box 8.3: Categories of Maker Manifestos

As a basis for summarising our propositions for Maker thinking, we have generated a list of prominent manifestos (including Dougherty's, quoted above). The concept of "manifestos" is in this case in its widest sense: A condensed document that attempts to summarise a movement, ideology, purpose or way of working. They are represented in Table 8.1.

The intention behind including the different manifestos in the mapping can be grouped into the following categories:

Seminal: Those are seminal writings that have contributed to and shaped the international discourse on the Maker movement in such a way that they have been used as a reference point to debate, communicate and to cement what it means to be a Maker.

Local: Those are manifestos for/by local or online sub-groups, Maker/hackerspaces and the like. These manifestos represent local voices (and thus nuances) in forming communities around the Maker movement. The manifesto of Danish interactive art collective illutron has been included because it has a central role in shaping the ideas behind the Fablab at Roskilde University and the manifesto represents one of the early writings about Maker intentions in Denmark. We would have liked to include other seminal Maker/hacker spaces like Sudo-Room (Oakland), Noisebridge (SF), TechShop (SF), Homebrew Computer Club (SF), C-Base (Berlin) and the local Danish hacker space Labitat (CPH). However, as much as we acknowledge their role in the development of the Maker/digital fabrication mindset, we have yet to be able to find public writings from them with clear enough intentional framings (manifestos/visions) to be able to include them.

Positional: This category represents manifestos that are not central Maker manifestos but serve to position more central Maker manifestos relative to related fields and schools of thought. Design thinking and the Manifesto for Agile Software Development are on the outskirts of what can be considered embedded in the Maker movement. They have been included for two specific reasons. Design thinking (Plattner et al., 2010) is the closest to a formal method within the Maker movement. Although design thinking has its roots in user-centered design (Simon, 1969) the possibility of prototyping through digital fabrication tools has been shown to be fruitful for engaging in design thinking methods. Similarly, the user-centered approach in design thinking creates a framework for arguing for empowerment and creativity in technology within a Maker mindset. The Agile Software manifesto (Beck et al., 2002) represents a seminal turn for software development within computer science in which more open-ended and exploratory approaches to problem solving were put at the forefront. Since a significant part of Maker spaces revolves around programming and hacking software for new possibilities, the two share similar agendas. Finally, Problem-oriented project learning is the overarching way of thinking within Roskilde University. It thus also serves as a basis for understanding the interest and role of Fablab RUC as a part of the university.

Table 8.1 Mapping of different maker manifestos

Name	Category	Ideation through problems	Ideation through materials	Freedom to fix	Freedom to create	Freedom to hack/ modify	The vision to experiment	Knowledge sharing	Learning through making/ hacking	From consumers to producers	Technological ownership
The maker mindset (Dougherty, 2013)	Seminal		x	x	x	x	x				x
The maker movement manifesto (Hatch, 2013)	Seminal		x	x	x	?		x	x		(x)
Maker's bill of rights (Torrone, 2006)	Seminal			x	x	x					x
The Fab Charter (2012)	Seminal				x			x	x		(x)
Open design manifesto (Kadushin, 2010)	Seminal		x		x	x		x			
Repair manifesto (ifixit, n.d.)	Seminal			x					x		x
Hacker's manifesto (Mentor, 1986)	Seminal					x	x				x
The creative commons (Lessig, 2004)	Seminal				x	x		x		x	x
Hacker ethics (CCC, n.d.)	Local	x	x	x	x	x	x	x	x		x
What is a maker mindset? (Myers, 2017)	Local		x					x	x		

Table 8.1 (continued)

Name	Category	Ideation through problems	Ideation through materials	Freedom to fix	Freedom to create	Freedom to hack/ modify	The vision to experiment	Knowledge sharing	Learning through making/ hacking	From consumers to producers	Technological ownership
Illutron manifesto (Hobye et al., 2009)	Local		x		x		x	x		x	
Design thinking (Plattner et al., 2010)	Positional	x	x								
Manifesto for agile software development (Beck et al., 2002)	Positional	x	x								
Problem-oriented project learning (Roskilde University, n.d.)	Positional	x						x	(x)	x	

(x) in parenthesis means that it is not explicitly stated, but they indirectly phrase similar arguments. In the case of knowledge sharing and PPL, there is an implicit sense that knowledge sharing in this case points towards academic knowledge more than Maker (practical) knowledge classified. The themes have been derived by comparing the manifestos and grouping common interests together in themes. The themes are:

Ideation through problems is the approach of project work, in which a problem serves as the basis and focus for collaboration, solving and knowledge gathering. Although not a common perspective within the different Maker movement manifestos, it is prevalent in the last three manifestos, which are closer to academic and business development traditions.

Ideation through materials is a prevailing approach to making, which considers exploration with and through the materials as a primary source of ideation compared to the more problem-oriented approach. This is especially the case with the more grassroots-like movements of Making and hacking, while the Fablab charter has in its tone more weight on the engineering and production approach to Making.

Freedom to fix: Similarly, this theme is about the political activist's right to be able to repair commercially available products. A common parole is "If you can't open it, you don't own it".[1] Most products today come with a voided warranty if you open the product. Some are designed to be unrepairable. This is a strong argument in the Maker's Manifesto and the Repair Manifesto, and has since found a strong resonance within legislative bodies in both the US and EU.

Freedom to create is similar to the perspective of ideation through materials. Still, it focuses on your right to create and learn through engaging with materials, production methods and technology, and the value of creation, even if not market-ready.

Freedom to hack/modify takes freedom to fix one step further. It highlights the importance of also being able to modify things to suit your needs—maybe even leaving behind the original intentions of the product from the company that produced it in the first place. To some extent, one can say this is where true ownership is brought back to the people/users, who are not just consumers (with maintenance rights to the products).

The vision to experiment puts focus on the potential of pushing boundaries and creating new knowledge beyond mere physical objects and products.

Knowledge sharing is the notion of having a communal practice of sharing (often technical) knowledge and inventive methods as openly as possible with other people so novel solutions and best practices can be disseminated across communities. 'Fix once fix everywhere' is a common parole. It is to some extent considered rude not to share because new knowledge often builds upon the experiments of others.

Learning through Making is the belief that only by actually doing and Making one can embody theoretical knowledge. By programming a robot to follow a line, one can understand programming in practice and discover and solve real problems instead of merely following theoretical logic. This is a central argument and justification for the American tradition of the Maker and Fablab movement because it

[1] What is known as a common phrase seems to be credited to: Bas van Abel, Waag Society Guimarães, Open City Project July 2012.

justifies integrating with STEM[2] education, especially at a middle and High School level.

From consumers to producers: This ties into other themes presented here. The reason we have chosen to make it a separate theme is to point out the relationship with Problem-oriented project learning at RUC, which emphasises exploration, research and learning primarily in the form of being consumers and producers of academic knowledge and not in the sense of actually producing, fabricating and Making (see Box 8.4 for a longer discussion on PPL in relation to academic production).

Box 8.4: The Problem-Oriented Project Learning (PPL) Model

Roskilde University draws reference to the pedagogical thinking of Dewey (as learning by doing or action-learning, see e.g. Dewey, 1916, 1925, 1938) as well as critical Marxism, in its foundation. The pedagogical model at Roskilde University is based on the principle of problem-oriented project work (PPL); this entails that 50% of the students' curriculum is anchored in their own projects and experiments. In doing so, the model explicitly positions students as researchers in their own right. In other words, students are expected to not only acquire knowledge, but to contribute to knowledge production. They are positioned as producers of knowledge rather than mere re-producers of knowledge.

Inherent in the PPL approach are the values of interdisciplinarity and engaging with practice; inviting students to think on related problems, challenges or phenomena in their complex and multifaceted nature, rather than strictly confine themselves within predetermined disciplinary borders, aim for short-term quick solutions, or fall prey to methodologism (see Gao, 2014; Thomas Teo, 2009). The key principle of problem-orientation connects to a basic tenet from Aristotle; namely that of phronesis, or the importance of science to engage in or with practical matters (Flyvbjerg, 2001; or see Thomas Teo, 2009 for a similar discussion). It also underlines an ambition for higher education to be both productive as well as personally and socially transformative (underlining the emancipatory heritage from critical theory and critical pedagogy), rather than merely accumulative and reproductive. The emancipatory agenda of critical theory implies that teaching and learning is indeed a political project of questioning and pushing boundaries and of co-creating novelty. It implies creating and opening new spaces of possibilities—for oneself and others—thus continuously engaging with a fundamental question of what society should and could be. Thus, a shared understanding in the Maker movement and PPL is the importance of expanding, rather than restricting, our knowledge, regardless of the form of knowledge. This is the case whether one works with design, geography, psychology or public health.

(continued)

[2] Science, Technology, Engineering and Mathematics.

Box 8.4 (continued)

"Education is not – or ought not to be – simple transmission of stable ideas across generations, but should be a way of reconstructing social relationships in ways that enable human beings to respond to the changing world in which they find themselves. In other words: Education is society's way of making sure that fruitful new ideas will be devised in the future, and this is achieved only through communication" (Brinkmann et al., 2010, p. 244).

The visions of PPL resonate with what Brinkmann & Tanggaard (2010) call an epistemology of the hand. With reference to William James and John Dewey, they call into question the purpose of education and challenge the dominant idea of an epistemology of the eye: The epistemology of the eye draws on Plato's philosophy stating that ideas are "out there", waiting to be discovered, or properly seen. This understanding has been rather persistent in Western thought and has had an enormous impact on the educational system and dominant conceptualisations of learning, conceptualising learning as something that (automatically) follows from visual confrontation with something. In this light, learning becomes a question of representation and transference. In contrast, an epistemology of the hand recognises that learning is an active and practical matter—it is a doing (which clearly resonates with the German word for science "wissenschaft" and the Danish equivalent "videnskab"—science is knowledge creation). Adopting an epistemology of the hand implies that knowledge is never neutral or decontextualised; it is done by someone, with the purpose of something, in relation to something. It is an active creation—a manipulation of the world.

Mastering knowledge cannot be a reproductive venture (see Ilyenkov, 2007): it is an active, collective, cooperative venture (see e.g. Johnson et al., 1999) of engaging with the world. Thinking with the framework of PPL means challenging the notion that one can teach abstract decontextualised knowledge to be applied later; knowledge is always embedded in a social practice tradition. Subsequently, thinking—and learning—is an action relating to a concrete worldly matter, to concrete forms and materials, to concrete problems and societal issues (see also Andersen et al., 2015; Illeris, 1974; Negt, 1975).

Technological ownership: Whereas the "freedom to hack/modify" ties to the concrete practice of doing, some manifestos argue for "ownership" in a more explicit sense.

The focus in the Maker manifestos seems to be on experimentation for the sake of knowledge sharing and development. Concerning identifying the core "methodological" principles of Maker thinking, this seems to be a central feature. In the following, we will try to elaborate on what this means for formulating the "mindset" of Maker thinking.

8.3 Condensing the Themes into Six Propositions for Actionable Maker Thinking

What has emerged from the mapping of the different manifestos are roughly ten themes that in some way or another underpin values, intentions and dreams around what it means to be a Maker. However, the themes do not not explicitly state an approach or way to engage with a Maker mindset. Although the Maker movement is often portrayed as an innocent movement whose primary goal is to inspire kids and teenagers to get an education within STEM (re Micro Bit, The Maker Festivals, Arduino, etc.), it is also clear from the different themes that a more serious political stance underlies the ideals behind it (see Sect. 8.2). There is a sense of ownership of technology and a belief that knowledge should be free and shareable—counterculture to consumerism of closed, proprietary corporate products.

In the following propositions, we have deliberately attempted to formulate the creative practice of Maker thinking with actionable intention to engage with a Maker mindset. The propositions are concrete approaches to moving from politics and visions into how we actually engage—into something explicitly actionable. We attempt to translate what comes naturally within the Maker community into more transparent framing for newcomers and outsiders wanting to engage and participate.

The condensation of the themes into six propositions has been done through multiple online workshops with stakeholders from Maker spaces around Greater Zealand, Denmark (see acknowledgements). The themes have served as a basis for discussing what would be considered a methodological mindset that stems from within the Maker community (instead of borrowing from, e.g. Design thinking, Design sprint, Agile software development and the like). The intention is to capture major current tendencies and voices within the community. There is no intention of a complete fixed-in-stone scripture, which would go against the fundamental principles of the mindset. Furthermore, the propositions seek to not only challenge, but also contribute to, other forms of design thinking and project-based education and research. In doing this, we will distil our own experience with the "Fablab RUC perspective" (Box 8.2) and extrapolate on the commonalities of the manifestos displayed in Sect. 8.2.

8.3.1 Proposition 1: Question the Foundation of a Problem

Put simply, design is problem-oriented. It is based on the assumption that design is an iterative and solution-directed approach within a problem-oriented framework. However, this often implies the misconception that each problem has a 'right' solution instead of accepting that an infinite number of solutions exist. Problems can be great facilitators for engagement, but can also quickly create tunnel vision towards an (oversimplified) solution. Often, the problem is poorly defined and only touches

Fig. 8.1 Question the foundation of a problem instead of jumping to a solution

the surface of the complexity that lies beneath. A brainstorm on solving a problem can quickly lead to superficial solutions without actually pushing the fundamental boundaries of what is possible. Problem orientation, therefore, risks only making evolutionary steps without revolutionising the fundamental way of doing and thinking. Maker thinking is about speculating on previously unknown possibilities. In this framing, understanding the problem is an integrated and ongoing part of the process instead of a point of departure (Fig. 8.1). This is a central aspect of differentiating Maker thinking from traditional design-driven strategies. Instead of thinking in terms of problems and their possible solutions, we propose the need to ask questions based on curiosity. This is where things like asking "what if" come in. For example, what if we can make a 3D printer that can 3D print itself? What is seemingly a strange question opens up a whole palette of potential technologies and ways of doing digital fabrication, as well as ethical and social issues. In that sense, design practices, rather than simply solving problems, become resources for asking questions, posing problems and solutions, and exploring new potentials.

8.3.2 *Proposition 2: There Is Not a Fixed Recipe for Success*

Making is not about creating a specific final solution, but about approaching design as a knowledge-generative process in which the aim is more about creating a multitude of variations and highlighting the nuances and possible solutions for a possible point of interest (Fig. 8.2). In many design disciplines, the method is intended to secure the quality of the process, and thus, ultimately the outcome. Through this way of thinking, the methodology risks becoming the justification for the design. If you follow the "right" process, then your design is good and justified. Maker thinking leaves this behind by thinking in terms of the mindset as the method in itself, expressed by the fundamental and fuzzy ways of doing and engaging with materials, technologies, ideas, people etc. Where others pick up the experiment and how it facilitates new conversations is the primary success criteria—not the process in itself. This still leaves room for methodological reflection, but the different elements in different methods become a repertoire to use pragmatically as needed, not a rule book or a catechism. User testing happens when we need to understand real usage etc. Taken to the extreme, this also becomes a clash with the idea of following a (design) ideology (and thus indirectly gurufying the inventor of the chosen method).

Fig. 8.2 Making is a messy process where linearity is disturbed

The community is too eclectic and too anarchistic in their mindset for this to be attractive.

We must move away from explicit utilitarian purposes and into a more exploratory potential-seeking mode to achieve this goal. Making is an exploration of possible outcomes—the Maker mindset enables thinking agility as an explorative strategy that does not necessarily have a utilitarian perspective, but more seeks to create a space for a multiplicity of explorations that take place on multiple planes, from problems to possibilities.

8.3.3 Proposition 3: Think Through the Material

Often, design approaches pay tribute to an alleged need for clarification before engagement, as if one can map out anything before engaging with it. Post-it notes are sorted, and brainstorming sessions are held to make sense of whatever one plans to engage with. It is as if the world we live in is untouchable before we clearly understand how and what we want to change. Maker thinking is about the freedom to allow intuition to guide one's engagement with materials (and thus the world) (Fig. 8.3). Through an inherent curiosity in technology, one can gain clarity about what good choices are and the technology's potential. In this way, clarity is about understanding the complexity of any situation or technology, the limitations and potentials, and how these constraints relate to each other. Asking questions and exploring possible solutions through materials and material exploration is key to Maker thinking. Encouraging following your curiosity rather than a particular problem or method thus becomes discursive thinking in which materials open up potentials and speculations as a way of creating knowledge. Through prototyping, one becomes aware of the complexity of the problem domain from which one may initially have started. But it is by expanding your understanding through engaging in designing a series of curiosity-driven solutions—at first primitive and later more evolved ones—that you start exploring the unknown, inventing new concepts, or nuancing existing solutions and formulations of problems. This starts from the inside of the technologies and materials themselves and takes on an explorative approach into the potential of different kinds of technologies or combinations thereof. Objects designed from this perspective are solidified products or

Fig. 8.3 By iteratively shaping the material, the form appears

prototypes. They are not even sketches. They are curiosities that work as morphing and travelling objects to understand their own role. Though this exploration appears to be speculation around future possibilities and potentials, it does not start in a particular imaginary future, but uses material exploration to point towards these possible and potential futures—leaving good or bad behind and instead just reflecting on them as they are, or may be, or could be.

8.3.4 Proposition 4: Making Is About Experimenting Again and Again

Experimenting is an often overlooked resource for creativity. By engaging with questions and failing, success materialises atop of a pile of failures. Each experiment is a new beginning for new questions—potential experiments (Fig. 8.4). Ironically, the knowledge required is first gleaned at the end of the process. This is a beautiful paradox in which idealistic methods meet reality. Since engaging with making is the primary conversation and knowledge sharing is a central aspect, quick and dirty is a primary objective. Create something that shows new possibilities and new ways. Don't dwell too much and make a solid plan—it will be obsolete moments later anyway. If multiple people engage with the same questions, you draw on multiple perspectives, experiences and possible solutions. Instead of refining your grand ideas, making big plans and elaborate conceptualisations, test them out, and see what happens! Show potential instead, and engage with an ever-changing world. So, to approach Maker thinking, you need to allow yourself to go into an endless series of failures. Each failure starts a bit further down the path. Each failure is not only valuable as a test of whether your ideas work in practice or to build competencies—but failure may also open up new ideas and horizons. Use playfulness and curiosity! Begin with "what if…?"—"What if we could make the robot look around corners?". Try it out early—even without being able to answer "why?". A road to new knowledge is failing fast and using an agile approach. Allow yourself to fail so that you can learn and be inspired by new ideas. This way of experimenting is a continuous engagement with trying out, learning, and looking for new potentials.

Fig. 8.4 Making is about experimenting (the stars) again and again

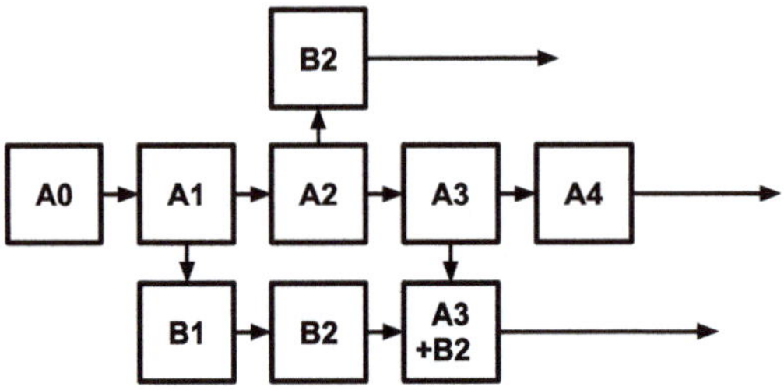

Fig. 8.5 When knowledge is shared, mutations appear

8.3.5 Proposition 5: Let Go of Ownership

A common saying within Maker communities is: "If you can't open it, you don't own it". A similar saying from the open source movement is "Sharing is caring". Both point to the fundamentals of sharing knowledge as a prerequisite for learning and innovation. While every Maker may have their own perception of what a good design for a 3D printer is, the community is constantly sharing new knowledge and making endless parallel strains of design solutions, each with its own qualities and limitations (Fig. 8.5). The Maker movement fosters those who help one another without intellectual property or exchange of monetary value. Ideas are shared with the intention of them being stolen. Ideas float—they are something you share with others, and you receive from others, and we have to respect the fact that creation is a collective enterprise that cannot exist without the others in the movement agreeing to share. Through the sharing of ideas and knowledge, single concepts turn into a multitude of variations. By actively participating in a practice that both shapes and is being shaped by everyone all the time, from a Maker's perspective, the Maker and the user are the same. So: build theories, concepts, methods and formulas. Share this knowledge through papers, presentations, videos, prototypes, open-source material, Instructables and demos: sharing builds communities and lets people build on each other's work.

8.3.6 Proposition 6: Accessibility as an Ongoing Endeavour

Making—and keeping—technology accessible requires ongoing effort. This effort is multifaceted and multilevel. This is the least observationally precise and most open-ended of our propositions, as it is partly an amalgamation of the earlier propositions and partly an emerging phenomenon relating as much to the future as to the observed past.

There is an ongoing effort to counter (and transform) commercial interests and inertia towards closed/blackboxed systems in which interface protocols are proprietary. The right to repair is only the essential starting point. As new technologies are rolled out, Maker culture actively de-black box technologies and enable more possibilities (Fig. 8.6). To create space for non-intended purposes, to enable maintenance when the technology is deemed obsolete, and to enable new experiments to piggyback on the current technological development. This effort is also emerging as politically successful, is coupled with climate, upcycling and e-waste agendas and has inspired multiple right-to-repair laws.

Accessibility is also about enabling and inviting diversity and different technological approaches. From the geek—to the activist—to the artist—to the entrepreneur—to a multitude of identifications of what different people deem valuable. While including different roles has already been quite successful, a comprehensive inclusiveness of people who may be less advantaged, more marginalised and have not traditionally been part of a tech enthusiast milieu is very much in need of further work. This is to challenge any communal bias knowledge bubble and echo chamber in such a way that it also creates a diverse set of voices to engage in the discussion.

In this light, accessibility is a multi-way street. It is about making technology accessible, but also about letting go of judgement and predefined intentions or purpose definitions of who has access to it and the community around it. This gives spaces for individual usefulness and purposefulness without judgement from set definitions and what is deemed valuable. It is about keeping the door wide open.

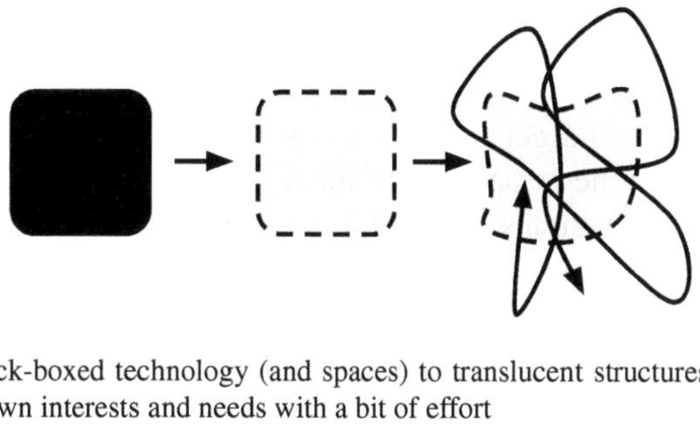

Fig. 8.6 From black-boxed technology (and spaces) to translucent structures where people can engage with their own interests and needs with a bit of effort

8.4 Closing Discussion: Potentials and Limitations

Gaining access to digital fabrication tools and environments that cater to non-technical practitioners is attractive for many disciplines. Interaction designers can argue that it is Sketching and Prototyping (Buxton, 2007), Design researchers can consider it Research Through Design (Frayling, 1993; Zimmerman et al., 2007 etc.) and a product designer can consider Making a part of the prototyping phase in Design Thinking (Plattner et al., 2010). The loosely coupled structure of the Maker movement gives ample space for such possibilities, and in this case, it is, in principle, okay. The Maker movement does not have a formal method, and the whole nature of it invites eclectic approaches to what it means to Make (and thus to design).

However, adding a design methodology with roots outside the Maker movement to the Maker movement comes with a risk of losing some of the tacit ways of doing and thinking embedded in everyday practices. More so, this becomes problematic (perhaps even unethical), as many practitioners in the Maker community are not academics and may not have the academic skills and institutional support to formulate and ground their perspectives in the academic discourse.

With this chapter, we have attempted to turn this tendency around. We have started with a diverse set of manifestos from some of the major stakeholders. We have distilled a set of themes that we think overlap in them. With Makers, we have formulated a set of propositions on what it means to think and act as a Maker. Finally, we have argued that this should be considered a methodology in itself. We hope this can be seen as an input to positioning the Maker movement's concept and ideals that go beyond the tacit practices already existing in the Maker space.

The way of thinking presented in this chapter also comes with limitations. Through our work in a university Fablab, we also observe how the Maker's way of thinking has its limitations and challenges. It is easy to say within a Maker/Fablab space, "here you can build anything". The reality is that making something that fundamentally challenges or moves the world often requires a strong commitment with serious engagement that often moves beyond the Maker spaces themselves.

From the perspective of STEM (science, technology, engineering and mathematics) and science education, a Maker thinking approach comes with a unique set of challenges. Within STEM, rapidly exploring and experimenting from a Maker perspective quickly comes across as naive compared to more structured approaches of calculating, planning, and constructing to exacting standards dominant in those fields. Similarly, from the perspective of the rigour of conducting a proper science experiment within the natural sciences, it is hard to acknowledge making and hacking as something that can be considered a knowledge contribution in itself.

However, what is lost in this segregational approach to dividing the fields is an acknowledgement of the potential of iterating between different modes of operation and thinking, depending on the intention of the process. First and foremost, from a problem-oriented learning perspective (e.g. PPL within RUC), where science is just as much about engaging with real-world problems. Moving out of the sterile lab and into a space in which it is possible to quickly experiment with a multitude of solutions can be an inspirational approach to understanding a problem (and its possible solutions) quickly. However, the final experiments in a Maker space may still need to be translated back into more traditional experiments and theoretical framings within their respective fields. Secondly, the mode of learning differs greatly from student to student. For some, theoretical knowledge is at the forefront, whereas for other students, theory is a necessary step when they have yet to figure out why the quick-and-dirty structure they created in the Maker space is not able to hold the load they expected. On the other hand, for some students, the open-ended mindset in a Maker space can destabilise their sense of direction and put them in decision paralysis. It is a challenge to—in an aware manner—shift between Maker thinking = practical doing, and classical study and research of existing knowledge (Haldrup et al., 2018; Svabo et al., 2020). The simplest answer is probably to see it as switching between different mindsets of operation. Based on the propositions put forward in this chapter, there is a need to educate students on the different modes of doing, to induce awareness, such that they know when to change mode, why and how they change mode, what the two different modes are good for, and how to increase synergies. This is especially the case for the ones who have a tendency to only be in one of them.

If we acknowledge that taking action is an integrated part of learning and solving complex problems, Maker spaces should be celebrated even if it is currently harder to measure, evaluate and systematise the "output" and knowledge contributions created in such a space. Creative Pragmatics, as outlined in the introduction to this chapter and in Chap. 1, shifts the perspective from trying to untangle the messiness of Maker spaces—to thinking about how we can value it and support it constructively. The six propositions in this chapter support this notion by moving beyond the lab as a technical production facility in which predefined design can be produced, emphasising instead the communities and complex entanglements which Creative Pragmatic learning and exploration are performed through. In doing this, we propose to see the Maker space as a vibrant space for constant negotiation of learning, development and exploration—in which the culture and the space itself serves as a hybrid hub for a creative and, as Manning (2016) formulates it, 'speculative pragmatics' that transgress the facilities themselves.

Acknowledgments This chapter has been created in collaboration with Makers primarily at Fablab RUC, but also involved the Maker community in Zealand, Denmark. Participants have been involved in illutron, Fablab RUC, Underbroen, Copenhagen Maker, Labitat, Fablab TI, Fablab Nordvest and more. A special thank you to students and staff working with and around the Fablab at Roskilde University for cultivating one of the main environments that inspired the above discussion.

References

Andersen, A. S., & Heilesen, S. B. (2014). *The Roskilde model: Problem-oriented learning and project work*. Springer.

Andersen, A. S., & Kjeldsen, T. H. (2015). Theoretical foundations of PPL at Roskilde University. In A. S. Andersen & S. B. Heilesen (Eds.), *The Roskilde model: Problem-oriented learning and project work* (pp. 3–16). Springer.

Anderson, C. (2012). *Makers: The new industrial revolution*. Random House.

Beck, K., Beedle, M., van Bennekum, A., Cockburn, A., Cunningham, W., Fowler, M., Grenning, J., Highsmith, J., Hunt, A., Jeffries, R., Kern, J., Marick, B., Martin, R. C., Mellor, S., Schwaber, K., & ThomasDave, S. J. (2002, February 3). *Manifesto for Agile Software Development*. Retrieved from https://agilemanifesto.org/

Birtchnell, T., & Urry, J. (2016). *A new industrial future?: 3D printing and the reconfiguring of production, distribution, and consumption*. Routledge.

Brinkmann, S., & Tanggaard, L. (2010). Toward an Epistemology of the Hand. *Studies in Philosophy and Education, 29*(3), 243–257.

Buxton, W. (2007). *Sketching user experiences: Getting the design right and the right design*. Elsevier/Morgan Kaufmann.

CCC. (n.d.). Retrieved from https://www.ccc.de/en/hackerethics

Cross, N. (1982). Designerly ways of knowing. *Design Studies, 3*(4), 221–227.

Davies, S. R. (2017). *Hackerspaces: Making the maker movement*. Wiley.

Dewey, J. (1916). *Democracy and education. An introduction to the philosophy of education. Teddington*. Echo Library.

Dewey, J. (1925). *Experience and nature*. Open Court.

Dewey, J. (1938). *Experience and education*. Touchstone/Simon & Schuster.

Dixon, B. S. (2020). *Dewey and design: A pragmatist perspective for Design research*. Springer.

Dougherty, D. (2013). The maker mindset. In *Design, make, play* (pp. 25–29). Routledge.

Dougherty, D. (2016). *Free to make: How the maker movement is changing our schools, our jobs, and our minds*. North Atlantic Books.

Dunbar-Hester, C. (2014). Radical inclusion? Locating accountability in technical DIY. In *DIY citizenship – critical making and social media* (pp. 75– 88). MIT Press.

Dweck, C. S. (2016). *Mindset: The new psychology of success (Updated Edn.)*. Random House.

Ehn, P. (1998). Manifesto for a digital bauhaus 1. *Digital Creativity, 9*(4), 207–217.

Ehn, P., Nilsson, E. M., & Topgaard, R. (2014). *Making futures: Marginal notes on innovation, design, and democracy*. MIT Press.

Flyvbjerg, B. (2001). *Making social science matter: Why social inquiry fails and how it can succeed again*. Cambridge university press.

Frayling, C. (1993). *Research in art and design*. Royal College of Art.

Gao, Z. (2014). Methodologism-methodological imperative. In T. Teo (Ed.), *Encyclopedia of critical psychology* (pp. 1176–1179). Springer.

Haldrup, M., Hobye, M., & Padfield, N. (2018). The Bizarre Bazaar: FabLabs as hybrid hubs. *CoDesign, 14*(4), 1–16.

Hatch, M. (2013). *The maker movement manifesto: rules for innovation in the new world of crafters, hackers, and tinkerers*. McGraw Hill Professional.

Hobye, M., Padfield, N., & Members. (2009, March 1). *Illutron manifesto*. Illutron. Retrieved from https://www.illutron.dk/about/manifesto-values

Hobye, M., Padfield, N., & Haldrup, M. (2016). Making makers make maker machines. *Fab12. Shenzhen*. Retrieved from https://forskning.ruc.dk/files/57331468/Fab12_Hoybe_et_al.pdf

ifixit. (n.d.). *Self-repair manifesto*. iFixit. Retrieved from https://www.ifixit.com/Manifesto

Illeris, K. (1974). *Problemorientering og deltagerstyring—oplæg til en alternativ didaktik*. Munksgaard.

Ilyenkov, E. V. (2007). Knowledge and thinking. *Journal of Russian & East European Psychology, 45*, 75–80.

Johnson, D. W., & Johnson, R. T. (1999). Making cooperative learning work. *Theory into Practice, 38*(2), 67–73.

Kadushin, R. (2010, September). *Open design manifesto*. Ronen Kadushin. Retrieved from https://www.ronen-kadushin.com/open-design-manifesto

Knoll, M. (1997). *The project method: Its vocational education origin and international development*. Retrieved from https://scholar.lib.vt.edu/ejournals/JITE/v34n3/Knoll.html?ref=psweboffice.com

Lessig, L. (2004). The creative commons. *Montana Law Review, 65*, 1.
Maker Media. (n.d.). The Maker Movement. Retrieved from https://makerfaire.com/maker-movement/
MakerSpace & Fablabs mm. (n.d.). Google Maps. Retrieved from https://www.google.com/maps/d/u/0/viewer?mid=1Sji0p_ENe1QbbNo9J85-4Wrb_Ls
Manning, E. (2016). *The minor gesture*. Duke University Press.
Mentor. (1986, January 8). *Hacker's Manifesto*. Phrack Magazine. Retrieved from http://phrack.org/issues/7/3.html
Mikhak, B., Lyon, C., Gorton, T., Gershenfeld, N., Mc Ennis, C., & Taylor, J. (n.d.). *Fab lab: An alternate model of ict for development*. Retrieved from https://cba.mit.edu/events/03.05.fablab/fablab-dyd02.pdf
Myers, N. (2017, April 3). *What is a maker mindset?* STEM MINDS; STEM MINDS Corp. Retrieved from https://www.stemminds.com/what-is-a-maker-mindset/
Negt, O. (1975). *Sociologisk fantasi og eksemplarisk indlæring: til teori og praksis i arbejderud-dannelsen*. RUC Boghandel & Forlag.
Nielsen, N. M., Blomhøj, M., Enevoldsen, T., & Haldrup, M. (2014). The bachelor Programmes and the Roskilde model. In *Roskilde model* (pp. 79–106). Springer.
Padfield, N., Haldrup, M., & Hobye, M. (2014). *Empowering academia through modern fabrication practices. Fablearn EU 2014.*
Plattner, H., Meinel, C., & Leifer, L. (2010). *Design thinking: Understand – Improve – Apply*. Springer.
Roskilde University. (n.d.). *Pedagogical profle at Roskilde University*. RUC.dk. Retrieved from https://typo3.ruc.dk/fileadmin/assets/adm/dokumenter_og_noegletal/vaerdier/Paedagogisk_profil_UK.pdf
Schelhowe, H. (2013). Digital realities, physical action and deep learning-FabLabs as educational environments. *FabLab: Of Machines, Makers and Inventors, 2013*, 93–103.
Simon, H. A. (1969). *The sciences of the artificial*. MIT Press.
Smith, A. (2017). *Social innovation, democracy and makerspaces*. https://doi.org/10.2139/ssrn.2986245
Smith, A., Fressoli, M., Abrol, D., Arond, E., & Ely, A. (2016). Hackerspaces, Fablabs and makerspaces. In *Grassroots innovation movements* (pp. 100–122). Routledge.
Svabo, C., & Borch, M. M. (2020). Oplevelseslæring og BioFabLab [Experiential Learning and BioFabLab]. *MONA*, 17.
Teo, T. (2009). Philosophical concerns in critical psychology. In D. Fox, I. Prilleltensky, & S. Austin (Eds.), *Critical psychology: An introduction* (Vol. 2, pp. 36–53). Sage.
The Fab Foundation. (2012, October 20). *The Fab Charter*. Retrieved from https://fab.cba.mit.edu/about/charter/
The Fab Foundation. (n.d.). Retrieved from https://fabfoundation.org/about/
Torrone, P. (2006, December 1). *The maker's bill of rights (PDF)*. Make: DIY Projects and Ideas for Makers; Make: Community. Retrieved from https://makezine.com/article/maker-news/the-makers-bill-of-rights/
Walter-Herrmann, J. (2013). *Fablabs—A global social movement. FabLab. of machines, makers and inventors* (pp. 33–45).
Zimmerman, J., Forlizzi, J., & Evenson, S. (2007). Research through design as a method for interaction design research in HCI. *Proceedings of the SIGCHI Conference on Human Factors in Computing Systems*, 493–502.

Mads Hobye PhD, Associate Professor at Roskilde University (Denmark) within the Department of People and Technology. His research is in interactive digital materiality as a hybrid between physical material and digital technology. It is research in the potential of aesthetic use of technology and how to democratise technology ownership through frameworks, open knowledge-sharing and digital fabrication as a maker mindset. The work has been published internationally and applied in practice through co-founding illutron (a collaborative, interactive art collective in Copenhagen) and as artworks presented at international museums, galleries, and festivals. For more see www.hobye.dk.

Michael Haldrup Professor at the Department of Communication and Arts, Roskilde University. Numerous publications on arts- and design-based practices including The Bizarre Bazaar (w. M. Hobye & N. Padfield in CoDesign 2018), Experimental Museology (w. M. Achiam & K. Drotner, Routledge 2021) and Arts, Heritage and Performative Politics (w. L. Nomeikaite & L. Frers, Edward Elgar (forthcoming)). In 2013 Michael was part of establishing FabLab RUC. Since then his research and teaching (as part of the B. A. program in humanities and technologies as well as the M. A. in sustainable digital transformation at Roskilde University) has often included multiple collaborations with art & technology practitioners from the maker milieux. For more see forskning.ruc.dk/en/persons/mhp.

Sara Daugbjerg MSc, Technology Consultant and Lab Management at FabLab RUC, Roskilde University. Sara has an interdisciplinary education in the fields of Computer Science, Urban Studies and Humanities & Technology. Sara works with IoT, environmental sensing, citizen science, guerrilla data collection, and digital fabrication. She also has a broad experience in teaching programming to kids, youngsters, and adults.

Sofie Pedersen PhD, Associate Professor in Social Psychology of Everyday Life, Department of People and Technology, Roskilde University. Sofie has extensive experience in researching, facilitating, and supporting developmental processes in various institutional settings, mainly targeted at young adults and people living with psychosocial difficulties. Her work is situated within a practice research tradition, contributing to both theory and practice development, with a special emphasis on participation, communities of practice, and the meaning of the environment for development and learning. She is particularly interested in how people develop and learn through their active relating to their environments, and what this means for how we design and conduct e.g., education or psychosocial rehabilitation. Recent publications include Developmental Dynamics and Transitions in High School (Bloomsbury Academic 2021), and several articles on the construction and meaning of enabling spaces.

Martin Malthe Borch M.Sc.Eng Biotechnology DTU. Master in interaction design CIID. PhD researcher at The Novo Nordisk Foundation Center for Biosustainability (DTU-Biosustain) at The Technical University of Denmark (DTU). Martin is a biological engineer, designer, consultant and artist. He is currently researching thermophilic mixed microbial cultures that convert CO2 into chemicals and fuels using bioreactors, metabolic, metagenomic and thermodynamic tools and models. He previously founded the BioFabLab at Roskilde University and worked as a management consultant. His strength lies in understanding and combining knowledge and practices from diverse disciplines for problem-solving and developing new ideas or concepts. He has collaborated with large international companies and architects, designers and artists on product innovation and concept development. He has organised and made several workshops, talks and exhibitions at museums, galleries, new media festivals, art fairs, universities and scientific conferences. He has appeared in newspapers, magazines, peer-reviewed articles and documentaries about biomedia art, biohacking, maker culture, innovation processes and citizen science, where his ideas, concepts and thoughts are described as "game changing" "disruptive" and "a new paradigm".

Nicolas Padfield MA, Lab leader, Fablab RUC, Roskilde University, Denmark. Nicolas leads the Fablab at Roskilde University and has extensive experience teaching programming, robotics, digital fabrication and making at university level. He holds a masters in Computer Science, Communication and Educational Studies. Has been instrumental in introducing new studies and hands-on methods at graduate and postgraduate level. Fablab RUC is acclaimed for assisting students, researchers, inventors and startups with building functioning prototypes of anything from programs to biotech to large scale 3D printed buildings. In his free time, he project leads large scale interactive fire, water, light and sound installations as co-founder of illutron artists' collective.

Schack Lindemann is a Technology Consultant, Project Lead, External Lecturer, and Teaching Artist at Roskilde University / FabLab RUC. Formally educated as an electrical technician in 1989, he later co-founded the Short and Documentary Film School and worked on editing and sound design for various projects within the Danish film industry. For over a decade, he served as a teacher and post-production coordinator at The National Film School of Denmark. Schack is also a co-founder of Halfmachine and later illutron Collaborative Interactive Art Collective, a Copenhagen hub that brings together artists, programmers, musicians, dancers, and other individuals working at the intersection of art and technology. His primary field of interest lies in exploring the interconnectedness of art and technology, along with the associated aesthetic, behavioural, and societal implications. At FabLab RUC, he teaches the use of modern digital fabrication, design methods and rapid prototyping as tools for aesthetic and artistic exploration.

Nikolaj Møbius is a renowned expert in rapid prototyping with extensive experience in digital fabrication machinery and electronics. His innovative projects span various domains, from constructing remote-controlled submarines and large-scale CNC machines to reverse engineering control systems of industrial robots. Møbius has made significant contributions to open-source hardware. His work includes educational initiatives and the promotion of technology hackability, empowering users to take control of modern technological systems.

Bo Thorning works and teaches at the FabLab at Roskilde University and is an active part of the Danish art and technology scene, as a light/sound artist and software/hardware developer. Recently he has built microscopes with AI vision, nano sheet machines, laser projection systems, video mapping and installations at major festivals and museums. Bo Thorning has toured with tesla coils playing concerts and has a wide range of knowledge within rapid prototyping and modern production. He is educated as a control and regulation technician from Danfoss A/S 2007 and is specialised in machinery, 2D and 3D with special knowledge about Virtual Reality (VR), Augmented reality (AR), Machine learning (ML), Computer vision (CV), Artificial intelligence (AI). Fablab RUC: fablab.ruc.dk; art: www.kobeactive.com

Chapter 9
Use of Spaces for Active Learning in University STEM Education

Rie Troelsen and Katrine Bergkvist Borch

9.1 Introduction: Creative Pragmatics and Learning Spaces

Creative Pragmatics offers an approach that can enhance active learning in STEM education, as discussed in Chap. 1. This chapter explores how teachers' engagement with Creative Pragmatics can strengthen active learning environments, facilitating the development of students' competences. Specifically, we focus on how educators in higher education can apply Creative Pragmatics in the design and use of physical learning spaces within STEM education to enhance student engagement and active learning.

Focusing on the physical learning space is not often viewed as a part of higher education teachers' pedagogical toolbox. It is, however, our argument that it should. Teachers face in every teaching situation a complexity of why, who, what, when,

---

We believe content and context are intertwined, and this is no less true when adopting a Creative Pragmatics approach. Creative Pragmatics holds that learning is always located. Written by a Danish associate professor and PhD student, this chapter shares the passion to understand how physical space and outdoor learning shape science education, in science labs, product-realisation labs, art studios, ateliers, workshops, fablabs, maker spaces, in the field. In Creative Pragmatics these are architectural or performative assemblages, stagings of people, artefacts (props), potential scenarios and dramaturgies (of learning, play, discovery). Rie and Katrine prompt questions of the design and use of learning space, with closely associated concerns of competence in such design and use. (Editors)

R. Troelsen (✉)
University of Southern Denmark, Odense, Denmark
e-mail: riet@sdu.dk

K. B. Borch
STEM Centre for Research in Science Education and Communication (FNUG), Department of Mathematics and Computer Science, University of Southern Denmark, Odense, Denmark

C. Svabo et al. (eds.), *Creative Pragmatics for Active Learning in STEM Education*, Contributions from Science Education Research 14,
https://doi.org/10.1007/978-3-031-78720-1_9

with whom, and how they are teaching, but also the *where* of teaching has an impact on the well-being and learning input in a teaching situation (Fraser, 2018). Through the lens of Creative Pragmatics, the physical space in which learning takes place is as important as the goals, content or evaluation of teaching because of the situatedness of knowledge building in a material context. To design learning situations which develop students' Creative Pragmatics in STEM means taking into consideration also the artefacts and environment of the learning process. Designing for active learning requires teachers' creative and pragmatic view of the learning space.

Active learning takes place in all kinds of spaces. If someone says the word 'school,' most people in the Western world will imagine a traditional square room where tables and chairs are lined up in rows one after another. But when addressing STEM subjects, such as chemistry, biology, physics, engineering and mathematics, the visual scenario of 'school' also includes laboratories, field excursions and workshop facilities as part of the learning environments for these subjects. These spaces are designed for the so-called *wet part* of STEM and serve a specific type of learning. One might say that active learning is an inherent part of the wet part of STEM education, considering the active experimentation and hands-on work occurring in labs and during field work, and surely laboratory and field work is a fundamental part of science education and the identity of scientists (Malm, 2020; Peasland et al., 2019). Several studies have explored the potential for learning in laboratories in STEM education. For example, Agustian et al. (2022) found that laboratory work has several potentials, such as the theory–practice connection, laboratory skills and techniques, higher-order thinking and transversal competencies, and that it influences the affective domain. However, when it comes to the laboratory as a physical learning environment, a space, little research has been conducted. Regarding the physical learning environment, Fraser et al. (1995) developed a questionnaire called The Science Laboratory Environment Inventory (SLEI) to assess the unique learning environment that characterises the laboratory. In their study, the authors found that science laboratory classes are dominated by closed-end activities. Ahmad et al. (2010) used a refined version of the SLEI questionnaire called the Physical Science Laboratory Environment Inventory in a study that explored how students and teachers perceive the physical laboratory learning environment. This questionnaire investigates aspects such as furniture, space, lighting, technology, air quality and safety but does not combine this with the pedagogical practices and teaching in the laboratory, for example how the use of physical space and artefacts in the laboratory affect the learning outcome as well as how the human and non-human agents and actors engage and enact to produce situated knowledge (Latour, 1999).

Although laboratory and fieldwork 'spaces' (and specific types of active learning) are important to the STEM subjects, in this chapter we focus on the spaces for the *dry part* of the subjects. In these learning situations, the physical space can facilitate active learning processes such as group work, collaboration and discussion, important learning processes for developing twenty-first century skills (Chism, 2006; Martin-Dunlop et al., 2018). Active learning often involves collaboration and interaction with others, and in higher education, the classrooms where the dry part of the STEM subjects are taught are lecture halls for many students and smaller flat-floored spaces for smaller groups of students. From an architectural point of view, spaces can be either *hard* or *soft programmed.* A lecture hall with fixed benches and an inbuilt one-directional focus of attention—and for that matter also a teaching lab with lab tables with fume cupboards—are hard programmed. They are optimised and locked to a specific use. The traditional lecture hall allows for visibility and audibility in the direction from teacher to student and to a lesser extent for the students to collaborate and interact. A flat-floored room with moveable tables and chairs is soft programmed and may not be optimal for anything in specific but can be transformed for many uses and is flexible. Most importantly for our purposes, flat-floored rooms are the most common teaching spaces at almost any university.

The argument that soft-programmed teaching spaces are the most common teaching spaces at universities and therefore interesting to pay attention to is not the only argument of focus in this chapter. These spaces are also the spaces in which active learning of dry STEM is supposed to be most immanent, at least as opposed to the learning in lecture halls, where teaching tends to be more monologic and passive for the students. As Chism (2006, p. 2.2) states: '*A room with rows of tablet armchairs facing an instructor's desk in front of chalkboards conveys the pedagogical approach 'I talk and demonstrate; you listen or observe. A room of square tables with a chair on each side conveys the importance of teamwork and interaction to learning.*' In a classroom, these interactions become easier, but it still takes an effort for the teacher to promote this type of active learning process in the classroom (Alfredson et al., 2022).

The inbuilt possibility of conducting active learning in flat-floored classrooms is also one of the reasons why many universities have experimented over the years with designing new innovative teaching spaces for the active teaching of group work and collaborating. The new and redesigned spaces may be called active learning spaces (ALCs), twenty-first century classrooms, student-centred active learning environments (SCALEs), next-generation learning spaces or other names to highlight that they are especially designed and furnished with the aim of enhancing active learning for students. One such example is the 'group process room' at University College Lillebælt (UCL) (Fig. 9.1).

Nevertheless, there are still more 'ordinary' classrooms at many universities than these exceptional, redesigned spaces, and teaching in these ordinary classrooms should also aim for active learning. However, teachers need help to develop their competences in using the learning space to enhance active learning (McDavid et al., 2018), no matter whether they teach in innovative learning spaces or in ordinary spaces. In specially designed rooms, they need to know how to use the room's affordances (Gibson, 1979), and in ordinary classrooms teachers have to act creatively with tables and chairs because regular rooms do not have the same possibilities as specially designed rooms. In both cases, it requires teachers to be aware of the space they are teaching in and how the space can support them in planning and teaching for active learning. In other words, Creative Pragmatics in terms of learning spaces means to *know* that space can support one's pedagogical ideas of active learning and to be able to *use* the space to meet one's pedagogical needs, *particularly* in

Fig. 9.1 This photo illustrates a 'group process room' at University College Lillebælt. In the room there are separate group tables, shielded by plants and felt walls, to facilitate group work amongst students. The room is equipped with different screens, such as smartboards, whiteboards and electronic monitor screens. (Copyright: UCL, 2023)

situations and spaces that are less than ideal, for example by manipulating the interior in ordinary classrooms to support active learning.

In this chapter, we focus on the role that the physical learning space in higher education settings in STEM education can play in fostering active learning. Our argument is that the physical learning space can be a valuable co-pilot in facilitating the teacher's pedagogical intentions of active learning. To underpin that argument, in the section below we present what research has to say about the effect of space on teaching and learning. Even though there is empirical evidence that space has a direct and positive effect on learning outcomes, the majority of the research points out that space works through the pedagogy employed in the room and thereby affects the learning outcomes. This puts the spotlight on the teacher being the one to read and use the space to enhance their teaching learning strategies to foster active learning. In other words, the use of the physical space requires the enactment of a spatial competence focused on teaching and learning spaces. Following the section on the evidence base for focusing on the relationship between space and learning, we will very briefly describe what it means to be spatially competent in a teaching and learning situation. We will then present cases where spatial competence is and may be played out. What does it look like when spatial competence is enacted in both the planning, conducting and reflecting phase of active learning, and how can teachers manipulate the space to invite active learning? Last, we give examples of how teachers can be helped to develop spatial competence by creating opportunities for dialogue on space in learning.

9.2 What Is the Effect of Space on Learning?

There is a common phrase used by architects, which is that 'space matters.' In this section, we try to explain why physical space matters, why it is so important for learning and why it is important for teachers to acquire the skills needed to use space, that is, to acquire spatial competence.

The physical spaces in which learning takes place are a crucial part of the learning environment. In general, the term learning environment refers to an environment experienced or perceived by students within an educational setting. '*Learning environment is a term widely used in educational and instructional psychology to capture the idea that learning always occurs in a social and physical context. Generally speaking, a learning environment can be considered as a particular place where individuals can learn by using a variety of information resources and tools that are designed and allocated in the pursuit of learning objectives*' (Seel, 2012, p. 1849). Thus, learning environments include both spaces inside and outside educational institutions, and inside they may be regular classrooms, spaces for group work or self-study in corridors, libraries, workshops, laboratories, etc. Historically, learning environment research has centred around three dimensions: relationship, personal development, and system maintenance and system change, covering different aspects of how students and teachers experience the environment and how positive classroom environments are likely to pay off in terms of improved student outcomes (Fraser, 2018). In the past, research on learning environments has primarily emphasised the psychosocial aspects of learning environments, that is the teacher–student and student–student interactions taking place there. And although there has recently been an increasing interest in investigating the physical environment of educational spaces and buildings (Fraser, 2018), the psychosocial aspects of being taught in a physical setting are still important factors in the learning process.

Only a small amount of the research on learning environments, including the research on the physical environment, is done in relation to higher education, and several reviews have stated that the question of the effect of learning spaces in higher education has been under-researched (Temple, 2008; Ellis & Goodyear, 2016; Leijon et al., 2022). However, some research has been done, and when it comes to research on formal learning spaces, it is mostly related to different types of reconstructed classrooms called innovative learning spaces, active learning classrooms (ALCs) and the like. Innovative classrooms like these are typically designed with group tables and moveable seating that allow students to face each other and the teacher to move around more and thus support small-group work. Some rooms also have screens or whiteboards attached to each group table to support joint problem-solving or collaboration between students (Baepler et al., 2016). These studies find that innovative classrooms can enhance students' engagement (Rands & Gansemer-Topf, 2017), motivation (Adedokun et al., 2017), the learning outcome (Brooks, 2011) and the academic performance (Bayers et al., 2014). Furthermore, a well-designed learning space can be motivating and engaging, sparking curiosity and interest in the subject matter (Oblinger, 2006). Engaged learners are more likely to work with the content. However, other (but fewer) studies show that there are no effects on student performance when comparing different learning spaces (e.g. Vercellotti, 2018). Martin-Dunlop et al. (2018) argue that innovative learning spaces such as ALCs are important for nurturing twenty-first century skills, such as developing creativity, communication, collaboration and critical thinking. However, they also state that there is a need for more research-based evidence on how, why and in what specific ways these spaces lead to more effective learning.

In general, the research on the effect of learning spaces on learning outcomes—no matter whether the learning space is an ALC or a more ordinary classroom—notes that when considering positive learning outcomes, we cannot attribute them solely to physical space. Space in itself cannot enhance learning outcomes, but it can encourage the pedagogy used in the room towards more active learning and thereby affect the learning outcomes (Lundahl et al., 2017). Rather, the outcomes are the result of the intricate interplay between individuals, their environment, interactions and the learning process itself; they are entangled (Leijon et al., 2022), as also indicated in the learning environment research (Baars et al., 2023).

This places the spotlight on the actors involved in the learning processes in a physical learning space: the students and the teachers. Being mindful that students and their behaviour in the space of course play a pivotal role in the entangled processes of space, pedagogy and learning, our focus in the chapter is nevertheless primarily on the teachers and their opportunities to use the space when aiming to plan and design active learning for the students. Therefore, we will next discuss the spatial competence of teachers as a means to acting in and with the physical space, leaving the spatial competence of students to a later work.

9.3 What Is Spatial Competence (in a Teaching and Learning Situation)?

In the entanglement of space, pedagogy and learning, it takes a certain awareness on the teacher's part to take pedagogical advantage of the physical space, whether it is specially designed or not. In this section, we describe spatial competence as the underlying awareness and ability necessary to utilise space to enhance active learning.

Spatial competence is part of our daily lives; we use it to read maps, solve puzzles, play video games and in other situations where we need to imagine or transform abstract concepts into concrete visualisations. Spatial competence has to do with being aware of one's surroundings, being able to navigate in space and being able to manipulate objects in space both mentally and physically. In a teaching setting, spatial competence translates into how teachers (and students) are aware of the classroom surroundings and how they can navigate and manipulate objects in the classroom with creative agency.

The awareness and manipulative components of spatial competence are useful to teachers in order to use space to underpin their pedagogical intentions because, as we started out in this chapter by stating, space matters. Every space—and in our case especially educational spaces—implies certain ways to behave and act in that space. The design of educational buildings and classrooms influences the behaviour and actions of teachers and students in those classrooms. One might say that educational buildings, classrooms and interiors embody the teaching philosophies of their designers. In architectural terms, this is called *built pedagogy* (Monahan, 2000), which means that the way a certain space is designed shapes the learning that happens in that space.

The spatial competence of a teacher requires an ability to read the signs of built pedagogy. Lackney (2008) calls this *teacher environmental competence* and explains it as '*the ability to understand and effectively use physical instructional space for a pedagogical advantage*' (p. 134). In this lies an awareness of the physical environment and its impact on the learning activities and the ability to manipulate the environment so that it suits the learning activities and their intended goals. Awareness of the physical environment means that a well-trained eye can read a space for the pedagogies it facilitates. A classroom with neat rows of desks embodies pedagogies or 'tacit curricula' of discipline and conformity, whereas spaces with an interior on wheels can be said to embody pedagogies of freedom and self-discovery.

It is one thing to be able to read and react to the inbuilt pedagogy of a learning space, but it is something else to be able to start with the space in mind and from there design the teaching. The latter ability requires the first ability, and as such spatial competence becomes a component of the teacher's didactic competence or didactic spatial competence (DiSCo) (Leijon et al., 2021). Having DiSCo means to be able to '*..plan and design teaching and learning in a variety of learning spaces; to implement teaching and learning, act and react in a variety of learning spaces; and to evaluate, reflect and transform both pedagogy and the learning space*' (p. 72). In terms of didactic questions, this means to ask questions about spatiality in the planning, implementing and evaluating phase:

- In planning and designing teaching
  - Why/with what purpose are you teaching, and how is this related to the learning space?
  - What/which content are you teaching, and how is this related to the learning space?
- In implementing teaching
  - How are you going to teach, and how is this related to the learning space?
- In evaluating and developing teaching
  - How did the intentions of your teaching play out, and how is this related to the learning space?
  - What did you learn from your teaching and about the learning space?

The DiSCo, in other words, helps the teacher to use the learning space as an extra tool to enhance the intended learning outcomes of the students.

Embedded in the DiSCo lies the experimental and reflective approach to using the learning space actively which aligns with the basis of Creative Pragmatics. Spatial competence in a learning and teaching situation is in other words the spatial component of a teacher's Creative Pragmatics.

9.4 How Spatial Competence Is Enacted

So far, we have dealt with space and spatial competence in theoretical terms. We now move on to the question of what it looks like when spatial competence is enacted in the planning, conducting and reflecting phase of active learning. This section will present cases where spatial competence is played out and examples of how spatial competences may be played out.

In a teaching situation, spatial competence is, in short, about being aware and using the space to underpin one's pedagogical intentions in the situation. How is this implemented in a teaching situation? Below, we describe three situations where spatial competence is enacted in various ways but all in learning spaces that resemble spaces where the dry parts of STEM are taught. The descriptions stem from a research project that aimed to investigate how teachers are in dialogue with the learning space they are going to use for teaching (Troelsen, 2016). In the project, a variety of teachers were interviewed and observed building on a threefold methodological framework inspired by Lefebvre (1991) and Kress and Leeuwen (1996):

- Perceived space—interviews with teachers describing actions and activities that can and will take place in the learning space
- Conceived space—teachers' sketching their perception of the learning space and analyses of these sketches as to which elements are drawn and in which order
- Lived space—observational studies of how teaching proceeds, focusing on how teachers and students use the learning space in a teaching situation

Together, the interviews, the sketches and the observations form the basis for a description of the enacted spatial competence of each teacher. The descriptions highlight an entanglement of space, people and pedagogy, as the use of space is related to the teacher's teaching philosophy, the class size, the teacher's previous experience and the physical possibilities of the room.

The cases below are only a selection of cases collected in the research project. They were chosen partly to show STEM-like teaching situations and partly to show spatial competence at its various levels, from a teaching situation where the teacher's spatial competence is high to a teaching situation where the teacher's spatial competence is yet to be fully developed.

9.4.1 Case A—Room for Fun

The class Andrew teaches is structured in the form of workshops where the students partly work on questions set out for them beforehand and partly do small experiments. The theme is ballistics and the students calculate trajectories and test out their results in practice. In the first part of a workshop, Andrew '*just gets everyone working and kind of walks around just helping people on the spot*' instead of employing '*lecture-type teaching.*' If he thinks that a question is very important, he will '*stop, perhaps grab a mike and tell everyone.*' However, he prefers to have the students work on the questions in a '*self-directed*' way. In the second part of a workshop, the '*lab component*', Andrew and his teaching assistant will set up a little catapult and a target plate at one end of the room. The students, in groups, are asked to model the process of a ball hitting the target plate using the catapult, and Andrew sets up the catapult according to the students' settings. The students '*get marked if they hit the target or not.*' The workshop is divided into two because '*it's kind of a long roll – if you keep asking questions they'll get bored.*' The lab component is simply more fun. And fun is important '*to keep them motivated.*'

It is important for Andrew to teach in this specific space, mainly because he needs the floor space—not just for the catapult experiment but also for other experiments during the semester that require floor space (e.g. Lego robots following masking tape tracks). He also emphasises the light and the colours in the room, which contribute to the 'fun factor' of the workshop.

This is a case describing a teacher who reads the built pedagogy of the room, that is, uses the room and its affordances. The floor space and the group tables underpin his pedagogical intentions and help illustrate to the students what kind of engagement the teacher expects from them and how he sees possible learning paths for them.

9.4.2 Case B—'We're Not Using the Room's Full Potential'

Daniel is tutoring an engineering class in soil mechanics where the students are supposed to put theoretical concepts into use through primarily calculation exercises. The teaching takes place in a space where the students sit at group tables that each have a screen attached. In the middle of the room, there is a desk for the teacher with a screen connected to the screens at the group tables. Before class, Daniel uploads some exercises on the learning management system that he expects the students to have solved. However, the students seldom live up to this expectation and come to class unprepared. He then gives '*a mini-lecture*' and goes through solutions to the exercises. He shows the solutions on his screen. However, '*my screen is supposed to be connected with the screens in front of them – often it doesn't work.*' If the screens are not working, Daniel just talks the students through the exercises, and in his experience the students do not ask any questions about the problem-solving process.

Daniel is aware of the space he is teaching in, but mostly he is aware of the possibilities that the space provides that he as a teacher does not take advantage of: '*We're not using the room's full potential; we're doing a standard tutorial workshop and this room is not designed for that.*' What it is the space is designed for, he is not sure; he just '*knows that something's missing.*' He recognises that the space is flexible, enables teachers and students to walk around and is probably designed for students to work in groups, but he has not been introduced to teaching methods relevant to the space. Daniel would like '*to really adapt to the room*' but does not know how. Instead, he teaches '*using old materials, the old paradigm*'—he teaches almost despite the space's potential.

This case illustrates a teacher who reflects on the design of a space but is unable to use it. The depth of his spatial competence makes him aware of a 'right way' of using the space, but he is not able to use or manipulate its objects. Instead, he knowingly teaches in spite of what the room offers.

9.4.3 Case C—A New Teacher in an Unknown Room

Arnold is new to teaching and teaches a discussion class in biology on population and evolution. The content for the class is very much cases in which Arnold guides the students in applying their theoretical knowledge. He teaches in a room where the seating for students is shaped in a half circle. It creates almost like a focal point in the middle of the room. Arnold lectures for a short while and then asks the students, in pairs, to engage with some reflective questions that stimulate discussion as to how theoretical concepts apply to the case. He then invites them to share their thoughts, and if no one is willing to share, he moves on with his own thoughts on the initial question. This procedure (short lecture, then buzz groups, and then perhaps plenary discussion) is repeated several times during the lesson. Arnold has planned all his teaching sessions before knowing which room he was going to teach in and thinks that his kind of teaching can take place in any kind of room. He would like to be able to place himself more in the middle of the room and thus closer to the students to '*minimise not only the physical space between them but also the mental space.*' He feels that this would create a safer space for the students to discuss the topic with him and ask questions. He is not that comfortable with this just yet but hopes he will learn as he gains more teaching experience. Arnold does not mention how the students are seated.

This last case describes a teacher who has made a rather fixed plan for his teaching before knowing which room he will teach in and who does not make use of the room as such. However, he is beginning to activate his spatial competence, as he feels uncomfortable and too inexperienced to stand too far away from the screen.

The above cases show in various ways how space and teaching are intertwined, how spatial competence can be enacted and how teachers use space to underpin their teaching intentions to various degrees. Throughout the cases, it is reflected how the space affords different roles for the students and the teacher. Space—and particularly open floor space—can facilitate teaching where students and teachers can have fun. However, space can also trigger almost agoraphobic tendencies and keep the teacher locked to one end of the room. Some rooms have a designated space for the teacher to act as a spider in a web of computer connections, and others signal by means of group tables that students are to work together. Some teachers, such as Andrew, read the room and are able to react to the possibilities, while others, such as Daniel, sense that the space has a built pedagogy but are not able to understand what they read and are not able to act upon it. To teachers such as Arnold, the space does not play as distinct a role; they would teach the exact same way regardless of the space they are assigned.

The learning spaces in the three cases are all designed in different ways and create different possibilities (and impossibilities) for the teacher to teach for active learning. Barad (2007) would say that the design of the learning spaces; chairs, tables, screens, carpets, lamps and all other non-human actors all have agency and shape the teachers capacity to reach the learning goals and intentions of the lesson. Translated into flat-floored classrooms with (somehow) moveable furniture, one can imagine different layouts for different active learning methods in teaching the dry parts of STEM for teachers other than Andrew, Daniel and Arnold to take up (see the figures below).

In teaching situations with high amounts of group work a fruitful layout will be one that underlines that students' primary attention points to each other rather than to the teacher as illustrated in Fig. 9.2. Here, the students (circles) sit in groups around tables that are placed in a kind of semicircle in the room. There is space in the opening of the semicircle for the teacher (the triangle) to move around to the groups (the dotted line). The teacher starts the session at the screen/board (the narrow rectangle). The introduction should be short, both to create active learning and as some of the students need to turn away from the group table to be able to see the screen (arrows). The joint attention to the screen can of course also take place at other times in the lesson, led by either the teacher or the students. This layout can be

Fig. 9.2 Group work with short introduction/ wrapping-up sessions

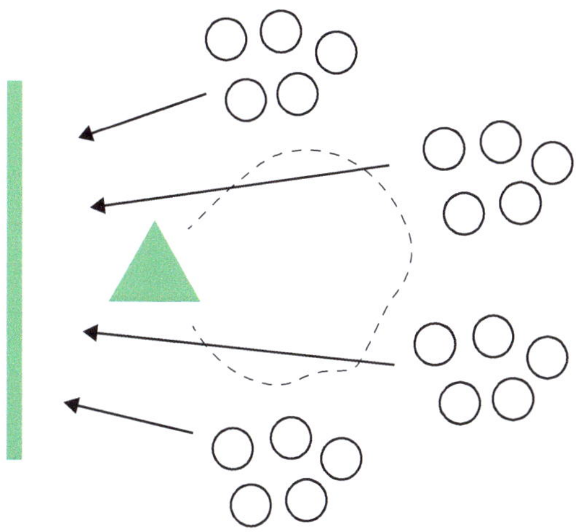

used for teaching situations where students work together on e.g. calculating concentrations, analysing graphs, visualising experimental data, proof-explaining, or interpreting theories.

Other teaching situations focus on students' communicative skills. In the layout shown in Fig. 9.3, students sit facing each other, possibly in multiple rows depending on the number of students. Students can see each other and focus primarily on each other. The teacher can move across the room, alternating between being close to the groups, but can also move between groups if necessary. The teaching method in this layout may be as in academic debates, with each group of students representing an academic point of view, like pros and cons on gene therapy, sustainability or surveillance. If students are seated in multiple rows, the back row can supplement the front row with data or good arguments in the debate or can be 'substituted' during the debate.

A third example of a teaching situation is collaborative project work supervised by a teacher. In Fig. 9.4, the tables are arranged at one end of the room so that the entire group of students sit together, and at the other end there is space for the entire group of students to stand in front of/view a shared whiteboard/screen. The

Fig. 9.3 Debate

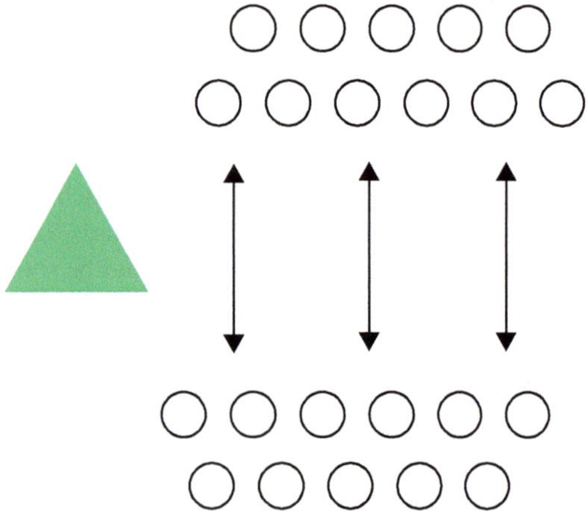

Fig. 9.4 A variety of working/teaching methods

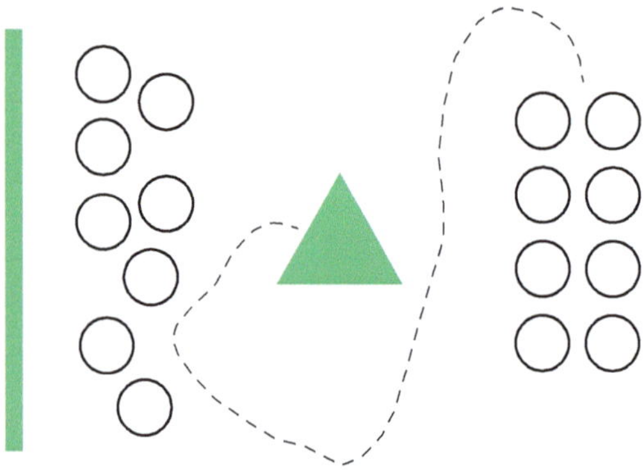

teacher—like the students—can move freely around the room and the working form and learning pace is driven by the progress of the project. The layout accommodates and supports teaching situations where the entire student group is working collaboratively on a project/assignment. This can be joint problem solving or designing a joint product, which is facilitated by alternating between working on a flat surface (a table) doing calculations, computing graphs, etc. and discussing in front of a vertical surface (i.e. a whiteboard or blackboard) while drawing, visualising or arguing for possible next steps in the project.

Manipulating a flat-floored room to accommodate the pedagogical intentions of active learning is, however, hard (and time consuming) work for the teacher. A suggestion to deal with this challenge is to involve students in the rearranging of furniture, which may serve several purposes. The obvious benefit is that it saves the teacher effort and time. But involvement can also help students understand that the layout is meant to do something special for them. In fact, involvement can actually support the teaching objectives because it gives the teacher the opportunity to justify the learning activities that the interior design is meant to reflect and to build their spatial competence. For many students, it can create a sense of calm and security to be introduced to the behaviour and activity expected by the teacher (Leijon, 2023).

9.5 Supporting Teachers to Become Spatially Competent

In the above-described research project, some of the teachers seemed to reflect on the role of the space in their teaching only as the research interview happened. Others seemed to have reflected on the relationship between space and learning before the research interview or even when designing the teaching situation. In both cases, the interview itself gave rise to the teachers' reflections and articulations of spatial competence. Even though the intention of the research project was not to develop spatial competence but merely to describe it, taking part in the interviews might have enhanced the involved teachers' spatial competence just by forcing them to think of how the space mattered to them. Taking this point further, we might ask ourselves how Daniel and Arnold and teachers like them can be helped to develop spatial competence by creating opportunities for dialogue on how space affects teaching and learning.

In general, there are multiple methods to make teachers become familiar with using and reflecting on a learning space (Knaub et al., 2016), and formal training is the most obvious one. However, we can also point to other methods that might support the process:

- **When refurbishing a space, involve teachers/users in the design phase**

Involving teachers creates ownership of a learning space and its resources by granting them the chance to engage in conversations about the space. When discussing and co-designing, the teachers increase their agency in relation to how learning spaces can be designed (Casanova et al., 2018). Ownership is seen as crucial for teachers to accept and embrace (new) learning spaces (Grummon, 2009; Lippincott, 2009). Of course, the involvement of teachers in the design phase can occur in various ways, from participatory designs and workshops to formal hearings.

- **Communicate/inform about the pedagogical intention of the space in different ways**

Just inside the door in many learning spaces, there is an overview of how tables and chairs should be placed in that particular room. Overviews like this can be supplemented by perhaps one or two variations of programming/design where the appropriate teaching method can also be described. Figure 9.5 shows an example from the University of Bergen. Having this inspirational information located in the room may facilitate an awareness in the teacher of other ways to use the space. It may also be illustrative if the teacher would like their students to help rearrange the furniture in a more appropriate design.

Another example of how the built pedagogy of space can be communicated is from UCL, referred to at the beginning of this chapter. In teams, teachers from the college were given the task of redesigning four different types of learning spaces: a room that facilitates group processes (Fig. 9.1), a room with a pleasant atmosphere, a room that aims at facilitating dialogue and a room that signals comfort for students and teachers (UCL, 2023). Using the 'dialogue room' as an example, the space has a large horseshoe-shaped table in the room so every student can see each other when sitting, which facilitates dialogue amongst the students. At each end of the room, there is a big screen if presentations are needed. At the far back of the room, benches are placed against the wall to create reading and immersion, with respect to the fact that active learning includes many different learning processes, including dialogue and solitary reflection. There are no whiteboards, but the students are encouraged to use the large windows in the room and mobile whiteboards can be placed in the room. To introduce the space, its intentions and recommended use, a video has been produced (https://vimeo.com/843147812) that can be sent to teachers who are planning to teach in that space. In the video, pedagogical arguments are used in explaining why the interior is designed, arranged and coloured as it is and which kinds of teaching methods a teacher can use in the space.

- **Lead by example through pedagogical courses in various spatial settings**

Many teachers participate now and then in staff development activities, not only relating to how to use space in their teaching but also to other pedagogical issues. As an educational developer, one may arrange these staff development activities to be conducted in spaces that are familiar to teachers and to use the spaces in innovative ways to lead by example.

New innovative learning spaces are being built at some universities and can also be used as exemplary spaces. One example is the Learning Lab at Örebro University (https://www.oru.se/english/about-us/centre-for-academic-development/perspectives/new-teaching-and-learning-environments/) where teachers and students can work together to design their learning environment and test new pedagogical approaches using the room's technology, furnishings and design. The awareness of

Fig. 9.5 Programming/design with three variations of how a room can be set up for different teaching methods (groups, lecture and plenary discussion). This is an example from the University of Bergen. (Copyright: University of Bergen (n.d.) Photographer Eivind Senneset)

space and teaching that teachers learn from conducting a lesson in the test environment hopefully follows them to their future teaching in more regular spaces. As such, teaching in a new, innovative learning space may be an initiator for reflection on the relationship between space and learning. Birdwell and Uttamchandani (2019) have described this as 'exploring classrooms with colleagues so that they can leverage any classroom for their teaching goals' (p. 20).

All the above-mentioned methods point to the need to make teachers aware of built pedagogy and the need to support them to become spatially competent. Institutions (including designers of buildings, building maintenance personnel and educational developers, who are the most obvious actors in the methods mentioned) need to work together to create opportunities for reflection and dialogue on space *in* learning, not just create new spaces *for* learning.

9.6 Conclusion: Space as a Co-teacher

In this chapter, we have offered some insight into how spaces may be used in higher education settings in STEM education to foster active learning. Why is this important? Over the past 20 years, physical space has received increasing attention as a factor that influences the teaching and learning that takes place there, and several studies show that the space can encourage learning activities that improve learning outcomes (Painter et al., 2013; Chiu & Cheng, 2017; Thomas et al., 2019). The space and its design stimulate a certain behaviour in the teacher, which stimulates a certain behaviour in the students (Brooks, 2012).

It is important to note, though, that the space itself does not guarantee the behaviour or activities in the space. It is not space itself that affects changes in teaching, but rather space can help changes in teaching to succeed. The space can only stimulate and encourage the people who use the space. This makes it even more important to consider how teachers use the space. The physical space, the didactic considerations (how to use the space and why) and the people involved are all significant for learning. It is in the interplay between these three perspectives that a good learning space is created, and it is the interplay that makes teaching complex.

The perspectives also point to the fact that all spaces have potential. In the above, we have focused on so-called regular classrooms, that is, rooms with a flat floor, rectangular tables usually placed in straight rows and a whiteboard/screen at one end of the room where most of the dry part of STEM teaching takes place. Rooms with this layout, where the students all face in one direction (towards the teacher) and the teacher has minimal room to manoeuvre, reflect a slightly old-fashioned way of teaching with a division of roles between teacher and students à la oracle and listener. So while waiting for more of these spaces at universities to be redesigned to align with modern, active approaches to learning, reflecting the development of twenty-first century skills, we argue that teachers in a creative and pragmatic way must convert the many regular classrooms to become active learning spaces in their own right.

Planning how an ordinary classroom can become an active learning space that can underpin the teacher's pedagogical intentions of active learning means enacting their didactic spatial competence. In other words, the teacher needs to use the space—use it as a co-teacher and as an additional resource alongside preparation activities and instructions that can help both teacher and students achieve the learning objectives. The teacher must use the spatial component of Creative Pragmatics to plan how the students are to be active learners in the physical classroom. Are they going to calculate, debate or present, and are they going to do that individually, in groups or in plenum? Teachers must also decide on how the physical space can help students become active learners. Do they sit in a particular way? Do they sit? Where is the teacher located? At whom or what do the students look? Who can they hear? In addition, teachers must communicate the pedagogical intentions through the layout of the space and collaborate with students in rearranging the furniture. But most importantly, teachers must experiment with space and teaching! Trying different layouts, experiencing the consequences, applying the experiences to other kinds of spaces and reflecting on how space was a driver or a barrier for the active learning the teacher had planned—these are all components of teachers' Creative Pragmatics when developing physical learning spaces that encourage and enhance active STEM learning.

References

Adedokun, O. A., Parker, L. C., Henke, J. N., & Burgess, W. D. (2017). Student perceptions of a 21st century learning space. *Journal of Learning Spaces, 6*(1), 1–13.

Agustian, H., Finne, L., Jørgensen, J. T., Pedersen, M., Christiansen, F., Gammelgaard, B., & Nielsen, J. A. (2022). Learning outcomes of university chemistry teaching in laboratories: A systematic review of empirical literature. *Review of Education, 10*, 1–41.

Ahmad, C. N. C., Osman, K., & Halim, L. (2010). Physical and psychosocial aspects of science laboratory learning environment. *Procedia—Social and Behavioral Sciences, 9*, 87–91.

Alfredson, V., Asker, N., Backman, C., & Uhnoo, S. (2022). *Använd Rummet. Högskolepedagogiska metoder för aktiva lärosalar* [Utilize the Room: Higher Education Pedagogical Methods for Active Classrooms]. Studentlitteratur, Lund.

Baars, S., Schellings, G. L. M., Joore, J. P., & van Wesemael, P. J. V. (2023). Physical learning environments' supportiveness to innovative pedagogies: Students' and teachers' experiences. *Learning Environments Research, 26*, 617–659. https://doi.org/10.1007/s10984-022-09433-x

Baepler, P., Walker, J., Brooks, D., Saichaie, K., & Petersen, C. (2016). *A guide to teaching in active learning classrooms: History, research, and practice*. Routledge.

Barad, K. M. (2007). *Meeting the universe halfway: Quantum physics and the entanglement of matter and meaning*. Duke University Press.

Bayers, T., Imms, W., & Hartnell-Young, E. (2014). Making the case for space: The effect of learning spaces on teaching and learning. *Curriculum and Teaching, 29*(1), 5–19. https://doi.org/10.7459/ct/29.1.02

Birdwell, T., & Uttamchandani, S. (2019). Learning to teach in space: Design principles for faculty development in active learning classrooms. *Journal of Learning Spaces, 8*(1), 19–27.

Brooks, C. (2011). Space matters: The impact of formal learning environments on student learning. *British Journal of Educational Technology, 42*(5), 719–726.

Brooks, D. C. (2012). Space and consequences: The impact of different formal learning spaces on instructor and student behavior. *Journal of Learning Spaces, 1*(2).

Casanova, D., Di Napoli, R., & Leijon, M. (2018). Which space? Whose space? An experience in involving students and teachers in space design. *Teaching in Higher Education, 23*(4), 488–503.

Chism, N. V. N. (2006). Challenging traditional assumptions and rethinking learning spaces. In D. G. Oblinger (Ed.), *Learning spaces*. Educause.

Chiu, P. H. P., & Cheng, S. H. (2017). Effects of active learning classrooms on student learning: A two-year empirical investigation on student perceptions and academic performance. *Higher Education Research & Development, 36*, 269–279.

Ellis, R., & Goodyear, P. (2016). Models of learning space: Integrating research on space, place and learning in higher education. *Review of Education, 4*(2), 149–191.

Fraser, B. J. (2018). Chapter 1: Milestones in the evolution of the learning environments field over the past three decades. In D. B. Zandvliet & B. J. Fraser (Eds.), *Thirty years of learning environments: Looking back and looking forward* (Advances in learning environments research series) (Vol. 11). Brill Sense.

Fraser, B. J., Giddings, G. J., & McRobbie, C. J. (1995). Evolution and validation of personal form of an instrument for assessing science laboratory classroom environments. *Journal of Research in Science Teaching, 32*, 399–422. https://doi.org/10.1002/tea.3660320408

Gibson, J. J. (1979). *The ecological approach to visual perception*. Houghton Mifflin.

Grummon, P. T. H. (2009). Best practices in learning space design: Engaging users. *Educause Quarterly, 32*(1).

Knaub, A. V., Foote, K. T., Henderson, C., Dancy, M., & Beichner, R. J. (2016). Get a room: The role of classroom space in sustained implementation of studio style instruction. *International Journal of STEM Education, 3*, 1–22.

Kress, G., & Leeuwen, T. V. (1996). *Reading images. The grammar of visual design*. Routledge.

Lackney, J. A. (2008). Teacher environmental competence in elementary school environments. *Children, Youth and Environments, 18*(2), 133–159.

Latour, B. (1999). On recalling ANT. *The Sociological Review, 47*(1_suppl), 15–25. https://doi.org/10.1111/j.1467-954X.1999.tb03480.x

Lefebvre, H. (1991). *The production of space* (D. Nicholson-Smith, Trans.). Blackwell.

Leijon, M. (2023). Flexibla lärandemiljöer med inkludering i fokus [Flexible learning environments with a focus on inclusion]. In M. Stigmar (Ed.), *Inkluderanda högre utbilding. Breddad rekrytering, breddat deltagande och studentaktivt lärande* [Inclusive higher education: Broadening recruitment, expanding participation, and student-centered learning] (pp. 207–238). Studentlitteratur.

Leijon, M., Malvebo, E., & Tieva, Å. (2021). It is time for DiSCo—A theoretical model for didactic spatial competence. *Journal of Learning Spaces, 10*(3), 72–77.

Leijon, M., Nordmo, I., Tieva, Å., & Troelsen, R. (2022). Formal learning spaces in higher education—A systematic review. *Teaching in Higher Education*, 1–22. https://doi.org/10.1080/13562517.2022.2066469

Lippincott, J. K. (2009). Learning spaces. Involving faculty to improve pedagogy. *EDUCAUSE Review*, (March/April), 17–25.

Lundahl, L., Cruise, E. G., Malmros, B., Sundbaum, A., & Tieva, Å. (2017). Pedagogisk rumtid och strategier för aktivt lärande i högre utbildning [Pedagogical space-time and strategies for active learning in higher education]. *Utbildning och Lärande/Education and Learning, 11*, 16–32.

Malm, R. H. (2020). *What is fieldwork for? Exploring roles of fieldwork in higher education earth science* (Publication no. 2349). Doctoral dissertation, University of Oslo. https://www.duo.uio.no/handle/10852/82828

Martin-Dunlop, C., Hohmann, C., Akers, M. A. A., Lewter, L., & Williams, I. (2018). Chapter 5: Evaluating the impact of a purposefully-designed active learning space on student outcomes and behaviours in an undergraduate architecture course. In D. B. Zandvliet & B. J. Fraser (Eds.), *Thirty years of learning environments: Looking back and looking forward* (Advances in learning environments research series) (Vol. 11). Brill Sense.

McDavid, L., Parker, L. C., Burgess, W., Robertshaw, B., & Doan, T. (2018). The combined effect of learning space and faculty self-efficacy to use student-centered practices on teaching experiences and student engagement. *Journal of Learning Spaces, 7*, 29–44.

Monahan, T. (2000). Built pedagogies & technology practices: Designing for participatory learning. In T. Cherkasky, J. Greenbaum, P. Mambrey, & J. K. Pors (Eds.), *PDC 2000 proceedings of the participatory design conference*. CPSR.

Oblinger, D. G. (Ed.) (2006). *Learning spaces*. Educause. http://www.educause.edu/learningspaces

Painter, S., Fournier, J. E., Grape, C. A., Grummon, P. T. H., & Morelli, J. K. (2013). *Research on learning space design: Present state, future directions*. Society for College and University Planning. https://www.scup.org/resource/research-on-learning-space-design-present-state-future-directions-report-from-the-recipients-of-the-2012-perry-chapman-prize/. Last accessed 2023/11/30.

Peasland, E. L., Henri, D. C., Morrell, L. J., & Scott, G. W. (2019). The influence of fieldwork design on student perceptions of skills development during field courses. *International Journal of Science Education, 41*(17), 2369–2388. https://doi.org/10.1080/09500693.2019.1679906

Rands, M. L., & Gansemer-Topf, A. M. (2017). The room itself is active: How classroom design impacts student engagement. *Journal of Learning Spaces, 6*(n1), p26–p33.

Seel, N. M. (2012). Learning environment. In N. M. Seel (Ed.), *Encyclopedia of the sciences of learning*. Springer. https://doi.org/10.1007/978-1-4419-1428-6_1860

Temple, P. (2008). Learning spaces in higher education: An under-researched topic. *London Review of Education, 6*, 229–241. https://doi.org/10.1080/14748460802489363

Thomas, C. L., Pavlechko, G. M., & Cassady, J. C. (2019). An examination of the mediating role of learning space design on the relation between instructor effectiveness and student engagement. *Learning Environments Research, 22*, 117–131.

Troelsen, R. (2016). What's space to learning? In A. Qvortrup, M. Wiberg, G. Christensen, & M. Hansbøl (Eds.), *On the definitions of learning* (pp. 213–228). University Press of University of Southern Denmark.

University College Lillebælt. (2023). *Motiverende fysiske læringsrum* [Motivating Physical Learning Environments]. https://vimeo.com/851640699. Last accessed 2023/10/27.

University of Bergen. (n.d.). *Fleksibelt klasserom* [Flexible classroom]. Copyright: University of Bergen, Photographer Eivind Senneset.

Vercellotti, M. L. (2018). Do interactive learning spaces increase student achievement? A comparison of classroom context. *Active Learning in Higher Education, 19*, 197–210.

Rie Troelsen, PhD, Head of SDU Centre for Teaching and Learning, University of Southern Denmark. With an educational background in science and science education, Rie's research has mainly focused on the meeting between student and study programme and the professionalisation of the university teacher. More recently, her research focus has been on how the physical space plays a role in the learning and teaching that both teachers and students can imagine taking place. Having been engaged in both research, teaching and consultancy in higher education over the years, Rie is now Head of the Centre for Teaching and Learning at the University of Southern Denmark focusing on staff development of teachers and educational leaders at university level.

Katrine Bergkvist Borch, PhD student at STEM Centre for Research in Science Education and Communication (FNUG), Department of Mathematics and Computer Science, University of Southern Denmark. Katrine holds a MA in Biology and has worked as a science high school teacher before starting a PhD in science education where she is doing research within learning environments in science education with a focus on field trips and outdoor teaching and learning.

Chapter 10
Design Studio for Active Learning

Michael Bell and Chris Ford

10.1 Design Studio as a Place for Active Learning

On any given day as many as 28,000 students are enrolled at United States schools of architecture (NAAB, 2019). By comparison approximately 700,000 students are enrolled in United States undergraduate and master's programs in engineering.

We encouraged our contributors to add perspectives that traditional STEM educators and champions might see as unconventional, yet are vital to "operationalizing" Creative Pragmatics. In this chapter two American architects-cum-academics explore how design studios, which are core in the discipline of architecture, can function as an active learning mechanism for STEM education. Studios, workshops, and ateliers have been key features of collaborative and participatory making and knowing since prehistory—Creative Pragmatics has deep time-proved roots. This is another way of grasping Creative Pragmatics—it's actually all around us and has been throughout history, embedded in the workplace and in relationships such as apprenticeship. Here the concept of Creative Pragmatics connects design studio with design community (makers and apprentice learners, professionals and clients) with credentials of competence (working one's way through projects pursued by an expert team), complementarity of learning and professional practice, transdisciplinary scope through STEM and beyond (integrating design philosophy, aesthetic principles, civil engineering, materials science), and the management of such a design-cum-learning process. So the concept of Creative Pragmatics offers a perspective on architecture, with further expansion of STEM through the experiential lifeworlds of physical structures, engineering, building, design. This wide scope is grounded by Michael and Chris in a case-study in studio-based learning focused on the design of future urban infrastructures, an epitome of complexity. (Editors)

M. Bell (✉)
Graduate School of Architecture, Planning and Preservation, Columbia University, New York, NY, USA
e-mail: mjb92@columbia.edu

C. Ford
School of Engineering, Stanford University, Stanford, CA, USA
e-mail: ford@infra-x.org

C. Svabo et al. (eds.), *Creative Pragmatics for Active Learning in STEM Education*, Contributions from Science Education Research 14,
https://doi.org/10.1007/978-3-031-78720-1_10

Architecture is a relatively small academic field in the United States. For architecture students the course that commands most of their attention is simply referred to as "studio". Often referred to as a place (*"I'll meet you in studio"*) as much as, or more than, a course based on active learning, it's difficult to overstate how central the design studio is within architectural education (Fig. 10.1).

As the time allocations burgeon into something more than a course, studio effectively becomes a 24-h accessible zone in which all other coursework is often completed. Studios in most schools provide each student with their own dedicated workspace, and are clustered with other students enrolled in the same semester, to form a studio-based unit. Studios are often grouped or are closely adjacent with other studios making the entire degree program visible—often in a continuous array. In this way, studio enrollments change on a semester basis, but students often cross with their colleagues from previous studios. These inherent exchanges help to build a school of architecture's respective culture. Historically, the spatial allocation per design student is typically the highest within a university. This is necessary as larger drafting tables were a provided university resource for each student's use. Whether for drawing, drafting, or model building, each architecture student outfitted their space according to their personal trajectory. Today this manifests at minimum as a single computer workstation but there are some schools with greater space allocations providing two to three large work tables, effectively providing 100 square feet (9.29 square meters) of space.

Fig. 10.1 Design studios in Avery Hall, Columbia University, Graduate School of Architecture, Planning and Preservation. (Photographer Hatnim Lee)

The time allotment is significant as longer periods of uninterrupted time are required for "project" conception, development, and execution, but studios also operate in an atelier model. A faculty member will often meet each student 1:1 to discuss the development of a student's specific project, and this interaction is also conducted in the setting of the studio.

At the graduate level, the number of credit hours is approximately half of a respective semester's full time enrollment. Design studio courses often meet three times a week; for 4–5 h per session.

The delivery of a design studio course is a significant financial proposition for a university when compared to a large format lecture course. Studio credit hours account for approximately ~66% of a student's course load. One can begin to imagine the financial and economic scale. For instance, 100 students enrolled in a design studio for two semesters per academic year might generate approximately $4 million per annum in tuition. The commitment has historically been a common feature of architectural education; what we imagine is the best use of that time and financial investment going forward is worth re-considering.

Far more in space, time, and funding, faculty contact time for studio exceeds the more common 2 or 3 h seminar course. For instance, at the University of Nebraska, design studios are scheduled to meet every Monday, Wednesday, and Friday afternoons for 4 h per afternoon, for a total of 12 h of contact time per week. At Columbia University, design studios are also scheduled to meet for a total of 12 h. The reality of studio hours usually far exceed the schedule and professors and students routinely meet for as many as many 15 h. The time is allotted to individual critiques of student work but also to collective and publicly shared presentations or review sessions with invited guest critics or public all-school presentations of studio works and goals. Studio life and learning is a public process with students working side by side in close quarters, and being asked to discuss and debate work in communal forums from day one.

From a pragmatic perspective, it makes sense to try to better understand what a studio is.

Across the history of architectural education, project-based learning in the form of the design studio has played a central role in the education of the architect. Whether in practice-driven ateliers, or in university-based academic environments, architectural education is routinely imagined both as a discipline and a project. Its institutionalization is hundreds of years old and based on what is at hand today—what we are designing now. The destabilizing nature of this framing is rarely challenged even if a school may momentarily fall towards one side of the equation or another—toward a discipline that has ancient foundations or is being defined today in the context of entirely new conditions, including economics, engineering, materials science, energy, or climate, and furthermore of new relations between people, society and power. The design studio is the site where these interrelations between these valences are enacted and an outsider might be very surprised by the scope and range engaged by architecture studios.

Academic work in architecture might give equal time to issues of proportion as found in the architectural orders of antiquity, or attempt to gauge applications of the emerging material science. Similarly, coursework may explore design outcomes with a physical range from a bus shelter to an infrastructure megaproject, or the most recent forms of socio-economics and power in the development of a neighborhood. If the divergence causes strife, one sees little concern to either reduce the anxiety; to fully modernize our focus or to narrow it historically—to imagine what we call the discipline was formed centuries ago. A large part of the architectural project is quite explicitly based in anxiety. Today's design studios are often framed with ambition—against a backdrop of immense social, economic and industrial change—but in our later-day post-structuralist era it is possible to leave the raw and fragmented organization of cities and architectural quasi-broken. To, in effect, reveal their inoperability even as the wider horizon favors resolutions.

Part of what drives this divergence is in part a recognition that the aspirational zenith of architecture is rarely achieved—the search for a discipline in this sense often is a proxy for a search for principles (of the field, of the profession), but a survey of the world at hand often shows little recognition of what we might imagine is success. Instead of principled buildings achieved at scale, most appraisals of the built world see a kind of market-based vernacular—buildings often realized with little, and at times no, professional engagement. Engineering and architecture are here not entirely absent, but instead embedded or carried by commodity items and regulated practices—a standard 2×4 tied to a sill plate with a commercially-available metal strap. Building in this sense are in themselves often projects; made incrementally between a dispersed set of stakeholders and achieved within a set of skill-based trades and established building techniques that require little discussion. Against such a vernacular economy of building and construction, the academic side of architecture often imagines both a better outcome (a real-politic) and also something far more ambitious and set within the most celebrated achievements in architecture. But whatever is imagined, the design studio is indeed today often highly speculative, and the project-based nature of its work embraces a very wide and even promiscuous understanding of who and what constitutes the team required for architectural production.

This chapter reflects upon the delivery of two graduate-level design studios in Architecture. It offers context for the curricular centrality of design studios in accredited programs of Architecture at R-1 universities, and provides insights into the planning, delivery, and execution of two particularly-themed design studios which challenge students to consider aspects of the built environment traditionally beyond the extents of disciplinary interest and scope. This chapter concludes with an identification of qualities and attributes seemingly suitable for constructing a successful design studio within STEM-related disciplines.

10.2 Case Study: Hybridized Urban Infrastructures

Who is best positioned to conduct the research and development of next-generation infrastructure? Who has the credentialed capacity to engage such wicked design problems? Who can best provide leadership in technological development, while also developing viable solutions with whole-system synthesis as similarly found on the front lines of professional practice? If it is to be any one, or a combination, of the sectors represented in the AECO acronym, then pedagogical adjustments are necessary today for delivering conceptually superior next-generation infrastructure solutions for operation in uncertain and wholly unknowable futures.

A twenty-first century pedagogy for STEM-related fields will place equal emphasis upon design and analysis combined. However, current pedagogical emphasis upon *analysis* will not alone induce the resilient infrastructural futures otherwise required by the public realm. Instead, the field of infrastructure design is presented with a timely opportunity for increasing the role of active, project-based learning within the education of tomorrow's infrastructural decision makers.

The pedagogical benefit of centralizing design studios in a STEM-based curriculum can be found, in part, in the duality of a student's convergence to problem-based research and divergence through an inquiry-led investigation. As a pedagogical model, it presents distinct advantages over more familiar STEM tactics including controlled experiments, calculation-based problem sets, and Pareto distributions. Agency is established and expanded through studio-based projects. Whether a student design consists of a singular, megalomaniacal infrastructure, or as a network of distributed nodes across a cityscape, in certain instances student proposals exceed curricular goals to produce feasible, viable, and desirable lifeline solutions that act as provocateurs for change in the larger public realm (Ford, 2010a).

This section describes how a funded research project led to a cross-university pedagogical collaboration, which then catalyzed a design studio of Master of Architecture students to pivot from the design of architecture, towards the design of next-generation infrastructure instead.

10.2.1 Research: REIs: Renewable Energy Infrastructures

The *Renewable Energy Infrastructure (REI)* funded research project was conceived in response to the demonstrated and forecasted demands for electricity at industrial scale. An REI generates megawatts (MW) of electricity at an industrial scale through the simultaneous operation of renewable wind, solar, and geothermal technologies within an integrated, holistic, and free-standing infrastructure solution situated in an urban environment (Fig. 10.2). An REI is not a retrofit of a pre-existing architectural condition, but rather is conceived as a new infrastructure typology to be owned and operated by an electrical utility for purposes of servicing users in high-population

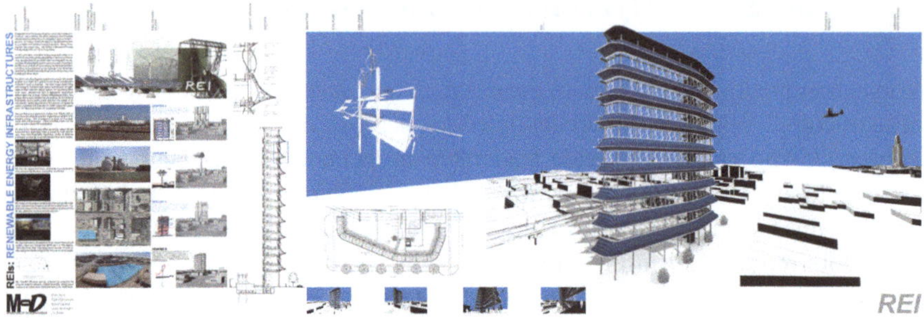

Fig. 10.2 Presentation panel for the *REI: Renewable Energy Infrastructure*, a research project

areas. The outcomes of this design research investigation have been published as a Final Report for the AIA Upjohn Research Initiative (Ford, 2011).

Across the REI timeline, the project PI recognized some key similarities and differences between architectural and infrastructural design problems, and found the structure of infrastructural design problems as sufficiently similar for engagement by an architecture design studio. The research team's quality of outcomes was proportional to the number of times it collectively crossed the threshold between the complementary yet distinct realms of *Research* and *Design*. These crossings contrasted with research behavior exhibiting sustained duration of time spent within one or the other. The experience of this dynamic plurality reinforced an interest to frame infrastructural design problems within upcoming teaching assignments, and to somehow pivot an architecture design studio to design infrastructure.

In April 2010, the in-progress REI investigation was presented by Chris Ford, then faculty in Architecture at the University of Nebraska, at *ECO-Architecture 2010: Third International Conference* in La Coruna, Spain (Ford, 2010b). Ted Shelton, faculty in Architecture at the University of Tennessee—Knoxville, also presented work titled "*Editing Eisenhower: Rethinking the Urban Segments of the U.S. Interstate Highway System*" (Shelton, 2010). After having discovered overlapping interests in infrastructural issues, these educators began to discuss options for a design studio collaboration that merged respective teaching and scholarship towards inventive infrastructure typologies.

The cities of Knoxville TN and Lincoln NE both host their respective State's flagship university, possess contrasting developmental histories, and have proximity to historically significant infrastructural achievements. Knoxville is near to Tennessee Valley Authority projects including a number of key hydro-electric dams that propelled a region out of economic poverty and generated an electrical supply for militarized nuclear research at nearby Oak Ridge TN. Lincoln NE, settled during the western migration due to its high concentration of salt deposits, houses multiple industrial-scale grain silos and has proximity to Union Pacific's first transcontinental telegraph line and railroad.

To pivot an architectural design studio towards infrastructural design problems, four questions were posed at the launch of the studio to contextualize the subject and drive student action:

1. *What are the most challenging problems urban dwellers will face in the foreseeable future?*
2. *What are the respective natures of these problems?*
3. *Can an inventive infrastructural solution address at least one of these problems?*
4. *How ought a hybridized infrastructure solution address multiple problems effectively in an inventive, pluralistic way?*

From a pedagogical perspective, it was undesirable for each respective studio to design for its own home city. It was equally undesirable for a whole studio to design for its reciprocal city, as the alien students would likely remain unaware of some locally demonstrated needs or features. Nonetheless, there is an interesting opportunity for creative discovery when designing for a place with which the student is unfamiliar. To this end, the specific city for which each student designed was determined by a draw, guaranteeing each studio section split between sites in downtown Knoxville TN and downtown Lincoln NE. Once drawn, the student was responsible for defining their own infrastructural design problem as informed by group research, shaped by the problem constraints of site, program, client/user, and determining their personal strategy for infrastructural hybridization. This produced a welcomed leveling effect between two different year-levels of students, and while some proposals would be advantaged through one's deep familiarity with a place, other proposals would be advantaged through one's fresh assessment.

The two university calendars across the 16-week semester shared similar first and last days. Our respective semesters began independently of one another, but with each studio engaging the same project assignments independently as lead-ins to the larger shared HUI project.

10.2.2 Project 01: Meta-Matics

Beginning in the mid-1950s, Swiss artist Jean Tinguely created a series of designed constructs titled *Meta-matics*. While these constructs included moments of compositional flourish within their physicality, they were not sculptures, despite being sculptural. The Meta-matics were themselves not the primary objects of interest. Instead, Meta-matics were machines that, when powered, then generated "art" as an outflow. Tinguely produced a series of Meta-matics in the decades to follow, and would reflect later to describe his Meta-matic works as either auto-destructive art machines, or auto-creative art machines. Although Tinguely generated a body of work consistent with, and deserving of the title of *Artist*, when working on the creative and technical development of a Meta-matic, Tinguely was instead working as a *Designer*.

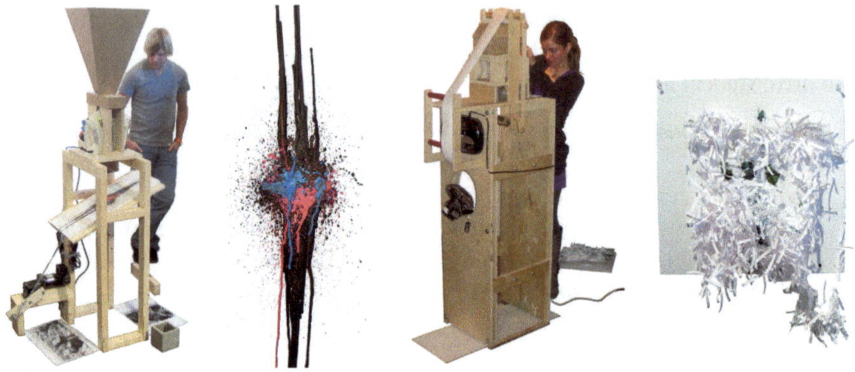

Fig. 10.3 Student-designed *Meta-matics*, or auto-creative art machines, with their respective *art*-ifact outcomes. (*Left*: Tyson Fiscus, *Right*: Kate Ankenbauer Hier)

For a span of 3 weeks, each student designed, fabricated and demonstrated their own *Meta-matic,* or auto-creative art machine (Fig. 10.3). Each *Meta-matic* would: Consider the art product yielded; Consider its source of power (hand, wind, electrical, gravity, or other); Consider its site condition; Consider its own mechanical operation; Consider its own construction methods.

Student Meta-matic designs included a machine that melted crayons between two facing electric clothing irons, which in turn dripped onto a teetering canvas below; a machine that shredded receipt tape from a spool, with shavings made airborne by an electric fan, which collided in mid-air with both melted hot-glue and sprayed green ink, all of which would ultimately fall and collect on a panel resting on the floor forming a two-and-a-half dimensional art product; and a spring-loaded machine capable of swatting paintballs in a split second onto an extruded foam canvas.

The pedagogical goal of this first assignment was to advance the range of architecture students' typical course outcomes for design beyond the representational, and into the demonstrative. By designing, building, and demonstrating an auto-creative art machine, students actively learn what it means to configure material for purposes of yielding a desirable resource unit outside of the designed machine itself. This important turn both mimics and foreshadows the responsibilities of a lifeline infrastructure system for yielding resource units for use within a service area. As the outflowing *art*-ifacts are of higher value, student designers were better positioned to observe and realize the physical Meta-matic designs were like infrastructure in that they too were indifferent to formal debate (Allen, 1999). This realization is important for pivoting from the disciplinary values of Architecture towards the disciplinary obligations of Civil Engineering in a meaningful way.

10.2.3 Project 02: Engaging the Essence

The second assignment "*Engaging the Essence of Technology*" consisted of both an in-field photo essay and graphite hand drawing. Each student designer artfully photographed an accessible type of infrastructure that captured the *essence* of that particular infrastructure type. Students were encouraged to shoot 20+ pictures of their chosen infrastructural subject, and only upon return to their screen and keyboard to then choose (9) finalist photographs for presentation (Fig. 10.4). Photographs presented considered their scale of subject(s), composition, lighting, shadow, exposure, and focus. Each photograph featured a 4 ¼" square proportion and utilized an Adobe InDesign template provided by the course instructors. The photos could be in any

Fig. 10.4 Examples of *Engaging the Essence* contact sheets, and 14″ × 14″ graphite drawings. (*Top*: Images of a recycling center by Kate Ankenbauer Hier, *Bottom*: Images of a wastewater digester by Taylor Nielsen)

combination of black and white or full color. After a pin-up and discussion of the photo essays, each student selected the single-most informative and visually-arresting photograph in their respective collection to then produce a hand-generated, fully-blended value scale (chiaroscuro) graphite rendering with extents of exactly 14″ × 14″ centered on a sheet of 18″ × 24″ paper.

The pedagogical goal of this second assignment was to position students to confront presumptions about the material configurations and assemblies that ultimately shape infrastructure design. Utilizing skills developed earlier in their foundational curriculum, the sustained engagement required to produce a 14″ × 14″ graphite drawing elevated personal awareness of an infrastructure's components and subsystems. The drawing exercise was actually less about the drawing artifact produced, and was instead a vehicle for students to develop greater awareness of the physical assemblies and overall configuration of infrastructure systems. The compositional relationships that may nominally be understood abstractly as *parts-to-whole* as described by Rudolph Arnheim (Arnheim, 1977), would be slowly superseded by an understanding of why and how a grain auger enclosure connects with a horizontal conveyor, or an understanding of how a high-voltage electrical transmission line connects to a fixed structure through an insulator. This greater compositional awareness could then be pedagogically leveraged to understand the contributing role of how prioritizing operational performance requires certain technical requirements, and affords opportunities for material assemblies, both of which configure infrastructure in ways that contribute to its overall, final aesthetic effect. The role of drawing in this assignment was a vehicle for slowing the viewing, for guaranteeing sustained consideration of infrastructural operation, and for provoking student contemplation of what other infrastructural configurations ought to come next.

10.2.4 Project 03: HUIs: Hybridized Urban Infrastructures

Both the *Meta-Matic* and *Engaging the Essence* assignments served as deliberately sequenced and scaled assignments which in turn served as lead-ins for the shared *HUIs: Hybridized Urban Infrastructures* project.

Capable of yielding multiple types of resource units through hybridization of its generative systems, a HUI is a context-dependent infrastructure solution positioned in an urban setting to best serve the demonstrated needs of urban dwellers. The quantities of a HUI's respective yields would be secondary compared against those resource units generated, transmitted and delivered by primary, regionally-scaled systems. However, these new infrastructure concepts would have the ability to generate yields *in-situ*, and if/when necessary, demonstrate operational independence from existing infrastructure systems. This independent "islanding" capability enhances its value during periods of primary infrastructure failure and thereby enhances a city's resilience. HUI proposals could occupy either public or private land, as it reflected a developed and consistent understanding of land ownership and urban land use. Some solutions may require the assertion of their own business model, if requirements for operation and ideas about ownership fall outside of a municipality's respective emergency management office. The HUI project seeks to empower architecture students as future leaders in infrastructure design, through conducting design research investigations of a context-specific proposal with holistic integration between its technological and tectonic systems.

Back-to-back field trips were orchestrated to kick-off the project period and guarantee an immersive experience. Nebraska students were hosted in Knoxville TN for a 4 day weekend in February, and Tennessee students were hosted in Lincoln NE for 4 days during the week immediately following. The motivation for these overnight academic trips was to create a shared, yet intense period of observation, documentation, research, idea generation, and creative exchange. Features of these trips included in-studio lectures by the course instructors, guest lectures by university colleagues including Jennifer Akerman (Tennessee), George Dodds (Tennessee) and Bruce Dvorak (Nebraska), reading discussions, visits to existing infrastructure types such as the Norris Dam and a Cargill urban grain silo, and intensive periods for mixed group research in a charette format. Across these 2 weeks, each student worked as a member of a larger research team to develop awareness of and share insight into the contemporary state of the art in one of five possible infrastructure sectors; *Agriculture, Energy, Transportation, Waste* or *Water* (Fig. 10.5).

In the interest of designing a *hybridized* infrastructure solution, individual students integrated at least one other type of resource unit during the development of their HUI concept, informed in part by team-based research work generated during the field trips. Incorporating multiple types of yields were also championed. Additional aspects for which students were expected to define their problem space included recognizing those constraints that shape design problems of the built environment. These included site, program, and client/user. At the time, it was believed the ownership model would play a key role in the conception of these solutions. Students determined if the owner/operator of the forthcoming infrastructure design was to be public (at either municipal, state or federal scales), private (corporate), or a third model of a public/private partnership. Ultimately, the contributing factor that most influenced final proposals was site selection. These decisions were informed by the student's respective expectations for infrastructural hybridization, and were usually identified in accordance with required resource inflows, or to optimize for preferred outflows.

Examples of selected sites include the Tennessee river, elevated freeways, and an existing parking deck. Most proposals can be understood as *infrastructural graftings*—the deliberate and coordinated superimposition of new regenerative technologies upon a pre-existing infrastructural asset with the structural capacity for forming a new hybridized asset, and without introducing operational compromises to either the pre-existing asset or new system(s) (Ford, 2023). For instance, the West "O" Street viaduct in Lincoln NE was selected for grafting new housing units, for use in a not-so-distant future in which battery electric vehicles proliferate. An elevated green park is used to cap the roadway and mitigate traffic noise. The roof scape and south elevations are used as substrates for architecturally-integrated

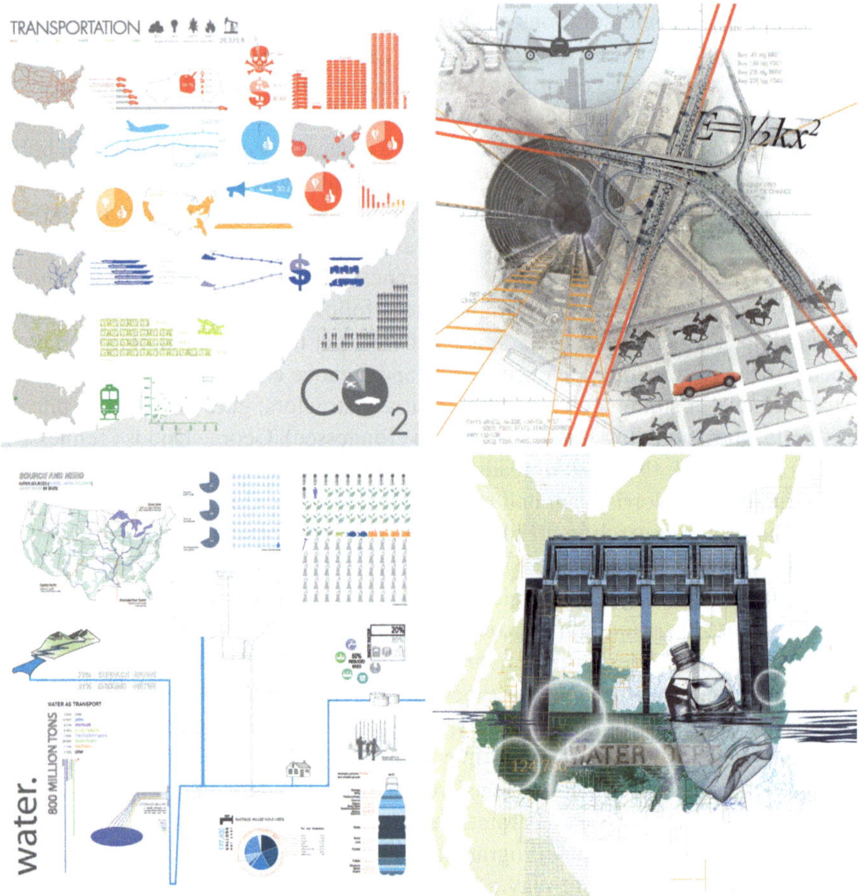

Fig. 10.5 Examples of research materials co-authored by architecture students from the University of Nebraska and University of Tennessee working in conjoined teams of four to five students per team. *Top*: Transportation. *Bottom*: Water

photovoltaics for meeting the facility's charging demand. For a proposal for Knoxville TN, a segment of I-40 was selected for the grafting of a new steel superstructure with two functional purposes, 1) to capture and redirect rainwater before falling onto the roadway, and 2) to suspend algae growth apparatus on south-facing elevations. The collected rainwater would be used to feed the algae tubes which in turn created high-nutrient food for Christmas tree farms enabled through a public-private partnership and grown in the typically inaccessible and undesirable in-between ground spaces created by crossing elevated roadways overhead (Fig. 10.6).

Beyond the fulfillment of course requirements, final presentations were informed by submission requirements to the annual ACSA I AISC Steel Competition. Final HUI project deliverables were to include an Abstract, a Composite Drawing (consisting of (4) 24″ × 24″ drawing panels), and the additional requirement for a ¼″ = 1′

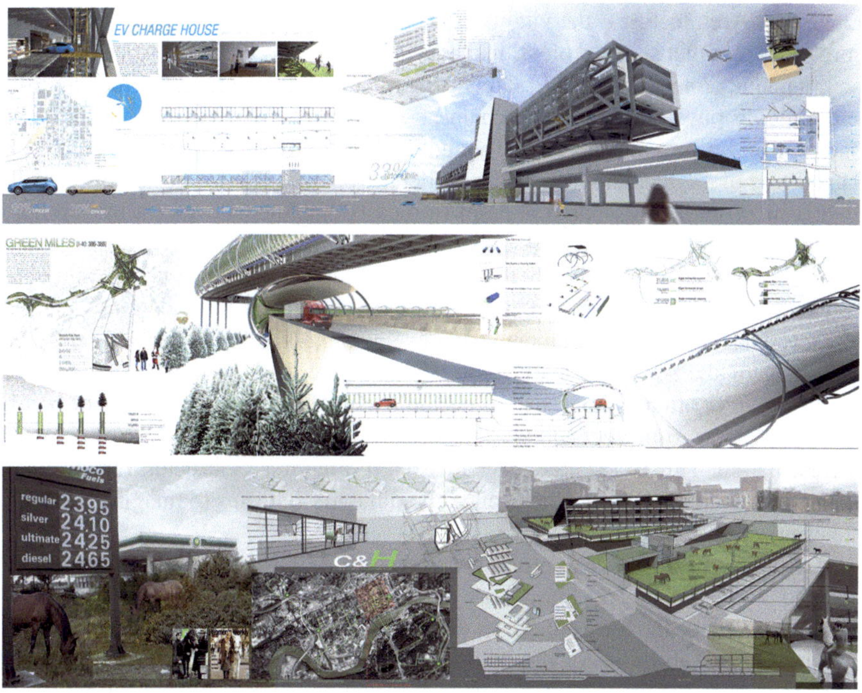

Fig. 10.6 *HUIs: Hybridized Urban Infrastructure* designs for Knoxville TN & Lincoln NE. (Figures by, from top to bottom: Taylor Nielsen, Kate Ankenbauer Hier, and Tyson Fiscus)

systems model. This systems model needed to be free-standing, and self-supportive in an upright position (not wall-mounted); be fully representative of technological and tectonics systems represented in full architectural materiality; represents either one column bay, two column bays, or ½ of one column bay (depending upon overall scale of proposed HUI solution). In ideal circumstances, such design studio projects would transcend curricular goals and would act as provocateurs for change in a larger and more inclusive public realm. If the unsolicited speculative designs resonated "with a particular community, or responds to a demonstrated need in an opportunistic way, then the student work can provoke a public dialogue about the feasibility of such a project beyond the semester extents in which it was issued" (Ford, 2010a).

10.3 Case Study: Everything Must Scale

Much of what is assumed as foundational to architecture and professions that address the built environment is routinely questioned inside the professional schools that prepare students for these fields. Architecture, Urban Planning, Urban Design,

Landscape Architecture—the vitality of these schools often relies on challenges to core principles that shape professional education. In fields that have a deep reliance on pragmatic concerns from life safety and structural engineering, thermal controls and building physics and climate change—indeed to access and cost of shelter—one will find a pronounced curiosity and drive to reinvent the field's parameters. This is true of architectural schools in particular, but in the wider design of the built environment there is a shared attempt to recast fundamental aspects of building. This extends beyond the actual construction of buildings to their patronage and ownership and would-be clients; to heritage and representation of people and constituencies. Schools routinely imagine inventive and novel interrelations between markets and capital, government and development—in particular an imagined reframing of the state and market forces that shape development. This nexus of development is often as contested as design itself. The pragmatic aspects of buildings can appear to be secondary—replaced by a discussion of power, equity and representation. This is a source of vitality but it often masks structural inadequacy in our perceived futures or more accurately in the mechanisms to imagine and realize even what is needed.

Architecture is manifest as property and real-estate. Designers are contracted and enabled as subsidiaries in the nexus of development but also routinely seek to expand or reshape the forms of power that hire them. If there is a routine revaluation of who shapes the future of the built environment there is also little actual legal work to modify the professional statutes and contractual relationships at play. The historic mechanisms that shape what is possible. Yet there is also little in the built world today that is not at risk of obsolescence. For example: As the United States economy began to emerge from the Covid19 pandemic, financial and real estate news reported that the southern areas of the nation were rapidly advancing in renewed housing starts. The motivations, described as access to larger homes, lower cost housing and lower taxes (than Census Northeast or West Regions, for example) portrayed a new version of a tendency that has been in place for decades. The Census demarcated South Region has been outpacing the North, Midwest and West Regions for decades in housing starts and the delta of completed housing units has increased virtually every year since 1968. If compared to the Northeast Region, the South has built between 238% and 712% more housing each year since 1968.[1] In fields that routinely imagine density of land use is a major integer in meeting sustainability goals the pragmatics of everyday development point to a nation that has recently renewed and accelerated its propensity to lower density urban development. The realpolitik of development, the scientific and social aspirations to convene a counter to climate change here meet a development field that builds

[1] The census map outlines the four regions that comprise the Census. The Census South Region is composed of: Alabama, Arkansas, Delaware, District of Columbia, Florida, Georgia, Kentucky, Louisiana, Maryland, Mississippi, North Carolina, Oklahoma, South Carolina, Tennessee, Texas, Virginia, and West Virginia. Alaska and Hawaii are included in the West Region (U.S. Census Bureau, n.d.-a, n.d.-b).

single-family houses; a form of design/engineering and economy that has little professional engagement at all. Professional schools are left to try to navigate private development forces that operate at scale; practices that replicate a building type (80 million single-family houses) that has little over professional design or engineering engagement.

10.3.1 Creative Leap—A Groundless Future

Over 12 academic semesters between 2017 and 2023 the phrase *Everything Must Scale* has served as title and protagonist for a series of graduate level architectural design studios offered at Columbia University Graduate School of Architecture, Planning and Preservation (GSAPP). The studios were each framed around immediate and specific pragmatic concerns but evolved as a series and built on each other and the input and creativity of an evolving team of collaborators, advisors and users. While design studios are often based in simulations of real-world concerns and work with would-be clients, needs, sites and social goals, this series was instead engaged in forecasting future economies, processes, and large-scale social change. For example, what would be needed in housing or seen as possible if a steep loss of jobs resulted from automation? In particular what are called *opportunity occupations*—jobs that do not require a bachelor's degree but pay more than the national median wage. In Phoenix, Arizona and much of the nation *opportunity occupations* accounts for 25% or more of available jobs. These are a foundation of housing security and of personal and national economic health but also are believed to be highly susceptible to automation. What would the effects be of an implementation of renewable energy at vast scale that effectively makes energy non-scarce? Renewable energy generation will by 2050 emerge from entirely different regions of the country than we currently procure energy. Will a less employed populace move to lower cost—less metropolitan—land and simultaneously find access to abundant clean energy? These are macro-economic and territorial questions that dramatically exceed architecture and urban planning but it is also impossible to imagine even the next two decades of building and professional education without imagining reframing professional schools aspirations within development—at scale. This is a project that exceeds sustainability and requires a reformation of how the very matrices of development occur and in particular the role of design and engineering in private development. The foundations of settlement are changing—practices that underlie settlement patterns in the United States since epochal changes in the nation's economy in the aftermath of World War II, but also in the era of globalization that ensued in the post-Bretton Woods 1970s. A global economy spurred newly hyper-financialized housing markets, in particular, as a newly global trade of mortgage debt and leverage. *Everything Must Scale* asked students to effectively levitate in an emerging future; to try to ascertain where it lands in 30 years, but more importantly, what might be possible in a future economy that could be glimpsed now. To imagine adjudicating what is possible—what is just and equitable—in an economy that has yet to arrive. *Everything Must Scale* strove to incorporate our partner's research but it more critically points to a new type of design faculty and new evolution in what constitutes design and collaboration. It imagined a new form of design studio.

Everything Must Scale was intended in part to be a warning. Invoking the business and economics term "scale" or "to scale". A concern that if applied to architecture asks how would architecture or a building scale but here phrased to say it must scale or possibly go extinct. Forecasting markets is not in itself new nor are obsolete (extinct) products and economies. Industry studies have also been very successful in forecasting markets—for televisions, (then) color televisions, washing machines and commodity items of all types. These studies emerged with sophistication in the twentieth century and became the realm of research in business schools and manufacturing. But another type of forecasting is less developed and today urgent: an attempt to model both a future and not yet known economy and its potential or downstream subsidiary markets and products. The matrix of climate change, automation and potential joblessness and housing make this urgent. Development is fueled by employment and consumption; it is all secured by housing. This is a forecasting on multiple fronts—a kind of groundless creative leap where what one considers pragmatic is hard to gauge, unless it is imagined at the crossing of many future vectors. How do you get to such a place? Who procures such design speculation? Who funds this kind of research and development? In our current startup venture capital culture we imagine maverick enterprises but to date development itself and building have not scaled in the ways products or services have—more so they are embedded in deep matrices of state and private interest that have often bypassed any deep engagement with design or speculation. *Everything Must Scale* attempted to engage both the design of the near-future but more importantly reconvene what we see as the networks and teams that would make such a future possible.

10.3.2 A World Without Scale

In economics the potential of scale drives investment in a project—scale promises a return on capital that not only exceeds investment, it accelerates growth at a rate faster than the acceleration of investment. Scale is exponential and nonlinear. Growth may increase revenue but alone it is a straight line. Scale is at the root of incentives to innovation and while it's a factor in any manufactured component and divided labor, it is difficult to imagine as widely applicable to the built environment as a whole. Buildings are routinely homogeneous and similar in construction—made of assemblies and subassemblies of scaled parts—but due to specifics of site, contractor capabilities, discrete ownership (goals) and access to design and engineering they are rarely wholly repeatable. Buildings on aggregate are, however, not immune to economic scaling—for example, repetitive housing units built in the hundreds or thousands address scale even if they vary by size or location. But the sub-components of building are far more universal—if mundane—and scale-able.

Design in the built environment is also based on *fee for services*—architects are paid a percentage of construction cost. Unlike *design for manufacturing* their work is not anticipated as replicated. More so, many of the methods and means in architecture have a rote quality—often as a hedge on litigation that may come from an experimental path in building that subsequently fails. In the United States, for example, none of the top ten homebuilders list expenditures for *research and development* in their annual statements. Apple routinely spends 6% or more on R + D and did so even in the years that neared bankruptcy in the late 1990s.

10.3.3 A Reality (Happily) Stranger than Fiction

Everything Must Scale took on building types that are built in large numbers; building types that serve as an apparatus for business and social need and that operate in the context of large and usually distributed populations at an economic and literal scale. We took on forecasting the design of the future "truck stop" (*Pilot/Flying J* operate more than 750 locations in the U.S.; there are in total approximately 120,000 fueling locations in the U.S.). With a team of two epidemiologists from the World Health Organization we devoted a semester to designing a rapidly deployable small-scale clinic urgently needed worldwide as Covid 19 spread. As the world began to shift to remote work and education in the spring of 2020 our studio was focused on designing a prototype teacher-less small-scale micro-school (Fig. 10.14). The micro-school was in part based in a U.S. Department of Education pilot program anticipating automation in learning and less teacher-centric schools. This work served as a basis to assemble a team to investigate the design of a manufactured deployable remote work space. A project made more feasible as corporations deployed employees and feasibly replaced investment in office buildings with a newly allocated private workspace. In each case the design work relied both analyzing the status quo but also imagining its potential near-future obsolescence. *Pilot and Flying J*, for example, had not only merged in recent years but also had grown nationally from less than 10 sites to 750 in 30 years. They were anticipating investing a billion dollars in renovating existing sites and building hundreds more. But the more compelling story that drove scale in our work was the fact that controlling interest in *Pilot/Flying J* was acquired by *Berkshire Hathaway*—the investment conglomerate founded by Warren Buffet. Buffet had first acquired approximately 40% of the company and then recently the remaining shares. *Pilot* and *Flying J* had in effect invented the modern "truck stop" and scaled it; they were now owned by the same conglomerate that also took controlling interest in *Precision Castparts*. A company whose singular precise metal tooling and casting provides means to the most advanced high-stress components in aircraft turbines. The future "truck stop" would have to adapt to eclectic vehicles; it would have to imagine entirely new temporal occupancy (a longer stay while charging), but also was now owned in the same economic umbrella that owned a global technological powerhouse. The seemingly pragmatic and everyday nature of the truck stop is increasingly a site of advanced technology. Many students noted that the interior of a new Tesla electric semi was more physiologically and optically advanced than professional architectural offices. The volumetric quality of pumping gallons of liquid fuels becomes the volumeless transmission of electrical currents. Does the new environment alter the pathos and social life of the trucker pausing at night to refuel and rest? Most think an entire social eco-system of transportation not only changes but becomes far more attractive as a life choice. This only becomes imaginable as the product of a future economy and design for scale (Figs. 10.7, 10.8, 10.9, 10.10, 10.11, 10.13).

Students of architecture are routinely asked to imagine budgets and pricing even if these are not feasible to detail until a project is fully designed. Budgets in construction are negotiated by architects but actual prices are set by contractors and approved by owners. The commitment to design and engineering prior to pricing is a deep uncertainty in current practices and it causes a retreat from unknown or untested goals or methods. That division between responsibility for design vs lack of control over cost is a major risk factor in design in the United States. Design for manufacturing—industrial production of commodity scale-able products offer far more integration of design and cost. They reduce the risk in design. Students in the *Everything Must Scale* series were asked to imagine not just the next "truck stop" but the economy and investment capital that might serve as its basin. Students analyzed corporate annual reports, explored where research and development might be possible that spread from one company sector to another—that might leak into the built environment itself. In the case of trucking and truck stops they imagined a conflation of shipping, distribution and warehousing becoming retail. In short: a entirely new interchange in the physical shape of money and the spaces it lands. In products, in mobility, in buildings and in energy. This all expands the design team.

Work on a rapidly deployable clinic (Fig. 10.12) was similar. In the case of Covid-19 the early need for low pressure patient spaces to keep the airborne virus from spreading was both a conceptual spatial concern and a relatively simple mechanical property to model. Low-cost software can allow a design student to model otherwise advanced computational fluid dynamic studies. Air and its viscosity is a fundamental quality in architectural discussions of space, for example in (Picasso's) Cubism. Its abstraction in art practices is here made technical. It's model-able inside an elemental digital architectural model in ways that would have required a PhD in engineering in prior years. The micro-school invokes a need to understand and speculate on computer interfaces and algorithms that sort what is delivered to the student's attention—but also these all quickly invoke historical readings of the window and its role in offering a controlled but likely attenuated and separate view to nature. Would a less teacher-centric and less hierarchical educational environment elaborate and expand access to outdoors. The classroom window is not a distraction here—it is the point. We felt prototypes around automation and education we have seen from major tech companies were not as experimental as they could be. But also that architects would by necessity need to learn far more about computer vision and human computer interfaces of all kinds.

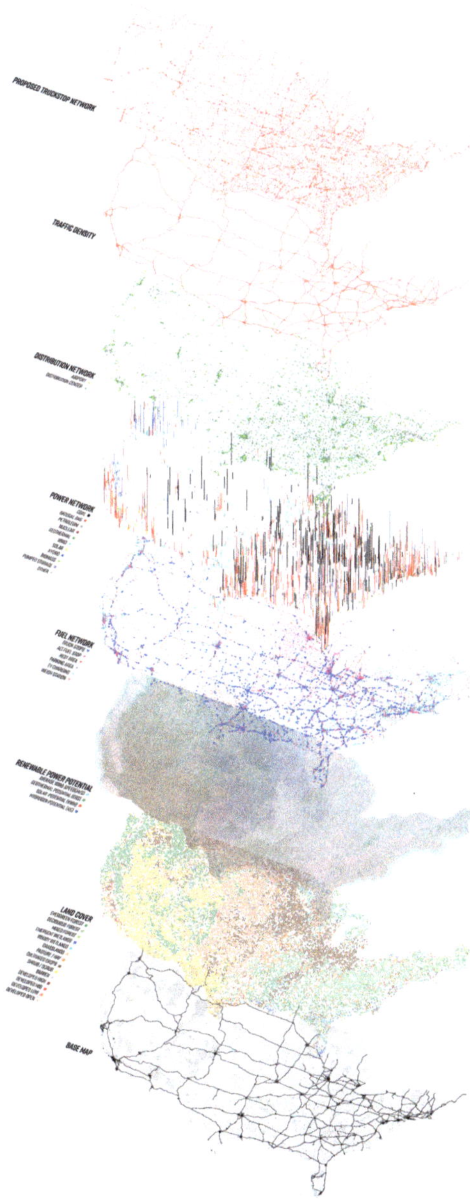

Fig. 10.7 A future truck stop design is preceded by extensive mapping using geocoded locations of the nation's existing fueling stations. Subsidiary maps chart emissions attributed to fuel sales, major roadways and power plants. Forestry and natural growth are shown as well. The student created this series of maps as a prelude to converging on design of a future truck stop. Michael Soderberg, GSAPP

In all these cases the user existed as an individual but also as a population—something that architects are often trained to approach with circumspection. Much of what drives architectural education is a faith in the unique experience of place and its forms of heritage. In approaching the realm of scale one has to be willing to imagine the user as millions of users. This reality of a truck stop is in this case a

Fig. 10.8 An automated truck employs computer vision. The cartography of Earth is quantified from above via satellite connectivity. The terrestrial roadway—the former navigation device is increasingly less definitive and could be less separate from the Earth itself. A truck might move on a modified tire, the roadway less a hardened and discrete surface allowing the pre-developed Earth to be more present. Jiyong Chun, GSAPP

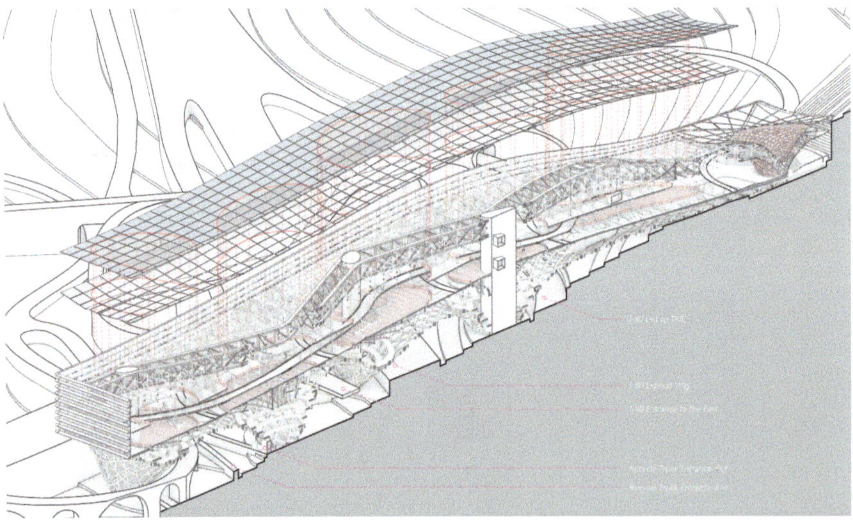

Fig. 10.9 A truck stop serves to transfer goods, fuel and energize truck batteries, offer rest and empathy for a driver but also to construct, manufacture and dismantle both vehicles and goods in a circular economy. All more closely aligned with sustainability and ecology/biodiversity it was titled by *Demolition from the Progress by the designer.* Seonggeun Hur, GSAPP

reality stranger than fiction. 10 p.m. in the middle of a sea of semis or bathing in a rented shower stall at *Flying J* is an experience few architects—few graduate students in a professional school—imagine designing. These studios were supported by a team that made the experience more immediate. Engineering, materials science, supply chain experts from *Tesla* hosted our team at the Fremont, California

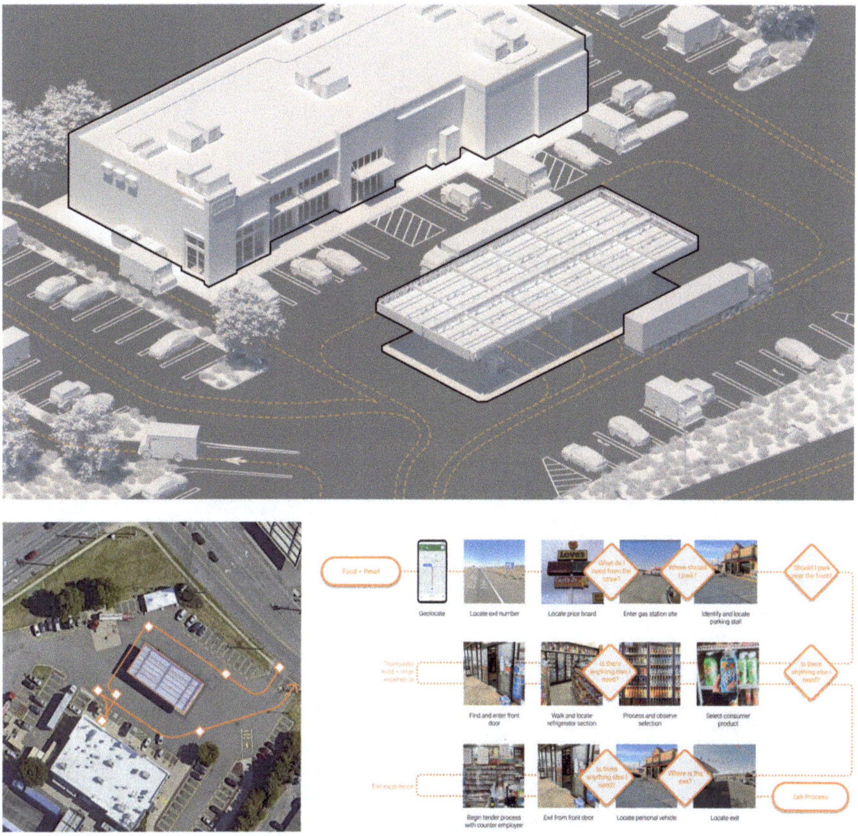

Fig. 10.10 Each consumer interaction and form of attention diagrammed atop the physical hardware and infrastructure. Most U.S. fuel stations earn more revenue from sales other than gasoline. Increasingly each step in the attention and behavior chain is mapped for potential increased sales. Electric vehicles change this entire experience. Jacob Kackley and Marcus Chan, GSAPP

Factory and Sparks, Nevada *Gigafactory*. We also were hosted at the Tesla Design Studio in Hawthorne, California. BMW *Designworks* and Mercedes Benz offered similar support as did several major engineering firms. Walmart opened their distribution hub at the Tahoe-Reno Industrial Center to us to meet the team leaders of the immense shipping and logistics center. The Center for Design Research at Stanford hosted workshops. Professor Michael Shanks, Classics and Archeology—a member of the d School contingent and founder of several urban and design initiatives at Stanford—hosted us over several years and traveled to New York to join us at Columbia over several days and visits. Chris Ford, my co-author here, did the same. We were also hosted by the Stanford Center for Design Research founding Director, Professor Larry Leifer with the goal of imagining the effect of CDR graduate-level research and design theory on architecture.

Fig. 10.11 The designers sought to reduce where and how often focused human attention was required for sales or fueling, opening the fueling stop to the broader possibly picturesque landscape. Jacob Kackley and Marcus Chan, GSAPP

10.3.4 Market Scale/State Scale

Levittown, New York, begun in 1947, reached a total of approximately 17,000 units over 4 years at its completion in 1951. As a forerunner to the industrialization of housing William J. Levitt built housing in the hundreds of units per year in pre-war years. Levittown, realized on 1200 acres in Nassau County, New York, brought this figure into the thousands. The New York City Housing Authority (NYCHA) —realized by the U.S. federal government, the state of New York and New York City—in 1950 was already far larger. NYCHA's 1950s Development Data Book showed a total of 75,952 apartments under NYCHA management. 30,730 were already completed, 34,967 new units were under construction or pending construction and intended as additional permanent housing. We ascribe to Levittown the seeds of the

Fig. 10.12 A Covid diagnostic clinic. A cellular monocoque shell structure manufactured on automotive assembly lines. Lightweight but structurally self-sustaining it is easily shippable. Its acrylic surfaces are easily treated and cleansed. Fan Liu and Haoran Xu, GSAPP

mass-produced suburb. NYHCA, while often seen as deeply successful, often faces a critique of being a state based housing system. Government competition with free markets; with the rights to property and its wealth. At its peak NYCHA housed as many as 1 in 8 residents of New York City (Marshall, 2015; New York City Housing Authority, n.d.). By the late 1960s leading architects were being increasingly asked to reconsider scale—in particular in regard to housing. Archival notes (Risselada & van den Heuvel, 2005) show George Candilis addressing an emerging, but overt sense of vulnerability in the face of large development. In the late 1960s his firm's immense housing project at Toulouse le Mirail, France—still under construction—forms a backdrop for Candilis reimagining his role in the design of works of such literal scale. Largely, but not fully realized, Toulouse le Mirail now could house 20,000 people—a significant portion of the 70,000 it was planned for. Candilis resigned from the project before completing it acknowledging a need to re-appraise the times and the "brutality and complexity" of the problems. His students and colleagues in the Technical University of Delft in part demanded a change formally declaring his team "lackeys of capitalism". Architecture often turns to the term empathy when facing scale—what or how do we better know or understand who we

Fig. 10.13 An evolution of the truck stop and warehouse become a new type of urban development. Emissionless electric vehicles, choreographed with short term warehousing and last mile transfers form an ensemble that is more integral to a natural landscape. Warehousing, shipping, commerce and the economies of each become more transparent yet also less intrusive and opaque to development. Shuyang Huang, GSAPP

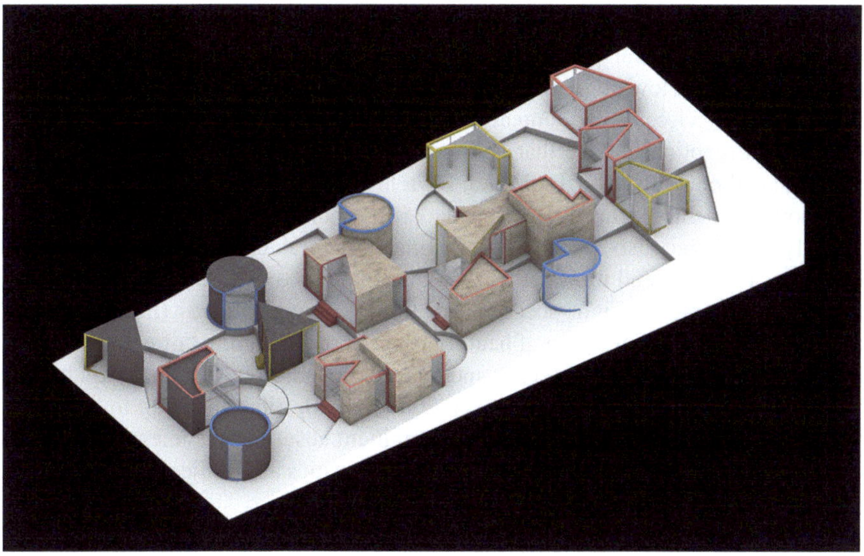

Fig. 10.14 A teacher-less micro school—small-scale private spaces for two or three students overlap forming a spatial and formal syntax. Titled Occupying Language it imagined a continually reforming syntax as architectural form, volume and duration. Ge Guo, GSAPP

design for. In the case of Candilis he imagined engaging the eventual occupant as more present in the design process. A massive turn after nearly single-handedly designing housing for 70,000 people. Studios grapple with this in unique ways but we often have instead overly simplified reality, oversimplified or reduced what a user or owner or occupant might imagine, see or desire. *Everything Must Scale sought to allow more complexity—more of the surrealism of everyday life.*

10.3.5 Absurdity, Pragmatism and Empathy

In his essay, "Visual Absurdity in Raging Bull" (2005), Todd Berliner described Martin Scorsese's, 1980 film *Raging Bull* in split second detail. He describes the film's sequence of boxing scenes as effectively temporally incoherent if imagined in the linear time of film. A left hook thrown by Jake La Motta (portrayed by Robert De Niro) as witnessed by the audience lands as a right hook all in less than a second of film. Under Scorsese's strobe lighting the scenes occur in linear time but also in split second periods of blindness. A blank screen. The audience believes they k*now* what unfolded and yet doesn't see a millisecond of the otherwise impossible (incoherent) blow. Berliner analyzing the collaboration of Scorsese and cinematographer Michael Chapman is unfolding the abstraction and unreality at the core of an otherwise mainstream and popular film. The *visual absurdity* is in his logic a testament to Scorsese's creativity but also ends up reflecting an audience that is not only participating in real time but is more than capable of a internalizing and normalizing an otherwise atemporal perception of time; a form imagination that pushes the film's experiments towards a form of empathy that effectively accepts an impossible sequencing of visceral but unreal events. These experiences are milliseconds borne—a perception of time that is dissonant to reality but capable of instigating catharsis and empathy—for La Motta, for a historical 1940s audience engulfed in the sport's violent spectacle. Berliner makes the point that *Raging Bull* was both visually absurd but also accessible—it's time structures unchallenged by an audience who are in the end more rather than less enabled to express and feel the emotion of the character. Fight scenes were filmed through flames (in front of the camera) to create the sensation of heat. Berliner describes a film "often on the point of falling to pieces" and fight scenes and images "so ludicrous that it's a wonder that viewers can make sense of it" (Berliner, 2005). For Berliner the goal: to show that "Raging Bull's visual incoherence and intermittent absurdity are integral to its success as a film and one of the primary reasons that critics and audiences find the film so compelling." (ibid.).

Architecture rarely approaches the level of experimentation that Berliner describes. Indeed Scorsese even imagined *Raging Bull* might "end his career." It instead accelerated his rise as a cultural figure. Scorsese's role as filmmaker allows creativity that is difficult to bring into the humanist aspects of architectural education but contemporary life is routinely filled with the same forms of discomfort if not alienation. A dissonance that we often see in our daily lives—the isolation that can occur from fueling a car to crossing a parking lot at night; to the hum and buzz of an air conditioner outside a thin plywood and stucco wall; to hearing a neighbor's television in a suburban apartment. In the case of a truck stop—a car driver taking an hour to recharge a vehicle at night in spaces shared with professional truck drivers—forms of vulnerability that are common in film and might drive a script are here also everywhere around us in real, everyday, life. Architecture relies on forms of empathy and with meeting public perceptions and goals; it struggles however with accessing the forms of experimentation that have allowed even a popular film like Raging Bull to communicate deeply even within abstractions of something as fundamental as time and perception.

10.3.6 The Client, the User, the Owner and Empathy

"Find what gave you the emotion; what the action was that gave you the excitement. Then write it down making it clear so the reader will see it too and have the same feeling as you had." Ernest Hemingway, acceptance speech, the Nobel Prize in Literature, 1954. (Hemingway, 1954).

A user in 1950s architectural education would likely have been imagined inside an industry resource such as *Architectural Graphic Standards*. The 11th Edition available today—a 1120-page tome updates the 1932 first published version—offers prototype design for thousands of architectural uses and program types. These orbit demands on use (the layout of a hotel lobby vs a school lobby) but also common construction details (bolt patterns needed to connect steel beams and columns or to the miter a bathroom ceramic tile corner). *Architectural Graphic Standards*' very title captures a source of creative angst in architecture school—to what extent are we a profession bound to norms and where are options to innovate. The issues of standards are by and large more acceptable in today's climate if applied to construction or materials but schools routinely imagine alternative building uses, functions and have dramatically expanded the would-be user or owner into a deeper concept of a political subjectivity. A person, imbricated within a series of power relations many of which they are unaware and unable to release or change. The architecture student is routinely imagining its service is to something that dramatically exceeds any professional contract. More so, even if such a contract were created one would have to ask if the building industry, the real-estate industry, and design firms that are legally formed to achieve such work could design or enact them. The average student is imagining a client that rarely if ever exists, and aspiring to physical design achievements that we might only imagine possible from industrial, capital intensive entities such as Airbus, Tesla, SEELE, Precision Castparts or Apple. In other words, what we imagine is needed socially and politically often exceeds our times but also requires immense capital and investment. The detail that carries architecture is often defined in the public realm—within shared language even—but the actual realization and description of building is carried in often arcane inaccessible drawings and specifications. That gap deserves far more study today and increasingly we can simulate and diminish the risk of misunderstanding between the user and the designers. Hemingway's statement above is a touchstone but the novelist was speaking of describing events past—design is a description of things and experience yet to occur.

10.3.7 Post-Scarcity

Architectural schools have been unevenly affected by calls to address climate change. If the term *sustainability* underpins our times it can also serve as a means to diminish speculative work. It can serve as a means to essentially replace design with engineering—to diminish the less quantifiable aspects of design. Sustainability is a term that would be impossible to try to isolate but one foundational source would have to be the United Nations, 1987 Report of the World Commission on Environment and Development titled *Our Common Future*. Written in an era when the cost of producing a photovoltaic cell had fallen from $77 to $10 per watt it offers only four pages on renewable energy but instead largely focuses on the inter-related nature of developed and developing economies in the post Bretton Woods era. Sustainable Development as a term assumes that inter-relationship is critical; it places a significant pressure on wealthy nations to advance sustainability goals in countries less able to do so. In the intervening 36 years the cost of producing a photovoltaic has plunged to less than 50 cents per watt. Architects are often taught to imagine energy as scarce; to impose restrictions and quasi-punitive limits on its use but the reality of renewable energy is that it is indeed not scarce in the same way as fossil fuels. The *Everything Must Scale* studios presented research on renewable energy and in particular 1970s National Science Foundation and Congressional Acts on renewable energy that describe solar energy, for example, as "inexhaustible." A form of energy that would nonetheless need industrial instruments—a PV panel, for example—but that was in essence abundant in source.[2] Design studios almost universally imagine building components as resources derived from capital intensive and industrial processes—from private companies who use technology to improve efficiency. These companies are the beneficiary of that wealth accumulation. In the context of renewing architectural and design education we have an opportunity to alter the equation of scarcity. This has arguably already occurred in relation to a host of digital services and devices and networks—entire communications industries essentially offer low-cost global communications; passenger automobiles increased fuel efficiency by more than 200% since the 1970s with even more dramatic gains in hybrid and electric drivetrains. Indeed, United States houses are as much as 40% more efficient on a per square footage basis than even 30 years ago (houses have countered this efficiency by growth in size). Scarcity is being altered into forms of abundance—but division in equity persists in extremes. For example: households that own their own home have a net worth that is on average 35 times those that rent and the recent financial crisis around housing costs have made more people renters. Design encompasses all these scenarios in today's architecture schools. We have come a long way from the discipline we understood in historical terms but we also have not reformed our schools to better enable the teams and real-time and parallel forms of design we need to shape a near future.

10.3.8 A Virtual School Run Between NYC and the Bay Area and Los Angeles

These studios were framed after having spent most of 2 years in the Bay Area, on a leave from Columbia University but also as a visiting professor at Berkeley and at the Center for Design Research at Stanford. This work is developed with Eunjeong Seong, partner in forthcoming 2025 book *8 Minutes 20 Seconds: Housing after Banking, Encrypting the Sun* published by Actar.

[2] Guyford Stever, Director, National Science Foundation, frames solar energy as capital that might replace "our use of the nonrenewable fossil and fission fuels." See: National Science Foundation (1975), also see statements by John N. Nassikas, Chairman, Federal Power Commission in: "Nassikas Backs Project Independence" (1974), also Nassikas (1974).

10.4 Conclusion

A shared feature between both the architectural academy and distinguished architectural practice is the central role of the design studio. The studio, as a place, as a course, and creatively productive culture, presents interesting characteristics that may be of useful import to STEM-based pedagogies and research methods.

Despite a millennia long disciplinary identity, architecture is only recently a legal profession and one that relies on professional schools for education. And yet it is indeed very much still an atelier model of education that relies on professor-to-student design studios as the core form of education and learning. These courses evolve over decades but by and large rely on students and professors together enacting the design of a building on behalf of a surrogate client. The design studio however also relies on a deep well of peer-to-peer learning that both shares knowledge and generates friendly yet anxious competition. Few schools of architecture have deep wells of research funding, but are nonetheless increasingly an exemplary model for active learning that is emblematic of what a university might strive for—An active and pragmatic form of design that often fuses a wide range of techniques and disciplines. They in fact are increasingly defining their goals not with problem sets or needs that are explicitly-stated, but instead seeking to refine and in effect discover the problems worthy of attention.

In conjunction with other available resources, the design student operates within a deliberately-designed physical and curricular environment in which the iterative development of their work is best supported.

A key pedagogical advantage of the design studio is the provision of a curriculum-based safe space in which students can diverge and converge repeatedly in the interest of architectural experimentation. Ultimately, each student synthesizes a range of constraints, requirements, and expectations into a workable solution for externalization to others. It is a daunting educational proposition. But unlike how a discrete chemistry lab is executed by a TA, design studios are course experiences requiring dedicated resources of space, time, funding, and faculty contact to best pursue the empirical development of solutions to design problems framed by the studio instructor. Focus and attention necessarily oscillate between scales, span across problem spaces and solution spaces, and shift often between the reflective and the projective. Task-switching is commonplace, and changing-up the media through which architectural possibility is explored is commonplace. The design studio makes this happen.

Little of what is designed inside professional architecture school studios today is imagined within a known commissioning protocol. There is often a deep desire to re-frame the way architecture is owned and commissioned; this can seem more critical than actual design work itself. A desire to free design future capacities from protocols and forms of power that leave vast amounts of the population underserved or indeed cut out of the benefits of development. The design studio—a vivid site of public learning, questioning and application of thought—often seeks to re-frame its own matrix and team. They model clients, communities, histories and the taxis of law and money that might shape a work. Architecture itself often remains central—but there are paths forged in studios that attempt to make the parties more equal and their engagement less linear and cumulative. With external funding pilot studio structures are tested that include an expanded teaching faculty. Housing experts, engineers, material scientists, legal, economic, climate and data/software specialists each have joined design studio teams offering engagement with relevant knowledge bases for engaging wicked design problems, and to prevent the premature convergence and resolution of a design proposal too soon. The persistent questioning of professional purpose inside architecture schools has we believe reached a crescendo; there is clear evidence that a new type of design school is needed.

A question that is asked inside design schools: is this unique to our time and does it constitute a demand to find a new way to learn? Architecture is deeply reliant on pragmatic concerns of every kind, yet for most of the last century the balancing between material facts (for example, the lifecycle of materials) and curiosity for change and experimentation has never been seen as unfortunate. Architecture and its partners across the AECO acronym have thrived in the kind of *active learning* environment of design studios where the framing and re-framing of scope is common.

Whether the prompts issued in a design studio emerge from questions *about* the world, or dynamics observed *in* the world, it is possible to see how the best attributes of studios found in Architecture might be applied to design studios launched throughout STEM-based fields.

References

Allen, S. (1999). Infrastructural urbanism. In S. Allen (Ed.), *Points + lines: Diagrams and projects for the city* (p. 55). Princeton Architectural Press.

Arnheim, R. (1977). *The dynamics of architectural form*. University of California Press.

Bell, M., & Seong, E. (2025). *8 minutes, 20 seconds: Housing after banking, encrypting the Sun*. ACTAR.

Berliner, T. (2005). Visual absurdity in raging bull. In K. J. Hayes (Ed.), *Martin Scorsese's raging bull* (Cambridge film handbooks) (pp. 41–68). Cambridge University Press.

Brundtland, G. (1987). *Report of the world commission on environment and development: Our common future*. United Nations General Assembly document A/42/427 https://sustainabledevelopment.un.org/content/documents/5987our-common-future.pdf

Ford, C. (2010a). The design studio as public provocateur. In *Proceedings of ConnectED 2010: 2nd international conference on design education*. University of New South Wales.

Ford, C. (2010b). REIs: Renewable energy infrastructures v1.0. In *Proceedings from ECO-architecture 2010: Third international conference*. Wessex Institute of Technology.

Ford, C. (2011). *REIs: Renewable energy infrastructures* (Final report, 2009 AIA Upjohn research initiative project) (p. 9). American Institute of Architects. https://tinyurl.com/u6vs3vl

Ford, C. (2023). *Resilient infrastructure futures: An Aetiological analysis of lifeline infrastructure system failures since 2000 for enhancing the resilience of next-generation infrastructure design* (p. 151). PhD dissertation, Stanford University.

Hemingway, E. (1954). Acceptance speech. Nobel Prize in Literature, December 10, 1954. https://www.nobelprize.org/prizes/literature/1954/hemingway/speech/

Marshall, C. (2015, April 28). Levittown, the prototypical American suburb—A history of cities in 50 buildings, day 25. *The Guardian*. https://www.theguardian.com/cities/2015/apr/28/levittown-america-prototypical-suburb-history-cities
NAAB. (2019). *NAAB annual report on architecture education*, National Architectural Accrediting Board. https://higherlogicdownload.s3.amazonaws.com/NAAB/21e8eae7-e532-47c0-bff1-4111ca0d4fb0/UploadedImages/PDFs/2019_NAAB-Annual-Report.pdf
Nassikas, J. N. (1974, July 30). *[Statement of John N. Nassikas, chairman, Federal Power Commission before the subcommittee on energy, committee on science and astronautics, US house of representatives]*. Digitized from Box 11, folder 1974/10/26 S3234, solar energy research development and demonstration of the white house records office: Legislation case files at the Gerald R. Ford Presidential Library, 10. https://www.fordlibrarymuseum.gov/library/document/0055/1668753.pdf
Nassikas Backs Project Independence. (1974, October 31). *The New York Times*. https://www.nytimes.com/1974/10/31/archives/nassikas-backs-project-independence-no-consensus-found.html
National Science Foundation. (1975). *Twenty-fourth annual report for fiscal year 1974*. U.S. Government Printing Office. https://www.nsf.gov/pubs/1974/1974-NSF-Annual-Report.pdf
New York City Housing Authority. (n.d.). *Archives 1950s development data book*. City of New York. https://www.nyc.gov/site/nycha/about/developments/1950archives.page
Risselada, M., & van den Heuvel, D. (Eds.). (2005). *Team 10, 1953–1981: In search of a utopia of the present*. NAi.
Scorsese, M. (Director). (1980). *Raging bull* [Film]. Chartoff-Winkler Productions, Inc.
Shelton, T. (2010). Editing Eisenhower: Rethinking the urban segments of the U.S. interstate highway system. In *Proceedings from ECO-architecture 2010: Third international conference*. Wessex Institute of Technology.
U.S. Census Bureau. (n.d.-a). *Housing vacancies and homeownership (CPS/HVS), [Current population survey/housing vacancy survey]*. Retrieved from https://www.census.gov/housing/hvs/data/histtabs.html
U.S. Census Bureau. (n.d.-b). *"Nation," and "Region," Maps*. Retrieved from https://data.census.gov/map?q=United+States&g=010XX00US&layer=VT_2022_020_00_PY_D1&basemap=basic&palette=Greens&mode=thematic&loc=37.7214,-99.7282,z4.4430

Michael Bell, Master of Architecture, Professor of Architecture, Columbia University, Graduate School of Architecture, Planning and Preservation (GSAPP), New York. Michael Bell founded and chaired the Columbia Conference on Architecture, Engineering and Materials; a research program in collaboration with Columbia's School of Engineering and Applied Science and the Institute for Lightweight Structures and Conceptual Design at the University of Stuttgart. Also at Columbia Bell served as a Director of the Master of Architecture Design Studios (2000–2014) and Coordinator for urban Housing Studios (2000 and 2011). Bell's work in housing has been commissioned by The Museum of Modern Art, and his design work is included in the Permanent Collection of SFMoMA. His *Binocular House* is featured in American Masterwork Houses by historian Kenneth Frampton. Books by Bell include *Engineered Transparency: The Technical, Visual, and Spatial Effects of Glass*; *Permanent Change: Plastics in Architecture and Engineering*; *Space Replaces Us*; *Slow Space*; and *8 Minutes, 20 Seconds* (Actar, Spring 2025) with Eunjeong Seong. Bell was a Fellow at the Joint Center for Housing Studies, Harvard University GSD and Kennedy School (2011–2013) and has taught at Berkeley and Rice University and served as visiting professor at Harvard University, the University of Michigan, Cornell and recently the Center for Design Research, Stanford University (2016). Bell founded his practice in Berkeley, California in 1990.

Chris Ford, PhD & AIA. Co-Founder of *Technology | Architecture + Design* (TAD Journal); Principal Investigator, INFRA-X (infra-x.org), Palo Alto CA. Chris is a design professional, design educator, and design researcher in the areas of both Architecture and Infrastructure design. He studies and shapes urban futures through design-actionable research utilizing human-centered methodologies. Chris is former faculty of Architecture at the University of Arizona and the University of Nebraska. Chris is a co-founding Editorial Board member of *Technology | Architecture + Design* (TAD Journal), a peer-review scholarly journal published by the Association of Collegiate Schools of Architecture (ACSA) and printed by Taylor & Francis. He served as Issue Editor for TAD: *Urbanizing* (Spring 2019) and TAD: *Engineering* (Fall 2022). In December 2023, Chris completed a PhD dissertation titled "Resilient Infrastructure Futures" in the School of Engineering at Stanford University. Chris is currently an Affiliate Researcher in the Stanford Center for Integrated Facility Engineering (CIFE), and is Principal Investigator for Infra-X, a research lab driven to enhance the operational performance of infra-systems, structures, and components used by the urban dweller.

Chapter 11
A Pedagogy of Enactment: Creative Rehearsals and Feedback Literacy in Science Teacher Education

Thomas R. S. Albrechtsen and Karina Kiær

11.1 Introduction

It is an ideal in higher education to make teaching more student-centered and promote active learning, but lectures and passive forms of learning are still often dominating and creating a need for staff professional development and better use of learning designs among other things to change the patterns (Børte et al., 2023). Some studies show that there is also a gap in teacher education between the intentions of active learning and the realization of it in practice (Niemi, 2002). In cases where active learning is part of teacher education it seems to promote the students'

The process of learning about learning matters in Creative Pragmatics because practice and experimentation are essential dimensions for those that frame and create a classroom experience. Two Danish senior researchers dig into this meta topic for us, investigating how science teachers benefit from ongoing processes of rehearsals and reflection. This case study foregrounds the fertility of the concept of performance in Creative Pragmatics and with respect to relationships between theory and practice. Rehearsal, enactment, scenario, improvisation are shown to be effective in conceiving of practice-based learning. The concept of "creative rehearsal" expresses the role of imagination in a reconciliation of tensions between theory and practice, while "feedback literacy" conveys a key pragmatic competency in the dynamic and iterative process of learning by rehearsing. (Editors)

T. R. S. Albrechtsen (✉)
University College of Southern Denmark, Esbjerg, Denmark
e-mail: trsa@ucsyd.dk

K. Kiær
University of Southern Denmark, University College of Southern Denmark, Haderslev, Odense, Denmark

C. Svabo et al. (eds.), *Creative Pragmatics for Active Learning in STEM Education*, Contributions from Science Education Research 14,
https://doi.org/10.1007/978-3-031-78720-1_11

professional competences (Darling-Hammond et al., 2017; Niemi et al., 2016; Virtanen et al., 2017). Active learning is typically defined as a learning or instructional approach but can mean different things (Hartikainen et al., 2019), and be practiced in different ways (Mintzes & Walter, 2020). In the context of teacher education, it is essential to differentiate between two educational arenas of active learning: teacher education on campus and in practicum or between 'coursework' and 'fieldwork'. In this chapter, we will discuss current research on the integration of the two arenas focusing on how the education of science teachers can become more active and creative. This involves integrating experiences from practicum but also creating scenarios or approximations of school practice allowing pre-service teachers to *enact* teaching situations to a higher degree than what they may be used to. The issue of integration in science teacher education follows a long tradition in teacher education in general of problematizing the relation between theory and practice (Rothland, 2020; Zeichner, 2012). Historically there has been a great focus on reducing the gap between what goes on in the classroom, and what teachers learn in teacher preparation programs. Ball and Cohen (1999) argue that knowledge is important to inform teaching practice. However, teaching knowledge *for* practice is not enough. Pre-service teachers must also learn *in* and *from* practice (p. 10) as they apply what they know and adapt as well as refine their knowledge in given contexts. As they say: "teachers must be actively learning as they teach" (p. 11). Pre-service teachers must learn about teaching as a part of teaching. The issue we will discuss in this chapter is also described by Darling-Hammond (2006, pp. 37–38) as 'the problem of enactment in teacher education':

> This problem often surfaces in complaints that teacher education is too theoretical, by which teachers often mean they have not learned about concrete tools and practices that let them put into action the ideas they have encountered. Teachers-in-training must not only understand how people learn and what teaching strategies may help them; they must also learn to present information clearly, lead discussions that really get at the ideas under study, manage discourse of many kinds, organize groups for learning and give them useful tasks they can do, manage student behavior, weigh difficult dilemmas and make quick decisions, plan well and alter plans for unforeseen circumstances, respond to children, and respond to questions about the material they are teaching" (Darling-Hammond, 2006, p. 37)

How to teach pre-service teachers in this way has been discussed intensively in the last 25 years under the heading "practice-based teacher education" (Matsumoto-Royo & Ramirez-Montoya, 2021). There is a greater interest in and emphasis on opportunities for pre-service teachers to engage in the practice of teaching during their teacher education time on campus. Sometimes the term practice-based has been used narrowly to refer to programs with more extensive opportunities for pre-service teachers to engage in actual classroom teaching (Forzani, 2014). Instead, we argue for the broader use of the term as it also allows for *approximations of practice* (Grossman, 2018; Grossman et al., 2009; Ochieng & Zoest, 2023; Schutz et al., 2018). This means that pre-service teachers engage in "approximate" teaching without necessarily having to provide it to children in a school classroom setting. How science teacher educators can arrange situations on campus to make this possible requires the formulation of *pedagogies of enactment* (Grossman & Mcdonald, 2008).

11.2 Method

Our contribution to the current research and discussions on doing *rehearsals* in science teacher education is to synthesize it with some new theoretical developments in creativity and improvisation in teaching and the concept of feedback literacy. The purpose of this chapter is to discuss how Creative Pragmatics can be part of a pedagogy of enactment focusing on the education of science teachers. We will argue for the fruitfulness of combining these insights forming a Creative Pragmatic pedagogy of enactment in science teacher education. To illustrate how this may be done we will present and analyze a case of pre-service science teachers rehearsing and enacting a lesson on modeling in a Danish primary school context (Thomas, 2021).

In the following we will first shed light on why it is important to develop a pedagogy of teacher education as part of being a science teacher educator. Next, we will introduce current research on rehearsals in practice-based teacher education with a special emphasis on the framework developed by Lampert et al. (2013). Inspired by this framework we will elaborate on the different aspects of the rehearsals and investigations of core science teaching practices giving pre-service teachers possibilities for active learning. We will analyze the case of modeling in science teaching in this elaboration of the framework. A central argument of the chapter is that to qualify this process the feedback literacy of both science teacher educator and pre-service science teachers is central and that the feedback on the teaching rehearsals can be formed in creative ways and thereby stimulate the improvisation and pedagogical imagination of the pre-service science teachers.

11.3 Pedagogy of Science Teacher Education

When we use the term "pedagogy" we follow the definition of Alexander (2009, p. 13) that pedagogy: "encompasses both the act of teaching and its contingent theories and debates. Pedagogy is the discourse with which one needs to engage in order both to teach intelligently and make sense of teaching – for discourse and act are interdependent". We also understand pedagogies of enactment—and active learning more generally—as a part of a *pedagogy of teacher education* (Peercy & Troyan, 2017; Russell & Loughran, 2007), which also characterize the professional knowledge and practice of teacher educators. The pedagogy of science teacher education concerns the knowledge and practice of teaching about teaching and learning science (Loughran, 2014). Science teacher educators are doing "second-order teaching". They are inducting the pre-service teachers "into the practices and discourses of both school teaching and teacher education" (Murray & Male, 2005, p. 126). There are different perspectives on what teacher educators in general should know and be able to do (Goodwin et al., 2014). Regarding science teacher educators Mork et al. (2021) distinguish between four essential knowledge domains. These knowledge domains are: (1) Natural science (2) Science education in school (3) Science teacher education and (4) Science education research. A task for the science teacher educator is also to assess and give pre-service teachers feedback on their *enacted pedagogical content knowledge*, that is how they transform science content in a way that makes it comprehensible for the kind of students they teach (Coetzee et al., 2022; Hume et al., 2019) What science teacher educators are supposed to know and do will also depend on the cultural context they are part of and what is part of the national curriculum for teacher education. For example, in a Danish context educating pre-service teachers for teaching in primary and lower secondary schools (the Danish *Folkeskolen*) is situated in university colleges (there are six university colleges in Denmark). The curriculum is *competency-based.* Pre-service teachers are taught general pedagogical knowledge and skills, and they choose to specialize in some subjects. Among the science subjects, Danish primary and lower-secondary teachers can teach Nature/Technology (1st–6th grade) or Biology, Geography and/or Physics/Chemistry (7th–9th grade). Common for these science subjects is that they all are oriented towards four main competence areas. Danish school students are expected to be able to: (1) Do inquiries (2) Do modeling (3) Communicate about science and (4) Take perspective/ relate content to different issues. We will later show a case of pre-service science teachers teaching modeling competences for 3rd and 4th grade students in the subject Nature/Technology.

The research on practice-based teacher education and the focus on teaching and learning of *core practices*, that we build on in this chapter, is primarily from a North American/US context and with a connection to the so-called *Next Generation Science Standards* for teacher preparation that emphasizes, among other things, teacher education pedagogies of analysis and enactment (Windschitl & Stroupe, 2017).

In their literature review Matsumoto-Royo and Ramirez-Montoya (2021) distinguish between the following core practices in teaching in general (and not specifically for science education) covered in the research in the field:

- Analyzing instruction, materials, and student learning
- Designing an appropriate assessment
- Eliciting and interpreting individual students' learning
- Explaining and modeling content, practices, and strategies
- Giving appropriate scaffolding
- Giving clear directions
- Implementing organizational routines
- Leading a group discussion
- Planning for teaching
- Positioning students as competent learners
- Pressing for evidence-based explanations
- Using methods to check or to assess students' understanding.

In the literature on practice-based science teacher education especially the core practice of "leading a group discussion" has been investigated (Davis et al., 2017; Firetto et al., 2023; Hume et al., 2019; Kademian & Davis, 2018; Masters, 2020). In the Danish school curriculum this could relate to the development of communicative competences in science and how teachers can arrange situations that support it. Science teacher educators have the task to find ways to promote the pre-service teachers' opportunities to enact and rehearse such skills (Hammerness et al., 2020).

A reform of the Danish teacher education system emphasizing among other things more coherence in the study to become a teacher began in 2023. The problem of coherence has also been addressed in teacher education research internationally that has to do with pre-service teachers experiencing a fragmentation in their studies to become teachers (Canrinus et al., 2017; Hermansen, 2020; Nordine et al., 2021). This has led to a discourse on the need for greater 'integration' in teacher education (Lehmann, 2020). This means first and foremost the integration between activities in school practicum and activities on campus in the university colleges to enhance the pre-service teachers' experience of connectivity and learning progression. A possible solution is to use more time in practicum or in other ways enable pre-service teachers to spend time in the 'real world' classrooms. This does not have to be the only solution to the problem of enactment, the theory-practice gap or the lack of coherence and integration in teacher education. We will argue that part of the solution is to develop new teacher education practices on campus using and cultivating the pre-service teachers' abilities to imagine different situations in classrooms and how to act upon them (Canrinus et al., 2019; Jay, 2023). We argue that this can be achieved through developing pedagogies of enactment. This implies on the one hand rehearsing core practices or teaching routines to become more familiar with them and ready to act in future similar situations. In other words, it will support the pre-service teachers' "science practice-readiness" (Bismack et al., 2022). On the other hand, giving rehearsals small variations or stimulating imagination, improvisation, and playfulness—like dealing with "what if"-questions—may also prepare pre-service teachers better for the unforeseeable and the uncertainties characterizing teaching. These approximations to teaching science in schools have some strengths regarding the enactment of components of practice as we will show in the following (Howell & Mikeska, 2021; Schutz et al., 2018). First, we will present a case showing how pre-service science teachers can engage in rehearsing and enacting modeling on primary school level.

11.4 Case: Pre-service Science Teachers Rehearsing and Enacting Modeling

The following case is inspired by an observation we made as part of our engagement in the national program *Danish Academy of Science Education* (https://nafa.nu/about-nafa/). The case concerns a group of Danish pre-service science teachers rehearsing and enacting modeling for children in 3rd and 4th grade in the subject Nature/Technology using self-made drawings (see also Auning & Auning, 2019; Joolingen et al., 2019). The pre-service teachers used drawings made by the children to model their understanding of what different human internal organs, such as the heart, look like. A central part of the pedagogy was for the pre-service teacher to have ongoing dialogues with the children. The pre-service teacher asked questions to understand the children's meaning making of e.g. the heart during the drawing process. Furthermore, the pre-service teacher took advantage of these dialogues to provide the children with feedback or assessment-on-the fly (see also Ruiz-Primo & Furtak, 2007). In their preparation the pre-service science teachers discussed how to plan the lesson and what pedagogical theory to base their interventions on. Part of the rehearsing of the lesson was to try to draw and get a feeling of how it is to engage in such an activity. It also included discussing possibilities of designing the lesson with peers and the science teacher educator. The pre-service teachers then enacted their plan in real classrooms. They took pictures of the drawings made by the children and they videotaped some conversations documenting how they asked the children questions and made them explain why they drew the internal organs in this way and how they could add new knowledge to the drawings. These video sequences were later used for analysis and discussion at campus reflecting on the pre-service teachers' pedagogical strategies.

11.5 Pedagogy of Enactment and Rehearsals

Ochieng and Zoest (2023) argue that: "Pedagogies of enactment are particular pedagogies used in teacher education that provide pre-service teachers with the opportunity to enact the practice of teaching and to learn from that enactment. Pedagogies of enactment are considered to be practice-based approaches to teacher education because they focus on pre-service teachers' learning of what teachers do" (p. 317). Pedagogies of enactment comprise things like microteaching, lesson study and rehearsals. In this chapter we focus solely on rehearsals.

According to Lampert et al. (2013) a pedagogy of enactment is focusing on certain *instructional activities* and can be perceived as containers for core practices that help to better prepare pre-service teachers for 'ambitious teaching', as they call it. Previous research on rehearsals provides insight about the decisions teacher educators make during rehearsals (Kazemi et al., 2016) and the interactions between pre-service teachers and teacher educators (Ghousseini, 2017) for example in the form of 'teacher time outs' (Wæge & Fauskanger, 2021). Additionally, studies on rehearsals have provided teacher educators with information about ways to improve the learning of pre-service teachers (Campbell & Elliott, 2015). Reviewing the literature Ochieng and Zoest (2023) point to a critical element of rehearsals when supporting pre-service teacher learning. One problem seems to be how the interaction plays out between teacher educators and pre-service teachers during the rehearsals. Ochieng and Zoest (2023) are primarily referring to how rehearsals are performed in mathematics teacher education, but the main ideas are also applicable to other subjects (Grossman, 2018) including rehearsals in science teacher education (Benedict-Chambers et al., 2020; Davis et al., 2017; Kloser et al., 2019; Lee et al., 2023; Masters, 2020). This is also a point made by Lampert et al. (2013, p. 227) that although they focus on the teaching of elementary mathematics the research has also "relevance for the design of teacher education more broadly". Lampert et al. (2013) call their framework for a pedagogy of enactment for the *Cycle of Enactment and Investigation* (the CEI-model). We will build on and expand this framework theoretically. Rehearsals are one of the stages in the cycle. The idea is that pre-service teachers do the same instructional activity across different settings.

11.5.1 Observation

The cycle begins with an *observation* of an enactment of a particular instructional activity from a classroom setting either live or on video (e.g. leading a group discussion in a science lesson). This observation can be informed by previous experiences the pre-service may have from their practicum, but it can also be a continuation of theoretical discussions on the course they participate in. In the case described above the observation could be in the form of previous experiences with teaching modeling in the subject Nature/Technology and using video recordings of the conversations with the children explaining their drawings of the human body. This observation sets the stage for the core practice in science teaching to be further investigated and discussed.

11.5.2 Collective Analysis

The second stage in the cycle is a *collective analysis* of the instructional activity the pre-service teachers have witnessed. This is a movement from watching the enactment of a teaching sequence to an investigation of it and makes it possible for the teacher educator to develop the professional noticing of the pre-service teachers by pointing out what may be of significance for the students learning particular science content (Baldinger & Munson, 2020; Benedict-Chambers & Aram, 2017; Chan et al., 2021; Luna et al., 2018). As illustrated in the case the form of talking about what the children understand as a model and how the use of their own drawings can help them to progress in their learning of the human body and how it functions. This is where the pre-service science teachers also give feedback to the children or assessment-on-the-fly.

11.5.3 Preparation

In the third stage of the cycle the pre-service teachers *prepare* for the rehearsal of the instructional activity in focus. This includes their choice of learning materials for the rehearsal. In the modeling case the learning materials in the form of drawings are central. The pre-service science teachers may use this preparation to find and collect the necessary materials to be able to carry out the lesson.

11.5.4 Rehearsal

The fourth stage is the actual *rehearsal* where selected pre-service teachers rehearse the instructional activity in front of their peers. The teacher educator acts as both a student and a coach and the other pre-service teachers who are not teaching play the role of students (Baldinger & Munson, 2020; Lampert et al., 2013). It is crucial to define and instruct how the pre-service teachers who play primary or secondary students—and function as a kind of audience in the rehearsal—should act and give feedback to the pre-service teacher rehearsing a core practice or competence. In the modeling case different moments in the lesson can be rehearsed e.g. in the form of how to introduce the activity for the children, motivating and encouraging them to make drawings and edit them as they get more knowledge on the actual details of the organs they are supposed to draw, how they can ask the children questions while they are modeling and so on.

We will expand this CEI-model by emphasizing its creative potential and the need of developing feedback literacy in the process. Lampert et al. (2013) stress the double purpose of the rehearsals. On the one hand, they are meant to give pre-service teachers the opportunity to learn important teaching routines. On the other hand, the rehearsals are designed to give the pre-service teachers opportunities to make good judgements of the instructional activity in focus when confronted with the unexpected. Building on research on expertise Lampert et al. (2013) argues for the use of 'deliberate practice' (Ericsson et al., 1993) and the development of 'adaptive expertise' (Hatano & Inagaki, 1984) creating a "balance between repeated practice to develop the efficiency of routines and the development of conceptual understanding to be able to innovate and adapt to new situations" (Lampert et al., 2013, p. 228).

In their study they identify four categories of roles played by the teacher educator during the rehearsal:

1. Providing directive feedback
2. Providing evaluative feedback
3. Scaffolding enactment (e.g. acting as a student)
4. Facilitating reflective discussion of instructional decisions within the rehearsal.

We will get back to the importance of feedback in the process of rehearsals.

11.5.5 Classroom Enactment Followed by Collective Analysis

In the framework presented by Lampert et al. (2013) the next stage of the cycle is a *classroom enactment* of the instructional activity followed by another *collective analysis*. This can be in a practicum setting where the pre-service teachers meet real children in school classrooms:

> Next, novices interact with students, doing the activity they have rehearsed and video recording their work. This is an opportunity for the novice to test in practice the results of the public and collaborative preparation that occurred in the rehearsal in a setting where many of the ways in which students participate were anticipated. (Lampert et al., 2013, p. 229).

The pre-service science teachers in the modeling case enacted their plan in a classroom setting and collected data from it. Through their enactment of the plan they gained new insights into what is possible to do with the children in practice. However, we will argue for the possibility of not necessarily doing the 'full circle' each time including the enactment in real classrooms. Resources can be scarce, making it difficult to visit schools in all cases of rehearsal although this may be the ideal to try out the instructional activities in real classroom settings. In some situations, the opportunity to rehearse the instructional activity only on campus and getting feedback from the teacher educator and the peers may be the only option. However, when it comes to both the enactment and the investigation or inquiry of the instructional activities it is crucial how the teacher educator manages it:

> Through teacher education programs, students of teaching commonly work in a context in which there is an expectation of learning through inquiry into practice. However, ensuring that such inquiry is meaningful and not just an assessment hurdle or program task is where the role of the teacher educator assumes even greater import (Loughran & Menter, 2019, p. 223).

As Loughran and Menter (2019) also point out, the quality of teacher education will be dependent on how teacher educators develop and learn in their work, strengthening their scholarship by critically interrogating and researching their own practice. In the following we will discuss how we can further develop the CEI-model focusing on creativity and feedback in the science teaching rehearsals.

11.6 Creative Pragmatics and Science Teaching Rehearsals

From a Creative Pragmatics approach, we suggest the concept of *creative rehearsals* in science teacher education, meaning the use and development of pedagogical imagination skills and the creation of scenarios simulating 'real' classroom issues or components of practice. This is especially thought of as an expansion and elaboration of the *improvisational* part of the rehearsals. As Kazemi et al. (2016, p. 18) say: "Rehearsals stimulate many different relationships between students and content that might emerge in teaching, requiring the rehearsing teacher to make both routine and improvisational decisions in practice".

As mentioned in this book's introduction, Creative Pragmatics is an approach "which involves iterative learning through practice, by doing, by designing, by making, by making art, by engineering knowledge, by collaborating, by observing and reflecting, through dialogue, participation and imagination" especially highlighting the importance of experimental enactment and feedback (Chap. 1, Sect. 1.5, "Creative Pragmatics Introdued"). This orientation fits well with the pedagogy of enactment described above. The selection of instructional activities—containing specific 'core practices' or 'areas of competence'—to rehearse in a science teaching course is a central aspect of the CEI-model. Selecting relatively easy routines to enact in a classroom might not invite the pre-service teachers to deep collective analysis and present different perspectives. But what might seem easy at first can also be found to be complex. "Keeping it complex" (Lampert et al., 2013) is a task for the science teacher educator orchestrating the rehearsals. The directive and evaluative feedback from the science teacher educator can open up more creative processes in the rehearsals. Creative rehearsals become guided improvisation in science teaching. Leading group discussions in classrooms about scientific issues or doing drawing-based modeling in primary science teaching, as in the case described above, are good examples of this. According to Sawyer (2021) a creative process has eight essential characteristics: (1) Iteration (2) Ambiguity (3) Exploration (4) Emergence (5) Failure and dead ends (6) It is deliberate and intentional (7) It involves conscious reflection and (8) Constraints are important. Also, building on pragmatism, Glăveanu & Beghetto (2021, p. 76) define creative experiences as: "novel person-world encounters grounded in meaningful actions and interactions, which are marked by the principles of: open-endedness, nonlinearity, pluri-perspectives and future-orientation". In a creative science teaching rehearsal, as we understand it, the rehearsing pre-service science teacher is challenged in different ways to judge how the instructional activity in question can be led in new directions if needed, because of changed circumstances. Ambiguity and dilemmas in science teaching may invite to discussions with different perspectives on the matter. Imagining different actions to take is one of the aspects of the creative rehearsal, but it is also important to point out the constraints and limitations in a scenario, since not all is possible, like time constraints or available learning materials in the classroom. In this regard, we find the so-called Perspective- Affordance Theory of Creativity developed by Glăveanu (2020) useful. According to the theory, differences of perspectives increase creative potential. Applied to the situation of rehearsals, the different perspectives on the science teaching component rehearsed may increase the creative potential of not only the pre-service teacher in focus, but also the rest of the group (the 'audience'). The exchange of perspectives between the participants fosters creative processes of the sort "what would happen if...?": "Differences of perspectives are particularly important in this regard given that perspectives are action orientations and, thus, being aware of or holding multiple perspectives means being able to potentially direct one's thinking and action in multiple, divergent ways" (Glăveanu, 2020, p. 343). Another part of the Perspective-Affordance Theory of Creativity is that the exchanges of perspectives can reveal previously unperceived affordances. "Affordances represent the meeting point between what is physically possible (material) and what the person perceives as possible (psychological), both shaped to a great extent by cultural norms and expectations (social)" as Glăveanu (2020, p. 342) explains it. Thinking Creative Pragmatics in this way, the science teacher educator can invite the peers of the rehearsing pre-service teacher to imagine different situations that could emerge in a classroom when students are engaged in the instructional activity e.g. doing modeling with children from 3rd grade. Not only as different responses from the children in the classroom, but also in the form of different uses of available resources, materials, artifacts, and spaces. This is also of importance in science subjects where learning materials, technologies and nature play a big role, since some of the rehearsals can be done outdoors and some in the science lab doing different kinds of experiments. These materials and spaces are some of the affordances pre-service science teachers may face and improvise on in their enactment of their prepared instructional activity.

11.7 Feedback Literacy and Science Teaching Rehearsals

Feedback is considered an important part of a practice-based teacher education in general as Matsumoto-Royo and Ramirez-Montoya (2021) also highlight in their literature review. Feedback plays an important role in rehearsals and deliberate practice. The idea with rehearsals is to promote deliberate practice, which is characterized by practicing and being given feedback from a more experienced practitioner (Ericsson et al., 1993). In our case the more experienced practitioner could either be the science teacher educator or peers of the rehearsing pre-service teacher. As mentioned, Lampert et al. (2013) argue for the importance of feedback as a part of their CEI-model where they distinguish between directive feedback and evaluative feedback. In the interactions between teacher educator and pre-service teachers the *directive feedback* can be in the form of "suggesting a next move or speculating possibilities for a next move" (p. 231) in the teaching rehearsal. Seen in the light of creative rehearsals, as described above, this is where the science teacher educator can stimulate the imagination and improvisation of the rehearsing pre-service teacher, by introducing new scenarios or constraints in the instructional activity. The science teacher educator can also give *evaluative feedback* like an evaluative comment highlighting "what was productive about a move or what could be improved on" (p. 231). In this case, the science teacher educator can also invite the peers to express their perspectives on the instructional activity in question. Peer feedback is not unproblematic and the possibilities and challenges in peer feedback activities are widely discussed (Mercader et al., 2020; Prilop & Weber, 2023; Sluijsmans et al., 2002; Wood, 2022). According to Lampert et al. (2013) this can lead to a discussion that the teacher educator facilitates.

The science teacher educator can *give* the rehearsing pre-service teacher direct or evaluative feedback, but how the pre-service science teacher in focus *receives, understands,* and *uses* the given feedback is also of great importance. The same thing can be said about the feedback given by the peers (or the 'audience'), but also their understanding of it as well as their reflections on giving and receiving feedback. Some researchers have therefore argued for the necessity of developing *student feedback literacy* (Carless & Boud, 2018). According to Carless and Boud (2018) feedback literacy consists of four interrelated features: (1) appreciating feedback; (2) making judgments; (3) managing affect; and (4) taking action. Sutton

(2012), who was one of the first to introduce the concept of feedback literacy, defines it in the following way:

> Becoming feedback literate is part of the process that enables learners to reach the standard of disciplinary knowledge indicated in module and programme learning outcomes, subject benchmarks etc.; that assists learners in forming judgements concerning what counts as valid knowledge within particular disciplines; and that helps them develop the ability to assess the quality of their own and others' work (Sutton, 2012, p. 33).

In a critical review of the conceptualizations of feedback literacy in the research literature Nieminen and Carless (2023) distinguish between two different approaches. The first approach understands the feedback literate student/teacher as a 'psychological capable individual'. The second approach understands the feedback literate student/teacher as a 'complex, contextual and social being'. In our case, we follow the latter understanding of feedback literacy that also is compatible with the Creative Pragmatics point of view. For example, the peers play the role of an audience or school children in the rehearsal and how they understand and give feedback is also significant for the creative process. According to Chong's (2021) ecological perspective on feedback literacy, the agency of learners should be included in the understanding of the concept. Chong distinguishes between three dimensions of feedback literacy: a contextual dimension, an engagement dimension and an individual dimension. The contextual dimension could in our case be the difference between giving, receiving and using feedback in the setting of 'coursework' on campus or 'fieldwork' in a classroom. This concerns the general 'problem of enactment' in teacher education as described in the introduction of this chapter. The sociocultural context forms the feedback and gives it meaning. The engagement dimension comprises the cognitive, affective and behavioral engagement of the pre-service science teacher receiving feedback from the science teacher educator and/or the peers in a rehearsal. As Chong (2021, p. 98) says: "With successful cognitive and affective engagements with feedback, feedback literate students can become behaviorally engaged with feedback by acting on feedback (e.g. revise their work or adjust learning strategies based on the feedback)". This raises the question of how creative rehearsals are contributing to engage the pre-service science teachers more in using the feedback given to them, or it may risk disengaging them if they feel negative attitudes towards the feedback. The individual dimension points to the differences between learners including differences in individual beliefs, goals, abilities, and feedback experience. For a science teacher educator to take this into account a good relationship, where the science teacher educator knows and appreciates the different backgrounds of the pre-service science teachers, will be of value when giving feedback to them. The same can be said about the relations between pre-service teachers on the science teaching course. To sum up, focusing on feedback literacy as part of the CEI-model the directive and evaluative feedback of the science teacher educator needs to be expanded further to understand the pedagogy of enactment and what is at stake here. The concept of feedback literacy makes us aware of the agency of the pre-service science teachers participating in the rehearsal

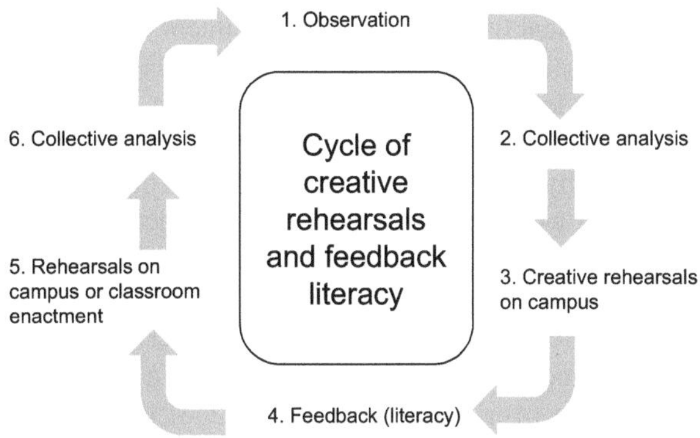

Fig. 11.1 Pedagogy of enactment: cycle of creative rehearsals and feedback literacy (developed with inspiration from Lampert et al., 2013)

and how they understand and make use of the feedback given by their science teacher educator and peers.

We exemplify the theoretical contributions of this article through the model presented in Fig. 11.1, elucidating how rehearsals can effectively occur on campus (or as real classroom enactment). Moreover, it extends the existing framework established by Lampert et al. in 2013, underscoring the critical significance of cultivating feedback literacy:

The model aims to provide a scaffold for both science teacher educators and their students, guiding and enhancing the instructional process.

11.8 Conclusion

In this chapter we have argued for an integration of different theories to form a pedagogy of enactment in science teacher education that is aware of both the importance of creativity, imagination and improvisation in teaching and the importance of giving, receiving and using feedback on such rehearsals. Our contribution to the field is therefore an argument for a theoretical synthesis that may be tested empirically and in practice be refined and nuanced further. Pedagogies of enactment are part of a broader movement arguing for a more practice-based or practice-oriented teacher education. This is an old issue in the history of (science) teacher education. From a critical point of view the risk of focusing too much on 'getting closer to practice' is ending in a 'tyranny of practice' (Imsen, 2012, pp. 112–118) reducing teacher education to be merely instrumental and downplaying the role of theory, research knowledge and reflection. In the circle of enactment and investigation proposed by Lampert et al. (2013) a balance is created between theory and practice by emphasizing the importance of inquiry and collective analysis of the rehearsals. We argue for doing more research on how the rehearsals can be further qualified in science teacher education using a Creative Pragmatics approach. Creative Pragmatics can support the development of a pedagogy of enactment and active learning in the teaching of pre-service science teachers. It offers a vocabulary for science teacher educators to design or redesign the way they arrange science teaching rehearsals on campus (see also Glăveanu et al., 2016). Although we argue that rehearsals can be fruitful without necessarily including an actual enactment of the instructional activities in real classrooms it would not be without consequences to decouple these two settings completely. Ideally, it would be preferable to do the 'full circle' by doing both rehearsals on campus and enact the rehearsed instructional activity in actual classrooms as also described in the framework by Lampert et al. (2013). An example of doing this in a partnership with schools is as part of a 'school adoption' where pre-service teachers 'take over' the lessons in a school in a short period of time teaching students something they have rehearsed on campus (Janssen & Wiedenhorn, 2021). Matsumoto-Royo and Ramirez-Montoya (2021) point in their literature review to the necessity of establishing organizational routines when developing a pedagogy of enactment (Feldman & Pentland, 2003; Kiær & Albrechtsen, 2024). When implementing a pedagogy of enactment in partnerships with schools we must pay attention to two central things. First, it is necessary to describe the new routine of this pedagogy. Second, it needs to be investigated as well as evaluated how this pedagogy is enacted in various practices paying certain attention to the specific artifacts used in these routines. This will provide science teacher education with knowledge on the dynamics in this new pedagogy of enactment which furthermore could contribute with knowledge on how to adjust the new routine as it emerges in science teaching practices at university colleges.

Acknowledgements This study was funded by Novo Nordisk Fonden and VILLUM Fonden, grant number LAVIN-18-6.

References

Alexander, R. (2009). Pedagogy, culture and the power of comparison. In H. Daniels, H. Lauder, & J. Porter (Eds.), *Educational theories, cultures and learning: A critical perspective* (pp. 10–26). Routledge.

Auning, C., & Auning, M. (2019). Students' explanations of a complex natural phenomenon using mathematical modelling as a design feature in a model-based inquiry unit. *Nordic Studies in Science Education, 19*(1), 62–77. https://doi.org/10.5617/nordina.8965

Baldinger, E. E., & Munson, J. (2020). Developing adaptive expertise in the wake of rehearsals: An emergent model of the debrief discussions of non-rehearsing teachers. *Teaching and Teacher Education*, 95. https://doi.org/10.1016/j.tate.2020.103125

Ball, D. L., & Cohen, D. K. (1999). Developing practice, developing practitioners: Toward a practice-based theory of professional education. In L. Darling-Hammond & G. Sykes (Eds.), *Teaching as the learning profession* (pp. 3–31). Jossey-Bass.

Benedict-Chambers, A., & Aram, R. (2017). Tools for teacher noticing: Helping preservice teachers notice and analyze student thinking and scientific practice use. *Journal of Science Teacher Education, 28*(3), 294–318. https://doi.org/10.1080/1046560X.2017.1302730

Benedict-Chambers, A., Fick, J. S., & Arias, M. A. (2020). Preservice teachers' noticing of instances for revision during rehearsals: A comparison across three university contexts. *Journal of Science Teacher Education, 31*(4), 435–459. https://doi.org/10.1080/1046560X.2020.1715554

Bismack, S. A., Davis, A. E., & Palincsar, S. A. (2022). Science practice-readiness: Novice elementary teachers' developing knowledge of science practices. *Science Education, 106*(2), 364–384. https://doi.org/10.1002/sce.21698

Børte, K., Nesje, K., & Lillejord, S. (2023). Barriers to student active learning in higher education. *Teaching in Higher Education, 28*(3), 597–615. https://doi.org/10.1080/13562517.2020.1839746

Campbell, M., & Elliott, R. (2015). Designing approximations of practice and Conceptualising responsive and practice-focused secondary mathematics teacher education. *Mathematics Teacher Education and Development, 17*(2), 146–164.

Canrinus, T. E., Bergem, K. O., Klette, K., & Hammerness, K. (2017). Coherent teacher education programmes: Taking a student perspective. *Journal of Curriculum Studies, 49*(3), 313–333. https://doi.org/10.1080/00220272.2015.1124145

Canrinus, T. E., Klette, K., Hammerness, K., & Bergem, K. O. (2019). Opportunities to enact practice in campus courses: Taking a student perspective. *Teachers and Teaching, 25*(1), 110–124. https://doi.org/10.1080/13540602.2018.1526171

Carless, D., & Boud, D. (2018). The development of student feedback literacy: Enabling uptake of feedback. *Assessment & Evaluation in Higher Education, 43*(8), 1315–1325. https://doi.org/10.1080/02602938.2018.1463354

Chan, H. K. K., Xu, L., Cooper, R., Berry, A., & Driel, V. H. J. (2021). Teacher noticing in science education: Do you see what I see? *Studies in Science Education, 57*(1), 1–44. https://doi.org/10.1080/03057267.2020.1755803

Chong, W. S. (2021). Reconsidering student feedback literacy from an ecological perspective. *Assessment & Evaluation in Higher Education, 46*(1), 92–104. https://doi.org/10.1080/02602938.2020.1730765

Coetzee, C., Rollnick, M., & Gaigher, E. (2022). Teaching electromagnetism for the first time: A case study of pre-service science teachers' enacted pedagogical content knowledge. *Research in Science Education, 52*(1), 357–378. https://doi.org/10.1007/s11165-020-09948-4

Darling-Hammond, L. (2006). *Powerful teacher education: Lessons from exemplary programs.* Jossey-Bass.

Darling-Hammond, L., Hyler, M. E., Gardner, M., & Espinoza, D. (2017). *Effective teacher professional development.* Learning Policy Institute.

Davis, A. E., Kloser, M., Wells, A., Windschitl, M., Carlson, J., & Marino, J.-C. (2017). Teaching the practice of leading sense-making discussions in science: Science teacher educators using rehearsals. *Journal of Science Teacher Education, 28*(3), 275–293. https://doi.org/10.1080/1046560X.2017.1302729

Ericsson, K. A., Krampe, R. T., & Tesch-Römer, C. (1993). The role of deliberate practice in the Acquisition of Expert Performance. *Psychological Review, 100*(3), 363–406.

Feldman, M. S., & Pentland, B. T. (2003). Reconceptualizing organizational routines as a source of flexibility and change. *Administrative Science Quarterly, 48*, 94–118.

Firetto, M. C., Starrett, E., & Jordan, E. M. (2023). Embracing a culture of talk: STEM teachers' engagement in small-group discussions about photovoltaics. *International Journal of STEM Education, 10*(1). https://doi.org/10.1186/s40594-023-00442-7

Forzani, M. F. (2014). Understanding "Core practices" and "practice-based" teacher education. *Journal of Teacher Education, 65*(4), 357–368. https://doi.org/10.1177/0022487114533800

Ghousseini, H. (2017). Rehearsals of teaching and opportunities to learn mathematical knowledge for teaching. *Cognition and Instruction, 35*(3), 188–211. https://doi.org/10.1177/1089268020961763

Glăveanu, P. V. (2020). A sociocultural theory of creativity: Bridging the social, the material, and the psychological. *Review of General Psychology, 24*(4), 335–354. https://doi.org/10.1177/1089268020961763

Glăveanu, P. V., & Beghetto, A. R. (2021). Creative experience: A non-standard definition of creativity. *Creativity Research Journal, 33*(2), 75–80. https://doi.org/10.1080/10400419.2020.1827606

Glăveanu, P. V., Tanggaard, L., & Wegener, C. (Eds.). (2016). *Creativity—A new vocabulary*. Palgrave Macmillan.

Goodwin, L. A., Smith, L., Souto-Manning, M., Cheruvu, R., Tan, Y. M., Reed, R., & Taveras, L. (2014). What should teacher educators know and be able to do? Perspectives from practicing teacher educators. *Journal of Teacher Education, 65*(4), 284–302. https://doi.org/10.1177/0022487114535266

Grossman, P. (2018). *Teaching core practices in teacher education*. Harvard Education Press.

Grossman, P., & Mcdonald, M. (2008). Back to the future: Directions for research in teaching and teacher education. *American Educational Research Journal, 45*(1), 184–205. https://doi.org/10.3102/0002831207312906

Grossman, P., Compton, C., Igra, D., Ronfeldt, M., Shahan, E., & Williamson, W. P. (2009). Teaching practice: A cross-professional perspective. *Teachers College Record: The Voice of Scholarship in Education, 111*(9), 2055–2100. https://doi.org/10.1177/016146810911100905

Hammerness, K., Klette, K., Jenset, I. S., & Canrinus, E. T. (2020). Opportunities to study, practice, and rehearse teaching in teacher preparation: An international perspective. *Teachers College Record, 122*, 1–46. https://doi.org/10.1177/016146812012201108

Hartikainen, S., Rintala, H., Pylväs, L., & Nokelainen, P. (2019). The concept of active learning and the measurement of learning outcomes: A review of research in engineering higher education. *Education Sciences, 9*(4), 276. https://doi.org/10.3390/educsci9040276

Hatano, G., & Inagaki, K. (1984). Two courses of expertise. In H. W. Stevenson & H. Azuma (Eds.), *Child development and education in Japan* (pp. 27–36). W.H. Freeman &.

Hermansen, H. (2020). In pursuit of coherence: Aligning program development in teacher education with institutional practices. *Scandinavian Journal of Educational Research, 64*(6), 936–952. https://doi.org/10.1080/00313831.2019.1639815

Howell, H., & Mikeska, N. J. (2021). Approximations of practice as a framework for understanding authenticity in simulations of teaching. *Journal of Research on Technology in Education, 53*(1), 8–20. https://doi.org/10.1080/15391523.2020.1809033

Hume, A., Cooper, R., & Borowski, A. (Eds.). (2019). *Respositioning pedagogical content knowledge in teachers' knowledge for teaching science*. Springer.

Imsen, G. (2012). *Hvad er pædagogik?* [What is Pedagogy?]. Akademisk Forlag.

Janssen, M., & Wiedenhorn, T. (Eds.). (2021). *School adoption in teacher education: Increasing pre-service teachers' responsibility during practice*. Waxmann.

Jay, P. L. (2023). Constructing imaginary classrooms: Teacher educators' use of representations to direct reflection about practice. *Teaching and Teacher Education, 132*, 104243. https://doi.org/10.1016/j.tate.2023.104243

Joolingen, W. R., Schouten, J., & Leenaars, F. (2019). Drawing-based modeling in teaching elementary biology as a diagnostic tool. In A. U. Belzen, D. Krüger, & J. Driel (Eds.), *Towards a competence-based view on models and modeling in science education* (pp. 131–146). Springer.

Kademian, M. S., & Davis, A. E. (2018). Supporting beginning teacher planning of investigation-based science discussions. *Journal of Science Teacher Education, 29*(8), 712–740. https://doi.org/10.1080/1046560X.2018.1504266

Kazemi, E., Ghousseini, H., Cunard, A., & Turrou, A. C. (2016). Getting inside rehearsals: Insights from teacher educators to support work on complex. *Journal of Teacher Education, 67*(1), 18–31.

Kiær, K. & Albrechtsen, T. R. S. (2024). *Dynamiske rutiner. Hvordan forandringer får fodfæste i organisationer.* [Routine Dynamics. How Changes Take Root in Organizations]. Dansk Psykologisk Forlag.

Kloser, M., Wilsey, M., Madkins, C. T., & Windschitl, M. (2019). Connecting the dots: Secondary science teacher candidates' uptake of the core practice of facilitating sensemaking discussions from teacher education experiences. *Teaching and Teacher Education, 80*, 115–127. https://doi.org/10.1016/j.tate.2019.01.006

Lampert, M., Franke, M. L., Kazemi, E., Ghousseini, H., Turrou, A. C., Beasley, H., Cunard, A., & Crowe, K. (2013). Keeping it complex: Using rehearsals to support novice teacher learning of ambitious teaching. *Journal of Teacher Education, 64*(3), 226–243. https://doi.org/10.1177/0022487112473837

Lee, D. T., Lee, C., Newton, M., Vos, P., Gallagher, J., Dickerson, D., & Regenthal, C. (2023). Peer to Peer vs. Virtual rehearsal simulation rehearsal contexts: Elementary teacher candidates' scientific discourse skills explored. *Journal of Science Teacher Education*, 1–22. https://doi.org/10.1080/1046560X.2023.2181505

Lehmann, T. (Ed.). (2020). *International perspectives on knowledge integration: Theory, research, and good practice in pre-servie teacher and higher education*. Brill Sense.

Loughran, J. (2014). Professionally developing as a teacher educator. *Journal of Teacher Education, 65*(4), 271–283. https://doi.org/10.1177/0022487114533386

Loughran, J., & Menter, I. (2019). The essence of being a teacher educator and why it matters. *Asia-Pacific Journal of Teacher Education, 47*(3), 216–229. https://doi.org/10.1080/1359866X.2019.1575946

Luna, J. M., Selmer, J. S., & Rye, A. J. (2018). Teachers' noticing of students' thinking in science through classroom artifacts: In what ways are science and engineering practices evident? *Journal of Science Teacher Education, 29*(2), 148–172. https://doi.org/10.1080/1046560X.2018.1427418

Masters, H. (2020). Using teaching rehearsals to prepare preservice teachers for explanation-driven science instruction. *Journal of Science Teacher Education, 31*(4), 414–434. https://doi.org/10.1080/1046560X.2020.1712047

Matsumoto-Royo, K., & Ramirez-Montoya, M. S. (2021). Core practices in practice-based teacher education: A systematic literature review of its teaching and assessment process. *Studies in Educational Evaluation, 79*. https://doi.org/10.1016/j.stueduc.2021.101047

Mercader, C., Ion, G., & Díaz-Vicario, A. (2020). Factors influencing students' peer feedback uptake: Instructional design matters. *Assessment & Evaluation in Higher Education, 45*(8), 1169–1180. https://doi.org/10.1080/02602938.2020.1726283

Mintzes, J. J., & Walter, E. M. (Eds.). (2020). *Active learning in college science: The case for evidence-based practice*. Springer.

Mork, S. M., Henriksen, E. K., Haug, B. S., Jorde, D., & Frøyland, M. (2021). Defining knowledge domains for science teacher educators. *International Journal of Science Education, 43*(18), 3018–3034. https://doi.org/10.1080/09500693.2021.2006819

Murray, J., & Male, T. (2005). Becoming a teacher educator: Evidence from the field. *Teaching and Teacher Education, 21*(2), 125–142. https://doi.org/10.1016/j.tate.2004.12.006

Niemi, H. (2002). Active learning—A cultural change needed in teacher education and schools. *Teaching and Teacher Education, 18*(7), 763–780. https://doi.org/10.1016/S0742-051X(02)00042-2

Niemi, H., Nevgi, A., & Aksit, F. (2016). Active learning promoting student teachers' professional competences in Finland and Turkey. *European Journal of Teacher Education, 39*(4), 471–490. https://doi.org/10.1080/02619768.2016.1212835

Nieminen, H. J., & Carless, D. (2023). Feedback literacy: A critical review of an emerging concept. *Higher Education, 85*(6), 1381–1400. https://doi.org/10.1007/s10734-022-00895-9

Nordine, J., Sorge, S., Delen, I., Evans, R., Juuti, K., Lavonen, J., Nilsson, P., Ropohl, M., & Stadler, M. (2021). Promoting coherent science instruction through coherent science teacher education: A model framework for program design. *Journal of Science Teacher Education, 32*(8), 911–933. https://doi.org/10.1080/1046560X.2021.1902631

Ochieng, A. M., & Zoest, V. R. L. (2023). How preservice teachers learn through a pedagogy of enactment in a middle school mathematics methods course. *Cognition and Instruction, 41*(3), 316–347. https://doi.org/10.1080/07370008.2022.2129636

Peercy, M. M., & Troyan, J. F. (2017). Making transparent the challenges of developing a practice-based pedagogy of teacher education. *Teaching and Teacher Education, 61*, 26–36. https://doi.org/10.1016/j.tate.2016.10.005

Prilop, N. C., & Weber, E. K. (2023). Digital video-based peer feedback training: The effect of expert feedback on pre-service teachers' peer feedback beliefs and peer feedback quality. *Teaching and Teacher Education, 127*, 104099. https://doi.org/10.1016/j.tate.2023.104099

Rothland, M. (2020). Legenden der Lehrerbildung: Zur Diskussion einheitsstiftender Vermittlung von 'Theorie' und 'Praxis' im Studium. *Zeitschrift für Pädagogik, 66*(2), 270–287.

Ruiz-Primo, M. A., & Furtak, E. M. (2007). Exploring teachers' informal formative assessment practices and students' understanding in the context of scientific inquiry. *Journal of Research in Science Teaching, 44*(1), 57–84. https://doi.org/10.1002/tea.20163

Russell, T., & Loughran, J. (Eds.). (2007). *Enacting a pedagogy of teacher education: Values, relationships and practices*. Routledge.

Sawyer, K. R. (2021). The iterative and improvisational nature of the creative process. *Journal of Creativity, 31*. https://doi.org/10.1016/j.yjoc.2021.100002

Schutz, K. M., Grossman, P., & Shaughnessy, M. (2018). Approximations of practice in teacher education. In P. Grossman (Ed.), *Teaching core practices in teacher education* (pp. 85–106). Harvard Education Press.

Sluijsmans, A. M. D., Brand-Gruwel, S., & Merriënboer, V. G. J. J. (2002). Peer assessment training in teacher education: Effects on performance and perceptions. *Assessment & Evaluation in Higher Education, 27*(5), 443–454. https://doi.org/10.1080/0260293022000009311

Sutton, P. (2012). Conceptualizing feedback literacy: Knowing, being, and acting. *Innovations in Education and Teaching International, 49*(1), 31–40. https://doi.org/10.1080/14703297.2012.647781

Thomas, G. (2021). *How to do your case study* (3rd ed.). Sage.

Virtanen, P., Niemi, H., & Nevgi, A. (2017). Active learning and self-regulation enhance student teachers' professional competences. *Australian Journal of Teacher Education*, 1–20. https://doi.org/10.14221/ajte.2017v42n12.1

Wæge, K., & Fauskanger, J. (2021). Teacher time outs in rehearsals: In-service teachers learning ambitious mathematics teaching practices. *Journal of Mathematics Teacher Education, 24*(6), 563–586. https://doi.org/10.1007/s10857-020-09474-0

Windschitl, A. M., & Stroupe, D. (2017). The three-story challenge. *Journal of Teacher Education, 68*(3), 251–261. https://doi.org/10.1177/0022487117696278

Wood, J. (2022). Making peer feedback work: The contribution of technology-mediated dialogic peer feedback to feedback uptake and literacy. *Assessment & Evaluation in Higher Education, 47*(3), 327–346. https://doi.org/10.1080/02602938.2021.1914544

Zeichner, K. (2012). The turn once again toward practice-based teacher education. *Journal of Teacher Education, 63*(5), 376–382. https://doi.org/10.1177/0022487112445789

Thomas R. S. Albrechtsen is a docent, PhD and leader of the Research Program on School and Teaching at the University College of Southern Denmark (UCSYD) and he is currently co-leading a longitudinal study on the Danish Academy of Science Education (NAFA). His main research interests include professional learning communities, science teacher education, teacher professional development, Routine Dynamics and Organizational Education.

Karina Kiær, PhD, is an associate professor at the University College of Southern Denmark (UCSYD) and affiliated with the University of Southern Denmark (SDU), and she is currently doing a longitudinal ethnographic study on knowledge ecology in science teacher education (NAFA). Her main research interests include Routine Dynamics, sociomaterial theory, Organizational Education, science teacher education, literacy coaching and data use.

Chapter 12
Innovating Science Education at a National Level

Gitte Miller Balslev

12.1 Understanding Education as a Complex Societal Phenomenon

On a broad and national scale, the Danish Academy of Science Education (NAFA) program can be understood in prolongation of the Danish National Science Strategy in education from 2018 (The Danish Government, 2018). This strategy envisions enhanced focus on motivating science teaching throughout the Danish educational system, improved capacity building of science teachers at all educational levels strengthening teaching skills and science didactics, renewing science courses and teaching material from a basis of newest and cutting-edge findings in research, science, and technology, besides network building and collaboration in local networks to unfold these key action areas (ibid.). The NAFA program is responding to these specific tasks, but also to more general agendas in the realm of education. Although the overarching and general objective of education seems clear—developing future

A Creative Pragmatics approach is not restricted to the classroom but can be applied broadly at all levels of learning. Gitte is a researcher-cum-research-administrator, and as the former head of Denmark's Center of Excellence in Science Education (CESE), she reflects on challenges at a national level. Creative Pragmatics is about navigating complex adaptive living systems. Gitte's is a case-study in policy, leadership, team dynamics, and intervention in large-scale complex systems that encompass science education and schooling. She advocates an holistic ecosystemic perspective working across administrative silos through abductive sensemaking applied iteratively, adaptively and collaboratively, especially in bridging large-scale and local initiatives, connecting science and society. (Editors)

G. M. Balslev (✉)
STEM Centre for Research in Science Education and Communication (FNUG), Department of Mathematics and Computer Science, University of Southern Denmark, Odense, Denmark
e-mail: gm@sdu.dk; https://findresearchersdu.elsevierpure.com/en/persons/gm

C. Svabo et al. (eds.), *Creative Pragmatics for Active Learning in STEM Education*, Contributions from Science Education Research 14,
https://doi.org/10.1007/978-3-031-78720-1_12

generations—the actual process can prove elusive and contingent, varying within different local educational frameworks. The complexity arises from multifaceted interactions spanning diverse boundaries, including sectors, institutions, cultures, and administrative domains. Research on collaborative innovation emphasises the value of transversal network-based cooperation as an effective approach to tackling and navigating intricate issues within the public sector in general (Aagaard et al., 2014) and in education specifically (Balslev, 2012, 2021; Daly, 2010; Díaz-Gibson et al., 2014; Díaz-Gibson et al., 2021). Current educational research highlights that educational development can be enhanced through learning communities (Albrechtsen et al., 2022) and understood as ecosystems rather than isolated institutions working in silos: one teacher, one subject, one class etc. Understanding education as ecosystems is far from new—as mentioned in Chap. 3 of this volume. Currently, much effort is put into transforming education in response to globalised cultures, economies and politics, and a networked, community based and systemic methodology holds a promising transformational approach (Díaz-Gibson et al., 2021; Kemmis et al., 2014).

To apprehend and eventually transform education at a large-scale, attention must be given to the understanding of key components within education—human behaviour. That is, behaviour of central stakeholders such as students, parents, teachers, managers, politicians etc. To understand human behaviour, individuals can be viewed as part of an ecological complex system (Bronfenbrenner & Ceci, 1994). With the metaphor of an ecosystem, Bronfenbrenner argues that human development is affected by the conditions to which the individual is exposed, and thus the interaction with other people and with the environment. In other words, human development always takes place in relation to and under the influence of the social circumstances and the context of which the individual is a part. The influences from the different system levels constantly affect human development, and results in a dynamic development model. Bronfenbrenner's ecological systems theory adopts a systemic way of understanding human development. Thus, the individual can be perceived as a system, and likewise can the social systems the individual is a part of. Both the individual and the social systems are in constant development and mirror themselves on multiple levels. Consequently, the systems understanding of education can be drivers for new orientations in education and change.

Bronfenbrenner's ecology model for human development (1986) consists of several systems. E.g., the Microsystem is closest to the child, the Mesosystem here representing the school, the Macrosystem the cultural and societal dimension. Bronfenbrenner's model is developed for analysis of children's learning and development and has inspired educational theory and educational change on a larger scale. In this chapter, the model will be used as inspiration to analyse the NAFA ecosystem and the potentials for educational change. According to Kemmis et al. (2014), schools as social forms are fixed in the nineteenth century. This static situation calls for a new balancing between reproduction of tradition and transformation into future education. Lifting schools and schooling out of fixed positions and into the twenty-first century requires transformation of educational practices (Kemmis et al., 2014). With Levin and Fullan's words, we must take a whole system approach

(Levin & Fullan, 2008) in the transformation endeavours and involve as many system levels as possible. Successful quality development and innovation occur when collaboration throughout the ecosystem—as well as within the management chain between municipality and institutions—is improved (Albret, 2010). When entire systems work together to common advantage, innovative results are more likely to occur (cf. Edquist, 2010; Van de Ven et al., 2008). However, collaboration across sectors, professional boundaries, and levels of governance is a challenge for public sector development (Balslev, 2012). Therefore, it is of the utmost importance that the collaboration is managed, facilitated, and aimed at common solutions for the core task of education. Hence, combined with Kemmis et al. (2014) transformation of education at different scales involves considering education more holistically as a complex ecosystem of real-world pragmatic practices.

Bronfenbrenner (Bronfenbrenner & Morris, 1998) proposed that to comprehend schools and learning with practical real-world applicability, rather than just in theoretical settings, it is essential to study the various interconnected subsystems that ecologically influence both youth, schools, and society as integral units. This perspective on educational innovation is increasingly adopting a relational systemic approach, recognising that each educational system operates within the context of near surroundings as well as the context of societal institutional settings. As a result, the emphasis on social relationships and sensemaking becomes crucial for school leaders to effectively mobilise resources in local communities as well as across schools and educational systems, drawing upon the concept of sensemaking and social capital.

Sensemaking is a central part of constructivist organisational literature, and is of pronounced importance in change processes (Ring & Van de Ven, 1994, p. 98; Van de Ven et al., 2008)—it precedes the decision-making process. Sensemaking initially involves an individual thought process where something surprising is attempted to be explained. In this understanding, meaning is created retrospectively based on a surprising event, aiming to make sense of the surprise and find plausible explanations and possible solutions. In this way, sensemaking helps provide structure to what we do not yet know or understand (Weick, 1995, p. 4). When a routine is interrupted by a surprise, the sensemaking process commences. Therefore, the study of sensemaking can shed light on how people react to disruptions, making the sensemaking process important when studying innovation in an organisational context. Since sensemaking contributes to the creation of meaning retrospectively based on a surprising event and gives meaning to what we do not yet know, it is helpful to draw parallels to the logical form abduction. Abduction, as first operated with by Aristotle and later Charles Sanders Peirce (1839–1914), is a way of understanding existence (Balslev & Jakobsen, 2000; Peirce, 1994). Abduction can be explained as "the expected" or "the hypothesis." Abduction is a pragmatic way of reasoning where an attentive observation of the surrounding world occurs based on an incomplete picture, from which people create meaning through a spontaneous confirmation or refutation of what they have previously experienced. In other words, it suddenly becomes clear what is happening without further contemplation. Abduction is an important type of reasoning in sensemaking, where people can see meaningful new connections and question what is taken for granted. In this way, abduction has an innovative property that is important for human perception of the surrounding world, particularly in an innovation context, where the goal is to establish and implement new understandings of a given phenomenon. Thus, abduction can be understood as a learning process and a pragmatic contribution to sensemaking, and sensemaking as a driving force when the goal is innovation—here of science education—in an organisational context at different scales.

The organisational context builds on social interactions of the people within and beyond organisational boundaries. Hence, sensemaking becomes a social activity where people discuss and create meaning. To this end, the ability of boundary spanning and establishing supporting structures for this (Balslev, 2012; Miller, 2009), becomes central. Here people work across institutions as flexible organisational navigators (ibid. p. 624), negotiating and creating meaning. Thus, using their social capital to move around is important. Social capital is generally defined as the resources embedded within social networks, encompassing resources that can be accessed or utilised throughout the network through intentional actions (Lin, 1999), and furthermore social capital constitutes high performing teams (Katzenbach & Smith, 2015). Theorists like Bourdieu and Richardson (1986) and Putnam (2001) agree that social capital resides in the resources found within relationships, rather than solely in the possession of individuals. These scholars underscore the importance of social networks in the concept of social capital. Social network theory is built on the idea that social resources, such as knowledge, information, and expertise, are exchanged through informal networks of relations between stakeholders in a system. A key aspect of this theory is the pattern of social ties among stakeholders, which collectively create an overall social structure (Scott, 2011). Accordingly, stakeholders with more connections are better positioned to efficiently distribute resources across the network due to their extensive links to many other stakeholders (Daly, 2010). On the other hand, stakeholders with fewer or no relational ties may have limited access to essential information and may not be as effective in disseminating information because their communication channels are less interconnected. Therefore, leaders must gain a clear understanding of the institutions' social networks and take a holistic approach to better grasp how resources flow within the broader ecosystem and to foster greater opportunities for all stakeholders. In prolongation, navigating complexity necessitates the cultivation of creative competencies, which include assessment of situations, the application of experiential knowledge, and effective communication with stakeholders (Zhou, 2016).

Summing up, the NAFA program can be understood as complex network-based innovation in public education to foster active learning and to motivate students and hinder dropout from science education and foster active STEM learning contexts, enabling students to learn to deal with real-world challenges effectively (Felder & Brent, 2016; Penprase, 2020). Presently the NAFA program is in the initiation phase (Fullan & Mundial, 1989; Fullan, 2007) and will be moving towards implementation and institutionalisation as the years go by. This chapter analyses the innovation of a NAFA as a national educational ecosystem crossing sectors and institutional borders and adopts a eco-systemic and whole system approach on science education for this purpose. The study focuses on the interplay of subsystems and social structures within the larger system on the national scale of the NAFA programme. Furthermore, this chapter will examine drivers and barriers for the NAFA national program leadership group to facilitate systemic educational change on a national scale, and how they juggle hopes and fears of change from multiple educational levels and organisational perspectives. After going through this study's research method, I will analyse how the leadership group navigates and struggles to create meaning of different elements of the program during the program-realisation process in the initiation phase.

12.2 Autoethnography as a Research Method

The empirical findings reported in this chapter are based on a qualitative case study of the NAFA program in its initiation phase, while I was the head of research in NAFA. The case-situations are carefully picked out to frame the NAFA program as an example of an attempt to innovate science education at a national level. This agenda on large-scale educational innovation matters especially in current times, where a rapidly changing society calls to prepare future citizens with the required knowledge, skills, and competencies to navigate increasing complexity. Innovating education is vital for achieving this goal, and this specific case focusses on pragmatic means of managing educational innovation as well as the formulation and implementation of strategy in this respect, as touched upon in Chap. 1. Even though NAFA is designed for science education exclusively, generic organisational and sector development is at its core. Therefore, NAFA can serve as an inspirational and relevant case of large scale educational development in general.

The aim of the empirical analysis is to explore the drivers and barriers for the national leadership group's facilitation of the NAFA program during the first years of initiation. The qualitative study was designed as a case study following Yin (2009). Robert Yin asserts that a case study is an empirical investigation of a contemporary phenomenon, where the goal is to study the phenomenon within its own context, where the boundaries between the phenomenon and context are not entirely clear, and where a multitude of sources are involved (ibid). In other words, the method suggests that the researcher should be close to the phenomenon studied (Maaløe, 1996). In this regard, autoethnography is a suited method to study a phenomenon close to the researcher (Alvesson, 2003), as is the case, where I—as researcher—am part of the investigated educational field. Thus, I draw on autoethnographic educational research allowing research application of existing field-data (Fujii, 2015; Levitan et al., 2017). The autoethnographic method allows the researcher-practitioner to utilise prior fieldwork through systematic analysis. In other words, using one's own work life experience pragmatically as research-data (Balslev & Raae, 2019). Furthermore, this case study draws on ethnography designed specifically for educational studies (Levitan et al., 2017), including a qualitative, retrospective analytic inquiry perspective (Brinkmann & Tanggaard, 2010). The data has been generated based on events occurring during the first years of the NAFA-programme from October 2021 to October 2023. I build the analysis on the three critical case-situations that are carefully picked out to frame the NAFA program as an example of an attempt to innovate science education at a national level. The data material consists of documents and logbooks as well as personal notes, thus founding my ethnographic work on first-hand experience. Using first-hand experience is an asset which the interview method may lack (Alvesson, 2003; Harrison, 2018). That being said, it could be interesting in the future to proceed with a mixed methods study, broadening the data field and supplementing the first-hand data.

I have used a thematic approach to analyse the data material, drawing inspiration from Braun and Clarke's method of thematic analysis in qualitative research (Braun & Clarke, 2006, p. 86–93). The process began with sorting all data material to ensure a comprehensive understanding of the data. Next, I identified overarching themes to explore within the material. In my analysis, I followed a holistic perspective, taking a whole system approach that treated the entire material as a unified entity, focusing on the interconnectedness of overarching themes and their possibility for opening the materials' potential for guiding further systemic change of science education on a national scale. For this chapter, I selected a central overarching theme:

- Drivers and barriers for the NAFA leadership group to facilitate systemic change across sectors and institutional borders—realising ideals/models of science education in primary and lower secondary education as well as in teacher education. And, more broadly in terms of systemic change in the understanding of sense-making in organisations at different scales—micro, meso, macro.

As a part of my autoethnographic inquiry, I have chosen to bring forward central narratives serving as critical case-situations pointing towards specific drivers and barriers for leadership and exemplifying the complexity of innovation leadership in the national leadership group.

The data analysis has been validated in collaboration with former colleagues in the NAFA leadership group. Thus, my former colleagues participated as co-researchers, providing critical and constructive feedback on the initial analytical findings. This open-ended dialogue was intentionally designed to confirm research results based on the reflections of the NAFA leadership group, leading to several refinements of the preliminary findings presented to them. It is essential to mention that I am a member of the NAFA leadership group myself, which is another reason for adopting an autoethnographic approach in this case study.

12.3 The Case of the NAFA-Program

NAFA is a national program designed to innovate Danish science education at a national level by promoting engaging science education and active learning in Danish schooling, and by enhancing the training of current and future science teachers. The vision is to enhance scientific literacy in Denmark by increasing children and young people's interest, motivation, and competence in science. The purpose is enabling a significant number of learners to translate scientific knowledge into actions that address societal challenges and dilemmas, thus being able to apply science knowledge and act with profound and discipline-based knowledge and thus bridge the gap between science and society. The program is funded by two prominent philanthropic Danish funds—Novo Nordisk Foundation & Villum Fonden. They share a common goal of setting high standards for Danish science education with the aim of developing the future work force in science and technology. The purpose of NAFA is addressed through three main goals:

- Science teaching in teacher training will be systematically based on scientific and didactic knowledge as well as current experiences from practice.
- New sustainable models for science teaching will be developed for improving skills in science subjects in primary and secondary schooling. These models will be designed in a way that makes upscaling possible with a nationally coordinated approach.
- The necessary knowledge about science didactics will be produced, which in interaction with relevant discipline-based science knowledge can be translated into improved teaching practice.

To fulfil these ambitions, the program consists of three nationally coordinated efforts:

- An effort in teacher education developing teaching through competence improvement of teacher educators in national and local Professional Learning Communities (PLCs) (Albrechtsen, 2013). All Danish university colleges are partners in NAFA.
- An effort in primary and lower secondary schooling aimed at developing and testing sustainable models for competence development to explore which development models meet both the demand of municipalities and schools, and at the same time ensure long-term anchoring of science related learning. More than 15% of all Danish primary and lower secondary schools are involved in the program.
- A research effort in the form of a Centre of Excellence in Science Education (CESE) which carries out research in connection with teacher training and school-oriented efforts to produce new and necessary knowledge. All research institutions at Danish university colleges and universities, that work with science education research, are involved in the research effort.

The NAFA program has a seven-year span (2021–2028), and the focus during the first years is to establish and start up the program activities. The last year of the program has special focus on completion and evaluation. The rationale for establishing NAFA draws inspiration from the Danish National Strategy for Sciences (2018), especially regarding the strategic areas highlighting the need for academically skilled and dedicated teachers updated with the latest knowledge in didactics. Therefore, teachers and university college teacher students should be "offered competence development and increasingly supported in incorporating and applying new knowledge in their teaching practices" (The NAFA Program, 2020, p. 11). However, only a few initiatives have been nationally anchored, resulting in some challenges (op cit., p. 4). With the central coordination of the NAFA program leadership these three initiatives are interconnected, and with Bronfenbrenner's ecology model (1986) the initiatives affect different system levels within the educational ecology on a national level.

The Teacher Education Initiative focuses on the competence development of teacher educators, organised in PLCs (Albrechtsen et al., 2022; Blonder & Vescio, 2022) both internally and across university colleges. The practice-oriented research and development activities in the Teacher Education Initiative contribute to generating experiences for the unique dual didactic situation in teacher education, which in broad terms can be described as the connection between teaching practices in teacher education and teaching practices in primary and secondary schooling. This connection can be explained through Bronfenbrenner's ecology model (1986) by using the concept of transfer between the microsystem closest to the learner and the mesosystem understood as the school organisation as well as the university college context. The concept of transfer can explain whether teaching and learning from one context, such as teacher education practice, can cross boundaries and be applied in another context, such as practices in schools. The work carried out in teacher education practice can, in various ways, lead teacher students or teachers in further education, in collaboration with others, to construct meaning and take place. Likewise, experiences from practice in schools can mediate the way work is conducted in educational or further education contexts. Hence, the Teacher Education Initiative is dedicated to enhancing the skills and expertise of teacher educators by fostering professional learning communities within and across university colleges at a mesosystem level.

The School Initiative focuses on exploring competence development models that address the specific needs, possibilities, and demands on the microsystem level of primary and lower secondary education. These models are designed based on current research, aiming to foster the integration of learning in the field of natural sciences. The primary objective is to generate interest and demand for these models at the microsystem level of a school, but also at a municipal mesosystem level, and the ultimate ambition is to establish a nationally coordinated selection of well-established competence models throughout the program period on a macrosystem level. These models will prioritise academic and science didactic development in natural sciences. The models should be sustainable within the educational sector, and attractive and feasible for the target audience in the municipalities. The School Initiative aims to develop and test research-based competency development models that are not only sustainable and economically viable but also scalable. Sustainability entails ensuring that the models can be supported both financially and organisationally, while being grounded in evidence-based learning. Scalability refers to the models' ability to travel between system levels and aim for significant advancement in teachers' competency development while remaining adaptable to accommodate local resources and practical requirements.

The vision of CESE is to establish and expand a comprehensive science didactics research environment across the Danish university colleges and universities, with a focus on primary and lower secondary schooling and teacher education. The research initiated through this effort is intended to be of the highest quality and of use to science education. To emphasise this point, it is referred to as a centre of excellence. According to the NAFA project description, CESE aims to conduct practice-oriented research in close collaboration between researchers and professionals (The NAFA Program, 2020, p. 44). Hence, the comprehension of practice-oriented research is related to how applicable and useful the research results are at various system levels of the NAFA program-initiatives. In other words, the long-term vision of research in CESE is to contribute to and strengthen science education in Denmark.

The three initiatives rely on each other for support, function, and success and can thus be understood as interdependent and embedded educational innovation from a systemic perspective. Furthermore, the design of the NAFA program follows leading educational research stating the importance of systematic capacity building among educational stakeholders to innovate quality education (Díaz-Gibson et al., 2021). According to the NAFA project description (The NAFA Program, 2020, appendix 9), the national program leadership's role is to establish a harmonious alignment between national goals and localised execution, working in cooperation with the steering group and the local administration at the partner university colleges. Within NAFA, the national program leadership comprises a program director (from one of the larger Danish university colleges), a head of research (from the same university college as well as from a university (25% employment)—this is my position in NAFA), a vice head (from another university college) and two program managers (as well from different university colleges) leading the two other program efforts. In this way, the governance of NAFA is anchored nationally and is geographically spread out. The program leadership group not only assists the NAFA steering group, but also leads and coordinates the program's efforts on both a sector-wide and cross-sector scale, bridging connections among the various participants.

Furthermore, the leadership group bears the responsibility of broadly disseminating and embedding the program's activities at a national level. The program director holds responsibility for the program's outcomes, financial commitments, and legal obligations within the NAFA program, reporting to the steering group. Additionally, the program director is entrusted with the professional and coordinating aspects of NAFA's other leadership functions. The primary role of the program managers is to ensure alignment among the various activities within the NAFA initiative, and support the development of infrastructure for research, teacher education practices, and the testing of models in the school initiative.

Although the general objective for NAFA seems clear, the actual processes are contingent and vary within different local contexts of the program. Thus, managing the program at a national level is complex, and the complexity arises from intricate interactions spanning diverse boundaries in the ecology of the system levels of the program. As mentioned earlier, research on collaborative innovation accentuates the value of network-based cooperation as a holistic approach to tackling intricate issues within the public sector (Balslev, 2012, 2021; Aagaard et al., 2014). With this in mind, and with the autoethnographic approach to this case study, it is possible to shed light on drivers and barriers for the NAFA leadership group's attempts to facilitate systemic change.

12.3.1 Case-Situations & Analysis

12.3.1.1 Case 1—Unspoken Expectations: Who Defines the Agenda?

Field-notes from an online NAFA leadership meeting, October 2023

> It's Friday and we have our regular online meeting with the NAFA program leadership. I present an agenda for an upcoming important meeting with an external expert group organised by CESE. Our program leadership group will participate in this meeting. During this agenda item I give an orientation about some changes to the agenda.
>
> My colleagues complain that the agenda has been changed once again. I argue that the changes do not have significant consequences for the program leadership group, but my colleagues believe otherwise. At this point it is not clear to me what the problem really is.
>
> The Program director mentions that it concerns the reduction of the part of the meeting related to the program leadership group. My colleagues try to address the issue in a conciliatory manner, acknowledging the program directors' perspective that it's important for them to be involved and at the same time expressing trust in CESE leadership's planning of the meeting.
>
> However, I still do not fully understand the problem, but I finally realise that it's all about unclear expectations. We haven't negotiated our different hopes clearly enough. The challenge is the session where the external expert group was supposed to share their experiences with the program leadership group. This session is no longer entirely exclusive to the program leadership group but includes several other participants. My colleagues still appear upset. Nevertheless, we move on to the other items on the agenda.

The core of this case is whether or not the external expert group will get the chance to provide supervision to the NAFA leadership team. The case can be viewed as an example of micro-level everyday interactions where the leadership group tries their best to communicate and collaborate. But even so, they fail to ask the fundamental communicative question of "what are we talking about?". The group fails to stay with the trouble and navigate complexity by engaging in a sensemaking reflecting dialogue on the different underlying expectations and perspectives of this apparently simple agenda point. We stay with each our taken-for-granted-perspectives and miss out on opportunities and hopes for spanning the boundaries of different understandings. The interaction in this meeting is taking place in relation to and under the influence of the context of which the individual is a part, and thus with the intentions and world views the individual meets the whole (Bronfenbrenner, 1986). And in this case, unspoken intentions collide in an unresolved dynamic manner. Even though there are good intentions, the leadership team fails with open and reflecting dialogue with potential to move the group further towards mutual understanding.

Hence, the group misses out on the opportunity for pragmatic understanding and development. The reflecting dialogue is, as mentioned above, one of the central elements of the work in the NAFA PLC organisation. Just to be clear, the NAFA leadership group is generally not expected to work as a PLC. But it is expected that the NAFA leadership team can collaborate on driving the NAFA program activities—including activities that transverse the three NAFA Initiatives—as the meeting mentioned in the case-situation. At stake here is the group's ability to collaborate through mutual investigation and being curious together, especially in a case like this where disagreements are experienced. It is the responsibility of every group member to be attentive and to pause or call for a reflective time out. This co-responsibility is especially important in groups like the NAFA leadership group where leadership is distributed and negotiated (Daly, 2010; Torfing & Diaz-Gibson, 2016). As Weick (1995, p. 4) puts it, sensemaking provides structure to what we do not yet know or understand. The potential for this to happen is during interruptions or uncertainties like the one described in this case. The sensemaking process begins within the individual. Furthermore, the individuals' sensemaking might result in new understandings. But in this case, the sensemaking does not reach the collective level, and thus does not result in mutual understandings with the potential of collective action, decision-making or change. While the main driving force in this micro-level case-situation is the NAFA leadership teams' constructive attitude and hopes towards moving forward in the meeting agenda, this drive also becomes the barrier for the groups' ability to negotiate and create meaning and mutual understanding. The meeting ends unresolved regarding the sensemaking process and leaves the matter of whether the external expert group can provide supervision to the NAFA leadership team, open-ended. Thus, the case-situation indicates that there is a strategic need for the NAFA leadership team to strengthen their focus on transversal collaboration across the three NAFA programme initiatives. And like twenty-first century STEM-learners, the NAFA leadership team must develop their team competencies to navigate complexity and to deal with unpredictability (see Chap. 1)—not the least in micro-level interactions with unspoken expectations.

12.3.1.2 Case 2—Research Collaboration: Who Does the Recruitment?

Field-notes from an online-meeting with participation of the School Initiative leader, both CESE leaders and two CESE-research project leaders, October 2022

> In CESE we have initiated several strategic research projects. One of the projects is working to establish a baseline for the NAFA program and will provide a foundation for evaluating NAFA initiatives as well as insight into the connection between teaching practices, actor perspectives, and students' acquired skills in lower secondary and primary schooling. Therefore, the research group needs to establish collaboration with several schools and their science programs. To this end, we (CESE leadership) have initiated this joint online meeting to coordinate recruitment of schools.

> Through the initial conversations it quickly becomes clear to me that we haven't properly coordinated understandings of the recruitment process. How many schools do we have now, and for how long a period? Furthermore, what is the process for CESE research projects to get access to schools in the NAFA project? It is clear that our views on these questions are quite different. And intuitively I think that it all comes down to the lack of communication, resources, and willingness to share access in this fragile initiation phase of the NAFA programme.
>
> The atmosphere is tense due to the underlying differences. The leader of the School Initiative expresses that he unfortunately can't assist with recruitment of schools for the research projects baseline measurement. The meeting ends unresolved, but with a clear message that the responsibility for recruitment of schools for CESE research projects lies with the research project group, which means that the schools recruited will not necessarily have connection to the NAFA program and thus risk decoupling the research results.

This case can be viewed as an example of a meso-level situation with different expectations between two of the three NAFA Initiatives: The Research Initiative, and the School Initiative. As described in earlier, the NAFA program is built on three pillars with very different goals and tasks described in much detail in the program application (The NAFA Program, 2020). And since this present research project is carried out in the first years of the NAFA program, the three NAFA initiatives are in their initiation phase (Fullan & Mundial, 1989; Fullan, 2007). This means that they are building up organisational routines and social structures with appropriate understandings and habits. In other words, the three pillars are grounding themselves and building up culture. Therefore, they are occupied with these inward oriented tasks. The case-situation involves two of the NAFA initiatives and it reveals a closure of the two. None of the Initiative leaders seems to be able to think systemically and leave the structures they are building in their own silos and span the boundaries between silos.

According to the education researcher Peter Miller (2009), the ability to span boundaries is crucial in successful cross-sector and cross-organisational collaboration, like the NAFA program. Boundary spanners are partaking individuals who possess necessary interpersonal skills to move in and out of different situations (Miller, 2009, p. 336) and accordingly contribute by their fluid participation.

As mentioned in Chap. 1, a central feature of Creative Pragmatics is for learning organizations and educational institutions to understand and engage with complexity (Nicholls, 1983; Gallagher & Savage, 2023), and thus building capacity to think and act systemically eg. by spanning boundaries.

Interestingly, it seems to be the NAFA project Initiative leaders themselves that become the obstacle for collaboration and not the NAFA stakeholders—here the research project members and the schoolteachers/leaders. This is because the initiative leaders remain in fixed their own silos conditioned by the NAFA structure as it is designed in the Program description (The NAFA Program, 2020) and are preoccupied with their own program silo-building and lack or fear the ability to look up and out towards synergies and the bigger holistic picture. Hence the lack of abilities or willingness to span boundaries, and the rigid NAFA program structures become barriers for the collaboration they initially were designed to create, and thus an obstacle for innovation through the practice-oriented baseline research project.

This calls for an attention to boundary spanning structures (Miller, 2009) in the NAFA program. Since the designed structures in NAFA seem to hinder collaboration, there is a need for identification and creation of such structures that can spur collaboration independent of specific boundary spanners—the role that the Initiative leaders ought to be acting out in this case. Creating such boundary spanning structures can, according to Miller (2009), prevent that neither physical, organisational, nor political boundaries limit collaboration. Furthermore, these structures can support the Boundary Spanner's work as flexible organisational navigators (ibid., p. 624) and enable them to "pierce holes in the silos" and allow the collaboration, interplay, and emergence of subsystems and social structures within the larger system of NAFA on a national scale. This case highlights the complexity of NAFA as large-scale systems innovation, and the notion and the need for a nature of professional practice that moves away from the idea that professionals, leaders, managers are in control (Davis & Sumara, 2006; Davis et al., 2008; Svabo, 2021), and thus to embrace the necessity to understand the concept of emergence.

12.3.1.3 Case 3—Institutional Politics: Who Gets the Money?

Field-notes from a series of partner meetings, October to November 2021

> To begin the establishment of CESE we (CESE leadership) arranged physical meetings with the science education research departments of all 10 partner institutions during the fall of 2021. It was quite a time-consuming nationwide round trip. But it was important regarding our collaborative take on creation of our first research strategy on science education. We discussed methodology and excellence and ambitions for CESE as the first National research centre for science education, but also equality and distribution of CESE resources among the 10 partner institutions.
>
> During all meetings I detected an implicit institutional-related tension. It seemed to revolve around the fact that university colleges in Denmark have the obligation to educate next generation teachers as well as competence-build current teachers from lower and secondary education. Furthermore, university colleges are obligated to engage in didactical research, but do not have the right to educate new researchers (PhDs). PhD-education in Denmark is exclusively undertaken by universities. The NAFA-program is targeted science education in primary and lower secondary education and in university colleges and is fundamentally a university college assignment. However, four universities are partner institutions in the NAFA Program because of their historical founded obligations and experience with science education research.
>
> The number of times and intricate ways in which equality was mentioned as problematic, made me fear that this issue could become one of the largest obstacles for success of CESE research and for my role as head of research. All in all, I sensed an underlying competition regarding prestige and possible research financing—especially from the university colleges. For example, I heard a person from a university college argue that an idea about a research initiative on informal science learning settings was too peripheral for a university college agenda—CESE research project implied. Furthermore, it was mentioned that the university in question could go right ahead and finance a research project like that on their own. In other words, the partner institutions play a political positioning game regarding prestige and resources.

This macro-level case-situation consists of several issues, institutional politics as well as questions about CESE-research topics' centre and periphery. Starting off with a focus on institutional politics, there exist quite different organisational framings and worldviews among the ten NAFA partner institutions. Firstly, there are very different legal bases and institutional tasks between university colleges and universities in Denmark, and despite the differences there exists an ambition of equality among partner institutions in NAFA. Thus, there is a built-in tension in the NAFA program which comes into play when research strategy and collaboration is designed in practice. Who gets the most out of NAFA regarding research financing and capacity building? There is no doubt that the main 'target group'—or the 'end users' with innovation lingo—are the students and their motivation, interests and competences within science in general. But the NAFA program's ambition is to unite forces between university colleges and universities, and thus affect the end users by lifting the sector as a whole—in other words, a whole system approach (Fullan, 2007). The question to be raised here is how the practices and procedures arise regarding determination of what becomes centre or periphery, and which rules and conditions as well as sociological and psychological mechanisms are in play? It would certainly be relevant to go through a thorough analysis of all elements of this complex question. However, here the rules and conditions will be the offset for analysis.

As mentioned, one of the set rules is the legal right to educate researchers. This rule or condition sets the scene for collaboration between the 10 partner institutions in CESE. When it comes to building future research capacity in the field of science education, the partner institutions are forced to collaborate, whether they like it or not. The first 2 years of NAFA and CESE demonstrates potential for this transversal collaboration through the many initiated research projects. However, collaboration must stand the test of time in the coming years with the launch of the CESE PhD program, where the collaboration among primary supervisors from universities and co-supervisors from university colleges is essential for success.

To continue, this case-situation provides a peek at the different perspectives that are at stake for the different partner institutions—here represented by different attitudes towards informal science learning settings, and the question about whether informal learning settings could become centre or periphery for CESE research. If a program like NAFA is to succeed with innovation of science teaching on multiple system levels—through a whole system approach—there seems to be an important level missing in the NAFA design: The broader society, and thus the informal learning settings. Fostering connections with the broader community and anchoring science teaching in local initiatives—municipal organisations, local companies as well as sport clubs and other educational institutions related to science in practice—is a central part of educational innovation from a network perspective (Díaz-Gibson et al., 2016; Díaz-Gibson et al., 2021; Longás Mayayo et al., 2020), and is furthermore at the forefront of the NAFA vision, which aims to enable learners to bridge the gap between disciplinary knowledge and transversal skills, as well as between science and society. The lack of focus on this element might become a barrier for the realisation of the NAFA vision.

In other words, the case-situation shows an important—and expected—power play of underlying hopes and fears that indicates a turn towards a 'diminished whole system approach' where the broader society and informal learning situations slides into the periphery of CESE research. The unresolved question is to what extent the NAFA initiatives, and in particular CESE, will prioritise incorporating specific and context-bound elements and expectations from local stakeholders in their science education research, and thus facilitate whole system educational change.

12.4 Crosscutting Conclusions

The three case-situations each on their level point to different types of drivers and barriers of large-scale educational innovation. The analysis of the first situation on unspoken expectations and agenda-setting, indicates that the cover ground at meetings can be a barrier for eventually more time-consuming, yet important processes of sensemaking on a micro-level meeting, with the potential to affect the other levels of the program. The NAFA leadership team should work with their team competencies to navigate complexity and constantly remind themselves that they are dealing with unpredictability while facilitating major large-scale change processes juggling multitudes of world views. Hence, there should always be time, space, and curiosity directed towards engaging in holistic dialogue revealing a multitude of perspectives. This simple outcome can be supported by former mentioned research findings regarding the organisation of PLCs as a driving structure for educational change and is furthermore backed up by the research on sensemaking processes in general (Weick, 1995) and on education specifically (Balslev, 2012).

Moving on, the meso-level case-situation on research collaboration and recruitment suggests that the NAFA program leadership should engage with complexity, and thus build capacity to think and act systemically. The leadership team could profit on strengthening the capacity of boundary spanning in general. They could beneficially begin with looking inward towards their own leadership group, and to a higher extent work with the transformation of this group into a high performing team (Katzenbach & Smith, 2015) with the willingness and abilities to span boundaries of practices, procedures, rules, and conditions—with Miller's (2009) understanding of change. Thus, they could work with their sociological and psychological fitness understood as social capital (Lin, 1999) that enables them to move in and out of different situations, so to speak. In this way they could become role models for the change processes and capacity building in the program on its different levels.

Turning to the last case-situation on institutional politics and distribution of research resources, this macro scale case implies that the whole system approach (Fullan, 2007) adopted by the NAFA program might not be as whole-system as assumed by the leadership and the partner institution stakeholders. According to front end research on educational innovation (Díaz-Gibson et al., 2016; Díaz-Gibson et al., 2021) people in the broader society, and thus the informal learning settings, play an important role in anchoring educational innovation—here aiming at bridging the gap between science and society. This implies a focus on local and informal micro educational practices with a possible long term and large-scale effect on student interests, and thus in this case-situation on the supposed effect of the NAFA program as whole. To this end the interconnectedness of people in large scale systems, as pointed to in Chaps. 2 and 3 in this volume, is important to take into account while navigating science education innovation at a national level. Creating and investigating science education collaborations with the broader community and anchoring science teaching in local initiatives, could be a beneficial model for long term and sustainable NAFA macro-scale impact. Thus, this discussion implies a much broader understanding of a whole system approach in NAFA, including not only formal education along the official educational chain, but including civil societal stakeholders and local organisations as well, and opening up to possibilities of individual and local actions leading to change in large scale patterns.

Furthermore, this collected volume's focus on Creative Pragmatics highlights the potential for enhancing STEM education through a Creative Pragmatics approach. This approach enables all stakeholders in the educational ecosystem to ensure a robust pipeline of diverse scientific talent and to cultivate future generations of adaptive thinkers capable of navigating the complexities of the real world. Thus further research could explore aspects of organizational change and human behaviour in relation to Creative Pragmatics. For instance, studies could examine how innovation management in STEM-related projects incorporates Creative Pragmatics.

12.5 Further Implications

Coming to an end of this analysis of large-scale innovation, it is noteworthy to turn to experiences from larger national science initiatives with similar elements as the NAFA program. For example, there are valuable insights from research-based science teacher education in Norway (Nilsen et al., 2021) and science education research in Sweden based on practitioners and teachers collaborating on defining research questions (Prøitz et al., 2021). These Scandinavian studies, including this present Danish study, contribute with knowledge on innovating science education at a national policy-driven level. Despite different research designs and scope, the studies draw some of the same generic conclusions—not the least about two areas: science education research and educational large-scale decision-making.

Regarding science education research, there is a need for integrating research traditions from various disciplines (Jorde & Dillon, 2012) which increases complexity, but also spurs possibilities for innovation and generating novel ideas and solutions that may not emerge within a single discipline. Regarding educational decision-making in the complex organizational context of innovating science education, sensemaking is marked by diverse preferences, unclear processes, and fluid participation, given that not everyone is consistently engaged in and overseeing all aspects (Prøitz et al., 2021). These features are typical of organisations at different times and levels, and carry significance for processes of change, innovation, and solutions. This is due to a heightened awareness of the context and different disciplines, in which various solutions or models are meant to function and make sense. As this study settles, large-scale science education innovation demands a specific type of holistic agency from those involved. Thus, sensemaking and boundary spanning are indicators for innovation and are therefore central concepts to be aware of in future studies of large-scale transdisciplinary initiatives like the NAFA program. Furthermore, studying science education as complex ecosystems of practices (Kemmis et al., 2014) will be important for future insight into transformational processes of education in general.

12.6 Closing Remarks

This chapter has addressed innovation of a national educational ecosystem across sectors and institutional borders and has analysed how the NAFA leadership group juggle hopes and fears of change from multiple educational levels and organisational perspectives—furthermore how they struggle to access central information to create meaning of different elements of the program during the program-realisation process. Through the analysis of three selected case-situations, different drivers and barriers have been highlighted. The situations point towards lessons to be learned involving the NAFA leadership group and their struggles with collaborative innovation of Danish science education. Thus, the ambition of unlocking possibilities for Danish science education innovation can be packed into the following implications.

Firstly, and most importantly, the drivers and barriers of large-scale innovation in science education in this specific case study results in 'being two sides of the same coin'. This means that a barrier—in its opposite form—represents a driver. This study points to the following pragmatic implications for leadership of large-scale innovation of science education:

- Go ahead and expand an innovation focus with structures that include the prioritisation of holistic reflecting dialogue that allows collective sensemaking (Weick, 1995; Weick et al., 2005) and invite a multitude of world views. Doing so could be a driving force for micro situations with the potential to affect the complex ecosystem as a whole (Bronfenbrenner & Morris, 1998).
- Work deliberately with the capacity to span the boundaries created and reinforced by general framing—e.g., project descriptions or institutional policies. Work with leadership-willingness and abilities of boundary spanning regarding practices, procedures, rules, and conditions (Miller, 2009). Thus, work with

leadership-sociological and psychological fitness understood as interpersonal skills allowing flexible organisational navigation (ibid. p. 624) and capacity to think and act systemically. Start drilling holes in the silos at the meso-level and work across initiatives. This will be an appropriate strategy with a changing effect on the educational ecosystem.
- Consider the appropriate span of the understanding of a whole system approach (Fullan, 2007). Build in the broader and informal learning communities (Díaz-Gibson et al., 2021) in a whole system approach, and thus lift the potential of anchoring science teaching in local initiatives to bridge the gap between science and society and spur long term and sustainable macro-scale impact.

All in all, take your time and engage in attentive, creative and pragmatic observation of the settings you move through, and co-create meaning of the incomplete pictures from the many perspectives you meet. Think abductively in your sensemaking, act pragmatically when creating innovation, and embrace the transformative potential of active learning in leadership and the interplay at all levels in the coherent ecosystem of science education. Now, you can create potential to unlock new possibilities for educational change in the buzzing, blooming and ever-evolving world of science education.

References

Aagaard, P., Sørensen, E., & Torfing, J. (2014). *Samarbejdsdrevet innovation i praksis* (Djøf/Jurist-og Økonomforbundet).

Albrechtsen, T. R. S. (2013). *Professionelle læringsfællesskaber: teamsamarbejde og undervisningsudvikling* (Dafolo).

Albrechtsen, T. R. S., et al. (2022). *Professionelle læringsfællesskaber: et forskningsoverblik (2018–2021)* [Professional learning communities: A research overview (2018–2021)]. NAFA.

Albret, P. (2010). New public management på retræte [New public management in retreat]. *Mandag Morgen, 7*, 46–48.

Alvesson, M. (2003). Methodology for close up studies–struggling with closeness and closure. *Higher Education, 46*(2), 167–193.

Balslev, G. M. (2012). *Omdannelse & uddannelse: innovation, læring og samarbejde* (FS&P Ph.D. afhandlinger, No. 72). Institut for Samfund og Globalisering, Roskilde Universitet.

Balslev, G. M. (2021). Learning throughout the innovation journey: A new dimension to learning as a search process. *International Journal of Innovation in Education, 7*(2), 86–101.

Balslev, G. M., & Jakobsen, R. M. (2000). www.kommunikationensvindue.dk: Refleksioner over det abduktive samfund. Internettet som eksponent for en erkendelsesteoretisk og kulturhistorisk analyse af et paradigmeskifte [Reflections on the Abductive Society: The Internet as an Exponent of an Epistemological and Cultural-Historical Analysis of a Paradigm Shift].

Balslev, G. M., & Raae, P. H. (2019). REFORMulering—Manipulation eller Metastyring? [REFORMulation—Manipulation or Metagovernance?]. *Samfundslederskab i Skandinavien, 34*(3), 207–228.

Blonder, R., & Vescio, V. (2022). Professional learning communities across science teachers' careers: The importance of differentiating learning, *Handbook of research on science teacher education*, pp. 300–312.

Bourdieu, P., & Richardson, J. G. (1986). Handbook of theory and research for the sociology of education. *The Forms of Capital, 241*, 258.
Braun, V., & Clarke, V. (2006). Using thematic analysis in psychology. *Qualitative Research in Psychology, 3*(2), 77–101.
Brinkmann, S., & Tanggaard, L. (2010). *Kvalitative metoder: en grundbog* (Hans Reitzels Forlag).
Bronfenbrenner, U. (1986). Ecology of the family as a context for human development: Research perspectives. *Developmental Psychology, 22*(6), 723.
Bronfenbrenner, U., & Ceci, S. J. (1994). Nature-nuture reconceptualized in developmental perspective: A bioecological model. *Psychological Review, 101*(4), 568.
Bronfenbrenner, U., & Morris, P. A. (1998). The ecology of developmental processes. *Handbook of Child Psychology, 1*(5), 993–1028.
Daly, A. J. (2010). *Social network theory and educational change* (Vol. 8). Harvard Education Press.
Davis, B., & Sumara. (2006). *Complexity and education: Inquiries into learning, teaching, and research* (1st ed.). Erlbaum. https://doi.org/10.4324/9780203764015
Davis, B., Sumara, D. J., & Luce-Kapler, R. (2008). *Engaging minds: Changing teaching in complex times* (2nd ed.). Routledge.
Díaz-Gibson, J., Civís-Zaragoza, M., & Guàrdia-Olmos, J. (2014). Strengthening education through collaborative networks: Leading the cultural change. *School Leadership & Management, 34*(2), 179–200.
Díaz-Gibson, J., et al. (2016). Networked leadership in educational collaborative networks. *Educational Management Administration & Leadership*, 1741143216628532.
Díaz-Gibson, J., et al. (2021). The SchoolWeavers tool: Supporting school leaders to weave learning ecosystems. *School Leadership & Management, 41*(4–5), 429–446.
Edquist, C. (2010). Systems of innovation perspectives and challenges. *African Journal of Science, Technology, Innovation and Development, 2*(3), 14–45.
Felder, R. M., & Brent, R. (2016). *Teaching and learning STEM: A practical guide*. Wiley.
Fujii, L. A. (2015). Five stories of accidental ethnography: Turning unplanned moments in the field into data. *Qualitative Research, 15*(4), 525–539.
Fullan, M. (2007). *The new meaning of educational change*. Routledge.
Fullan, M., & Mundial, B. (1989). *Implementing educational change: What we know* (Education and employment division, population and human resources department ...).
Gallagher, S. E., & Savage, T. (2023). Challenge-based learning in higher education: An exploratory literature review. *Teaching in Higher Education, 28*(6), 1135–1157. https://doi.org/10.1080/13562517.2020.1863354
Harrison, A. K. (2018). *Ethnography*. Oxford University Press.
Jorde, D., & Dillon, J. (2012). Science education research and practice in Europe: Retrospective and prospective. In *Science education research and practice in Europe* (pp. 1–11). Brill.
Katzenbach, J. R., & Smith, D. K. (2015). *The wisdom of teams: Creating the high-performance organization*. Harvard Business Review Press.
Kemmis, S., et al. (2014). *Changing practices, changing education*. Springer.
Levin, B., & Fullan, M. (2008). Learning about system renewal. *Educational Management Administration & Leadership, 36*(2), 289–303.
Levitan, J., Carr-Chellman, D., & Carr-Chellman, A. (2017). Accidental ethnography: A method for practitioner-based education research. *Action Research*, 1476750317709078.
Lin, N. (1999). Building a network theory of social capital. *Connect, 22*(1), 28–51.
Longás Mayayo, J., Riera Romaní, J., & de Querol Duran, R. (2020). Management and leadership of networks for educational success: Analysis of emerging networks in Spain. *School Leadership & Management, 40*(2–3), 183–202.
Maaløe, E. (1996). *Casestudier af og om mennessker i organisationer* [Case studies by and about people in organizations]. Akademisk forlag.

Miller, P. M. (2009). Boundary spanning in homeless children's education: Notes from an emergent faculty role in Pittsburgh. *Educational Administration Quarterly, 45*(4), 616–630.
Nicholls, A. (1983). *Managing educational innovations* (1st ed.). Routledge. https://doi.org/10.4324/9781351040860
Nilsen, T., et al. (2021). Et kritisk og konstruktivt blikk på naturfaget i norsk skole [A Critical and Constructive Perspective on Science Education in Norwegian Schools]. In Nilsen, T. & Kaarstein, H. (eds.) Med blikket mot naturfag: Nye analyser av TIMSS 2019-data og trender 2015–2019 [Focusing on Science: New Analyses of TIMSS 2019 Data and trends from 2015 to 2019]. Universitetsforlaget, 207–260.
Peirce, C. S. (1994). *Charles Sanders Peirce: Semiotik og pragmatisme* [Charles Sanders Peirce: Semiotics and Pragmatism]. Anne Marie Dinesen og Frederik Sjernfelt (Eds.), Lars Andersen (Trans.), København: Samlerens Bogklub.
Penprase, B. E. (2020). *STEM education for the 21st century*. Springer. https://doi.org/10.1007/978-3-030-41633-1
Prøitz, T. S., et al. (2021). *Utbildning, lärande, forskning: En evalueringsstudie–delrapport 3* [Education, learning, research: An evaluation study—Partial report 3].
Putnam, R. (2001). Social capital: Measurement and consequences. *Canadian Journal of Policy Research, 2*(1), 41–51.
Ring, P. S., & Van de Ven, A. H. (1994). Developmental processes of cooperative interorganizational relationships. *Academy of Management Review, 19*(1), 90–118.
Scott, J. (2011). Social network analysis: Developments, advances, and prospects. *Social Network Analysis and Mining, 1*(1), 21–26.
Svabo, C. (2021). Living world dynamics—Or: What Brian Eno can teach us about knowing in a complex world. In E. Brandt, T. Markussen, E. Berglund, G. Julier, & P. Linde (Eds.), *Proceedings of Nordes 2021: Matters of scale* (pp. 384–389). Nordes. https://doi.org/10.21606/nordes.2021.42
The Danish Government, Ministry of Education. (2018). *National Science Strategy in education.*
The-NAFA-Program. (2020). *The Academy of Science—NAFA. A National Program for Quality of Science Education in Denmark*, (Copenhagen), 60.
Torfing, J., & Diaz-Gibson, J. (2016). Transforming governance to enhance social and educational innovation. *Pedagogía Social, 28.*
Van de Ven, A. H., Polley, D., & Garud, R. (2008). *The innovation journey*. Oxford University Press.
Weick, K. E. (1995). *Sensemaking in organizations* (Vol. 3). Sage.
Weick, K. E., Sutcliffe, K. M., & Obstfeld, D. (2005). Organizing and the process of sensemaking. *Organization Science, 16*(4), 409–421.
Yin, R. K. (2009). *Case study research: Design and methods (applied social research methods)*. Sage.
Zhou, C. (2016). Fostering creative problem solvers in higher education: A response to complexity of societies. In C. Zhou (Ed.), *Handbook of research on creative problem-solving skill development in higher education* (Vol. 1, pp. 1–23). IGI Global.

Gitte Miller Balslev BA, MA and PhD, is a transdisciplinary scholar and associate professor in Educational Innovation and a part of the management team at FNUG—STEM Centre for Research in Science Education and Communication—at the Department of Mathematics and Computer Science at SDU, University of Southern Denmark, SDU. She has experience with research management as former head of the national research center CESE—Center of Excellence in Science Education. Gitte holds a PhD in collaborative innovation in public sector education, and develops her research and publications around STEM-related educational innovation through national reforms, educational leadership in reform realisation, collaborative innovation and educational ecosystems from a network perspective. Building on long practical experience and strong bonds to the Danish public educational system, Gitte builds on Engaged Scholarship and forms research projects as collaborative endeavours between multiple practitioners, professions and researchers. Gitte has long experience in the areas of education, innovation and organization and has facilitated major changes stretching from institutional, municipal to national level. At SDU Gitte also works with developing university teaching with connection to the UN goals of sustainability and has developed new master programs on pedagogical change management in and around public schooling as well as courses on project management in science. Gitte is associated at Center for Primary and Lower Secondary Education Research, SDU investigating organization and leadership of Eco-schools in relation to green societal transition. Furthermore, she is co-initiator of The Global NetEduProject, a transnational developmental research project exploring the potential of educational networks.

Chapter 13
A Call to Action for Embedding Creative Pragmatics in STEM Education

Tamara Carleton, Connie Svabo, Michael Shanks, and Chunfang Zhou

13.1 Taking STEM into Visionary Futures

Now we come to the last chapter in our book, where we aim to bridge thinking and doing, insight and action, just as advocated in our concept of Creative Pragmatics. We consider multiple takeaways across the mix of case studies from our various contributors. We explore different implications for the future of STEM education. And we ask where the dialogue and work on Creative Pragmatics can go next and what roles different stakeholders can take to embrace active learning more *actively* through Creative Pragmatics.

We have been considering how skills and competencies for addressing urgent contemporary and future challenges might be taught and learned. This question lies at the heart of research, policy, and initiatives in STEM education. Creative Pragmatics functions as an umbrella term that encompasses and can coordinate multiple active learning approaches—including problem-oriented-project learning, project-based learning, the Roskilde Model, and design thinking, among other pedagogical models. The concept's range and scope takes in so much beyond the classroom, curriculum, discipline-centred bodies of knowledge and skills, and the plans of educational policy makers. The concept takes us into how we understand the very

T. Carleton (✉)
Department of Mechanical Engineering, Blekinge Institute of Technology, Karlskrona, Sweden
e-mail: tamara.carleton@bth.se

C. Svabo · C. Zhou
STEM Centre for Research in Science Education and Communication (FNUG), Department of Mathematics and Computer Science, University of Southern Denmark, Odense, Denmark

M. Shanks
Archaeology Center, Stanford University, Stanford, CA, USA

C. Svabo et al. (eds.), *Creative Pragmatics for Active Learning in STEM Education*, Contributions from Science Education Research 14,
https://doi.org/10.1007/978-3-031-78720-1_13

building of knowledge, our relationships with the environment, with other species that share our complex living ecologies, our experiences of making and embodied engagement with materials and materialities. Our argument is that active learning, under this holistic vision offered by Creative Pragmatics, is a means of furthering STEM skills and competencies, and much more. This is the added value of the concept of Creative Pragmatics—it offers deep foundations for *future-oriented vision* for learning communities, (re)injecting choice, relevance, and joy back into learning models. As such, this book serves to inform and to advocate, to invigorate, and to inspire the need for educational reform.

13.2 Change Agents in Learning and Education

At its heart, active learning is situated learning that empowers educators and students, managers and policy-makers, to create their own meanings in dynamic learning environments. Creative Pragmatics directs our attention to the agency of stakeholders and to potential change agents in the reform of STEM education.

Higher education has a vital role to play in developing student's knowledge, skills, competencies, attitudes, and values that enable them to contribute to and benefit from a sustainable future (Miller, 2015; Wamsler, 2020). As the European Qualifications Framework (2017) urges, students should be able to manage and transform work or study contexts that are complex, unpredictable, and require new strategic approaches. Students should take responsibility for contributing to professional knowledge and practice and/or for reviewing the strategic performance of teams. As mentioned in Chap. 1, this implies developing competence in working in complex and unpredictable situations. Education should strive to enable students in structuring their own work, to develop into autonomous and self-directed individuals as well as competent collaborators. Students are expected to be able to collaborate across disciplines, to assume professional responsibility and to be able to formulate goals, interests, and concerns in self-directed ways. It is also a goal that students develop competencies in working with people that are different than themselves, to be able to work across organisational levels, and to be able to make informed decisions on situations of insecurity, risk and change an on the basis of incomplete information (Méhaut & Winch, 2012; Elken, 2016; Mikulec, 2017).

In a similar vein, as the Organization for Economic Co-operation and Development (OECD) (2018) addressed in the learning framework for 2030, students who are best prepared for the future are change agents; they should not only learn knowledge for their future work but also skills needed to become active, responsible, and engaged citizens. This also calls higher education to respond to challenges of developing twenty-first century skills (Tan, 2021).

Beyond students as individual change agents, higher education should function as a systemic change agent too. The development of higher education should be described as a pattern of 'complex responsive process', which involves ever-changing reforms, reflection of practice, and evaluation of changes (Davis & Sumara, 2006; Zhou, 2016).

Contrary to the prevailing tendency of education as a reactive institution, particularly in response to societal demands, the evolution of STEM education necessitates an innovative and proactive paradigm, underscored by a commitment to meaningful pedagogical approaches (Kennedy & Sundberg, 2020). This is reinforced by the advocacy for future-oriented strategies, emphasising the alignment of student learning with skills essential for future professional efficacy, in contrast to a conventional focus on predefined knowledge that is quickly out of date (Zhou, 2012, 2016; Kennedy & Sundberg, 2020; Dondi et al., 2021). In navigating complexity, the cultivation of creative competencies becomes imperative, encompassing the adept assessment of situations, drawing on experiential knowledge and effective communication with stakeholders (Zhou, 2016). Such creative competency extends to the facilitation of sense-making in the present and the envisioning of collective trajectories for the future (Smith & Ashby, 2020).

Most notable here are major initiatives, undertaken on a global scale, that advocate skill sets, tools and techniques that come under various labels such as Futures Studies and Strategic Foresight. As an editorial team, we note the exponential growth of this field over the last decade, and we have all been actively involved in this development. Much of this growth has been in policy and business applications, but research centres in futures studies are now common. Here we draw attention to the UNESCO initiative in Futures Literacy launched by our colleague Riel Miller (Miller, 2018; UNESCO, n.d.) that aims to promote skills and competencies for anticipating different futures, decolonizing the future by freeing us from unexamined premises and unchallenged agencies. UNESCO, as an organisational change agent, has aimed to empower educational institutions to establish centres and programs that challenge us to build better futures with appropriate skills and competencies. Futures Literacy is what we can call an initiative in Creative Pragmatics because it places students and local community stakeholders at the heart of collective learning and shared decision making.

13.3 Chapter Themes and Takeaways

In Chap. 1, Sects. 1.5 and 1.8 we outlined some key features of Creative Pragmatics. These sketches were expanded in Chaps. 2 and 3. We have used six metaphors or schemes in describing Creative Pragmatics, its features, and contexts of application, in order to capture how it is a synthetic concept, dynamic and suggestive of intervention and action.

- **Intersecting fields** of interest, research, study, concern—science and technology studies, arts and design practice, philosophical pragmatism. These are conveyed in the Venn diagram that is Fig. 1.1.

- A **constellation** of concepts, of bright points of enlightenment with different characteristics—like stars, dwarves and red giants, occasional planets and comets—perceived to be connected in loose and often partially-defined arrangements in the dark sky that prompt efforts to wonder and sensemaking, storytelling and mythmaking. Concepts, with reference to subjects of interest such as complex social phenomena, as stars arranged in the sky, are distinct but can only be fully appreciated in the context of their arrangement. In this metaphor, philosophical concepts and their relation to social phenomena should be understood not in isolation but in relation to other concepts and contexts. Such relationality allows for a more nuanced and dynamic or dialectical understanding, where the relationships generate insight and significance. In Sect. 1.5 we pointed to the six brightest concepts in the constellation of Creative Pragmatics: process and relationships (in relation to knowledge and dynamic systems), design and performance (in relation to making and competence), experience and creative agency (in relation to embodiment and (im)materialities).
- **Making**. Table 1.1 summarises key premises, frames, and modes of practice of Creative Pragmatics, broadly following a model of making. This involves establishing infrastructures and premises, adopting a mindset or design philosophy according to which one will work, and then organising and getting on with making through competencies and tool kits. As in any building or making, there is no one way to proceed in Creative Pragmatics—what one does in making anything depends on what one perceives and wants, one's resources, competencies, the feasibility of the project, opportunities and challenges.
- **Mapping and navigation**. Throughout our presentation of Creative Pragmatics, we have resisted a tight definition of the concept and have instead referred to Creative Pragmatics as involving means by which one might navigate the waters, archipelagos, the landscapes of complexity and uncertainty, when changing weather and currents, falling mist and storms, and ever-imperfect knowledge call forth quick judgement and hunches, improvisation, rules of thumb, collaboration and sharing knowledge and know-how, following faint tracks, pathfinding, work-arounds, border crossings and escapes. No one map can suffice for all purposes, but a map can direct our attention and aid navigation, suggesting without determining where one might go. In this, the map of emerging terrain is performative and eidetic, augmenting processes of coming to know in a changing world, aiding action rather than merely serving as a representation.
- **Policy frameworks**. Chapter 3 expanded on the features of Creative Pragmatics that relate pragmatically to current policy and planning in STEM education and serve to offer lines of action and implementation. Creative Pragmatics in this particular context focuses upon competence in the performance of knowledge, in processes of coming to know, in relation to policy specifications of knowledge, skills, attitudes, responsibilities and autonomy in learning.
- **Rules of thumb and shorthand guidelines**. Chapter 2 explored the foundations of Creative Pragmatics in an understanding of knowledge as situated performances of knowing. It was appropriate therefore to end the chapter with five

suggested summary rules of thumb that might guide how one pursues Creative Pragmatics.

We should not expect any of these metaphors or schemes to be perfect in fitting all purposes. Our schematics are suggestive, intended to be more provocative than prescriptive. If we turn now to the case studies presented in Chaps. 4, 5, 6, 7, 8, 9, 10, 11 and 12, we will see different ways that the features of Creative Pragmatics might be explored, mobilised, and made manifest.

Above all, we witness in the case studies how the authors have adopted and adapted particular features of Creative Pragmatics, mixing and matching according to need, purpose, intention, and aim. We might group these mobilizations into four categories:

- Participation, agency, responsibility and autonomy—student-centred creativity
- Thinking through making—hands-on active learning
- Collaboration and co-creativity—teamwork
- Foresight, imagination and worldbuilding—framing the future.

Firstly, Creative Pragmatics emphasises participation, agency, responsibility and autonomy. The learner's creative agency is a key topic in Chap. 4 (Niss), and runs through many case studies of this book. Having this focus on agency is crucial in order to enable learners to develop the necessary competencies to navigate complexity. This principle is most importantly directed at students' agency and participation, with its core connection to the student-centred, active learning approach. On a broader educational system level this can translate to creative agency of teachers as well, who are not just implementing the curriculum, but are active designers of the learning process with a voice in a systemic context. Although student agency and active participation can often be seen as a goal, and even brought forward through various teaching methods, it can be challenged in existing educational systems because of the lack of holistic and systemic approach. For example, it can be challenged by inflexible assessment systems and methods (Chap. 6, Svendsen & Larsen), or by the question of priority setting and resources in educational institutions. As Niss points out in Chap. 4, open-inquiry and problem-oriented project work is sometimes criticised as requiring too many resources, but this critique overlooks the importance of educational goals these methods are aimed at. It is thus inspiring to see examples in the case studies of this book of educational institutions that dedicate time, adapt spaces and provide other needed resources for learning that empowers agency and autonomy.

Second is a group that exemplifies a practical orientation towards hands-on thinking-through-making: materially, physically embedded, tangible learning through the use of artefacts, spatial awareness and material engagement. Chapter 5 (Eich-Høy & Zhou) explores how creative artefacts enliven mathematical learning. Chapter 6 (Svendsen & Larsen) considers the importance of emotion in educational assessment and offers a holistic view on learners' experience: cognitive, sensory, emotional, material and affective. Chapter 9 (Troelsen & Borch) discusses the influence of space, further showing how physical space may be manipulated to enhance active learning. This principle has close ties to art and design, and maker thinking, their connection to STEM education is analysed in Chaps. 8 (Hobye et al), and 10 (Bell & Ford). Creative Pragmatics is as embodied as it is intellectual, but as in traditional educational settings cognitive dimensions tend to dominate in expense to other aspects, Creative Pragmatics draws attention not just to cognitive, but also sensory, emotionally engaging experiences and materiality of learning. It is important to move beyond the segregational, either-or approach, and integrate both aspects into learning experiences and, as Chap. 8 (Hobye et al) emphasises, learn to switch between these modes: practical Maker thinking, and classical, more structured knowledge acquisition.

The third group of case studies is collaboration and co-creativity—teamwork. Chapter 10 (Bell & Ford) applies the design studio model from architectural education to STEM-related disciplines as a type of safe and experimental instructional space, where the learning relies on ongoing interaction and critical inquiry between students and faculty. Then Chap. 12 (Balslev) shares personal reflections on organisational change from her time observing and leading a research leadership team at a national science program in Denmark. By understanding team dynamics better, she realises the link to organisational learning. As it is vulnerably shown in Chap. 5 (Eich-Høy & Zhou), it can be challenging to move from content or designer-centred towards a user or learner-centred view, but this inclusion of the perspective of learners is crucial in the process of learning.

In a fourth group of case studies, Creative Pragmatics invites us to engage in foresight, imagination and worldbuilding. Chapter 7 (Carleton & Uebernickel) investigates the project brief as a common tool used by educators and student teams for corporate-sponsored projects. The authors propose prompts that guide students to innovate on four different horizons, from near term to visionary, and suggest a two-question guide for crafting project briefs more closely aligned with desired innovation outcomes related to active learning. Imagination and worldbuilding is closely interlinked with play, and in this book, play-infused learning by students as well as educators is explored in Chaps. 5 (Eich-Høy & Zhou), 8 (Hobye et al) and in Chap. 11 (Albrechtsen & Kiær), describing a performative pedagogy of enactment based on a case study of training Danish science teachers. If we want to develop transformative competencies, it is important to apply multiple horizons of perspective: the multilayered experience in the now, and the long-term view.

These four groupings by no means exhaust all that the case studies offer in the way of a rich exploration of the features of Creative Pragmatics.

13.4 Creative Pragmatics by Stakeholder Group

This first set of chapters is only a start to bringing Creative Pragmatics to life. We recognize that it takes courage to challenge and especially to change any existing paradigm, let alone how the sciences and STEM topics are taught. Navigating these challenges requires a concerted effort from multiple groups—that span educators to administrators, policymakers to funding bodies—to recognize the value of STEM approaches informed by a visionary Creative Pragmatics approach. In many ways, active learning as a shared academic practice is still at a primitive stage. Our call to action for Creative Pragmatics raises implications for all the various stakeholders involved in STEM-related education. We turn now to discuss an initial set of implications by stakeholder group.

13.4.1 Implications for Teachers and Educators

We use the term teachers loosely to refer to anyone responsible for instructing, guiding, or training students at any level, including tenured faculty, adjunct instructors, lecturers, graduate students, and teaching assistants. We see this book as a way to inspire anyone to adopt Creative Pragmatics wherever they are in their journey of active learning. Those already teaching with active learning methods can widen their pedagogical strategies by considering the core principles of Creative Pragmatics to create more participatory and student-directed learning experiences. Moreover, teachers who have hesitated to identify with a single pedagogical model, such as design thinking, or tried to evolve a well-known approach, such as problem-based learning, can feel encouraged to draw from a much broader philosophical outlook embodied in a Creative Pragmatics stance.

Our approach emphasises connections and relations, which challenges the traditionally siloed, even deeply conservative, nature of education—especially at higher education levels. A Creative Pragmatics stance deliberately fosters cross-disciplinary teaching, both in terms of considering multiple modes of hands-on learning within a classroom and bridging the learning transition between different education levels. For example, students often find that how they learned mathematics or science in high school (gymnasium) differs from their experiences at a university, becoming more demanding without becoming more relevant to what they see as real-world science or engineering. Teachers adopting a Creative Pragmatics approach can infuse more interaction and practical building that augments classic instructional models, helping students to feel more prepared advancing to the next level in their education.

From a student's perspective, Creative Pragmatics more closely aligns a set of twenty-first century skills with how they actually learn because students are required to take a more active role in their own learning. As such, we feel that students gain an emotional platform that aims to empower them, giving them more chances to develop self-initiative and reduce any fears of continued math-heavy programs. As we hear more young people raise concerns about how they can address the growing complexity of societal problems, such as climate change or the challenges represented by the UN's Sustainable Development Goals, we might ask—How does a Creative Pragmatics approach help them practise being change agents as they learn STEM-related material? We argue that not only does this learning approach increase their knowledge in school, learners also carry confidence and skills past graduation into their careers and future learning goals.

13.4.2 Implications for Educational Institutions

Beyond teachers and students, how can Creative Pragmatics offer lasting benefits at the institutional level? We believe that this approach can help universities and other schools more effectively attract and retain students in STEM-related fields, especially as many schools are funded based on student headcount. As European nations and the United States face a demographic cliff—based on dropping population forecasts of the next generation of students—a Creative Pragmatics approach offers a more compelling way to connect with young people and be more emotionally sensitive to their learning needs, plus counter growing dropout rates across science fields. For example, in Denmark, we know of several gymnasiums that have ended their physics programs due to fewer students and low interest, all of which impacts the next stage in the talent pipeline of universities seeking to sustain their physics programs. Moreover, Creative Pragmatics can influence the broader ecosystem of institutions supporting schools and universities, sparking more collaborative and student-centric partnerships.

13.4.3 Implications for Research Groups and Scholars

This book introduces a manifesto for Creative Pragmatics with several empirically backed examples, yet there are many other areas of research that will help build bodies of knowledge and good practice. Most of the empirical cases in our book focus on the university level, so more research is needed to describe and compare how Creative Pragmatics functions at other education levels—including elementary, secondary, apprenticeship, professional development, and executive education. As one line of inquiry, in what ways does Creative Pragmatics help foster interdisciplinary thinking among students by integrating elements from the humanities, art, design, social sciences, and other "non-technical" fields into science and STEM education? In addition, the examples and cases in our book predominantly feature Northern Europe with some American counterparts, so it would be good to examine how schools and groups respond to and adopt Creative Pragmatics in other states and regions, such as India, Latin America, Asia, and across the African continent.

Considerations related to measurable impact and systemic structures should be studied too. For instance, how do student engagement, retention rates, and performance outcomes change when Creative Pragmatics approaches are integrated into the curriculum? What mix of incentives and deterrents can encourage more training and experimentation for teaching Creative Pragmatics within the classic science research model?

In terms of space, how does an understanding of Creative Pragmatics influence the design (or redesign/adaptation) of STEM labs, school buildings, and even academic campuses? How do classrooms and other learning environments configured for Creative Pragmatics serve as affective learning spaces that support individual agency beyond current models of active learning?

Additional research studies could investigate aspects of organisational change and human behaviour regarding Creative Pragmatics. For example, how do the different academic disciplines for STEM-related fields implement Creative Pragmatics; which disciplines excel most in experimentation and why; and what lessons in cultural obstacles and pedagogical change can be shared across disciplines and regions? Also, does a certain profile exist of a new type of teacher trained in Creative Pragmatics delivery, especially given the high levels of ambiguity and collaboration required? What factors are critical when training and assessing effective teaching in Creative Pragmatics? How do levels of emotional intelligence influence the ability to teach Creative Pragmatics? Another set of questions should investigate how conventional power structures and learning dynamics change when a Creative Pragmatics approach is used, particularly between teachers and students and also between students themselves.

A different direction for future research might analyse the connection between Creative Pragmatics and emotional wellbeing, especially as more studies report rising problems in student mental health. Studies could explore any correlations between emotional wellbeing and Creative Pragmatics as seen in the process of learning and learning culture. In fact, high performance culture has been considered to negatively impact students' mental health (e.g., Högberg, 2023), so by emphasising more active and engaging pedagogical strategies, as seen in the practical examples in this book, additional evidence can indicate how Creative Pragmatics strengthens learning flexibility and enjoyment, as well as reduces performance burdens.

13.4.4 *Implications for Funding Agencies and Policymakers*

Funding groups—including government agencies, private foundations, family offices, active donors, and other investors—play an important role in the educational ecosystem because they recognise that investing in the next generation(s) of talent requires capital funding as much as those willing to push change forward. Creative Pragmatics is no different. Which funding models are more effective in launching and delivering Creative Pragmatics in schools and with youth? What types of funding partners are needed to establish new or transitional reskilling programs to train teachers in Creative Pragmatics at primary, secondary, and tertiary education levels? How might investors set funding milestones to accelerate the development of implementation materials for Creative Pragmatics, including at the tactical level of teaching manuals, playbooks, case studies, and scripts?

Equally important are policymakers who work to set requirements for accreditation, regional delivery, legal incentives (and disincentives), and other measures of academic performance. For example, how can European governmental leaders help to champion how a Creative Pragmatics approach acts on and contributes to the implementation of the European Qualifications Framework, plus directly builds competence in essential twenty-first century skills? Moreover, by taking a Creative Pragmatics stance, what role can policymakers play—especially in nations with declining youth populations, such as Europe and the US—when facing the growing talent gap in education? How might they encourage more young learners to stay in STEM-related fields?

13.5 Putting Creative Pragmatics in Practice

13.5.1 Long-Standing and Time-Proven Models of Learning

Our case study chapters offer rich examples of putting Creative Pragmatics into practice. We reiterate that particular features of Creative Pragmatics involve long-standing and time-proven models of knowledge and understanding, thoughtful practice (praxis), and learning (we have been particularly concerned to connect to the pragmatist educational philosophy of Dewey, for example). Active learning by doing is as old as humanity, and we have mentioned in Sect. 1.5 that the constellation of concepts that illuminate Creative Pragmatics takes us back to the very beginnings of recorded philosophical reflection upon ontology, epistemology and ethics—how we might come to know and act appropriately in a world of particular being, substance and form, and of which we are a part. It is appropriate to make the point that in addressing the challenges of navigating complexity and the competencies involved, we might aspire to be more aware of history and genealogy of our concepts and concerns, and to establish the kind of holistic connections we offer in Creative Pragmatics.

Closer to the experiences that motivate and inform this book are developments since the 1970s in student-centred and project-based learning offered in Denmark and at Stanford. Student-centred learning was initiated at Aalborg and Roskilde in the 1970s and the "Roskilde Model" (Andersen & Heilesen, 2015) represented a strong case for the value of project-based problem-oriented transdisciplinary learning. The design group in Stanford's School of Engineering pioneered project-based making-to-learn from the 1960s. By the foundation of the Hasso Plattner Institute of Design (the "d.school") in 2004, the programs had come to be called design thinking (see Tamara's book on the graduate course ME310, Carleton et al., 2023). As we mentioned in the Preface, this book started with the aim of connecting the project-based experiential learning of Roskilde with Stanford's design thinking. It was early editorial work and reflection upon their grounding in pragmatist science and technology studies, in art and design practice, that led Michael and Connie to coin the term Creative Pragmatics.

13.5.2 Personal Experiences of Intervening into Institutional Practice with Creative Pragmatics

Let us therefore expand a little more on these experiences of putting into practice what we are now calling Creative Pragmatics.

The STEM Centre for Research in Science Education and Communication (FNUG—Forskningscenter for Naturvidenskabelig Uddannelse og Formidling, https://www.sdu.dk/en/forskning/stem), was established by Connie Svabo and colleagues in 2021 at the University of Southern Denmark. The centre involves 26 faculty members and affiliates, post-docs and doctoral researchers, and exemplifies the principles of Creative Pragmatics in STEM education. As a transdisciplinary centre, FNUG unites education and communication scholars with scientists from various fields to create transformative educational experiences, actively remodelling science education for engagement and impact.

FNUG serves as a dynamic hub where Creative Pragmatics is actively manifested in practice. The centre's approach is rooted in adaptability, openness, and a performative understanding of knowledge. A signature method at FNUG involves spanning from research-based creative production to practice, developing open-ended, participatory activities that spiral into formalised education and pedagogy.

One example is FNUG's public-facing events and pop-up exhibitions, which serve as educational interventions within traditional academic settings, offering creative, pragmatic learning opportunities for a diverse audience, including educators, students, and researchers. These curated activities demonstrate how science can interface with society in meaningful ways, making disciplinary knowledge accessible to a broader audience and embedding small-scale interventions into everyday practices, as well as expanding our understanding of the kinds of activities that take place at universities.

A recent FNUG education flagship initiative is the course on project leadership for science students. This course unites students from various disciplines, equipping them with essential transversal skills and competencies. It aligns with Creative Pragmatics by emphasising transactional relationships, contextual sensitivity, and fostering agency and autonomy. Complementing this course are other inter- and transdisciplinary courses, such as teacher training and communication courses, all rooted in FNUG, aimed at supporting STEM students in developing disciplinary knowledge, and transversal skills and competences.

FNUG's supervision and facilitation of student project-based learning are another example of Creative Pragmatics in action. Researchers and students work together in design thinking inspired prototyping poster seminars. Taking place once every semester, these seminars involve faculty and students collaborating to rapidly prototype and visually communicate emerging research projects, facilitating co-creativity and challenging conventional academic practices. This fosters a performative understanding of knowledge that is iterative, dynamic, and creative. FNUG's educational offerings also extend to innovative Ph.D. seminars, co-created with an emphasis on transdisciplinary engagement in the learning process.

FNUG affiliates, faculty, and students together advance a Creative Pragmatics approach to education, where knowledge is actively constructed, shared, and developed in dynamic learning environments. This FNUG way of working is validated by its central role in a consortium with Copenhagen University and Aarhus University, supported by an historic 125 million DKK grant from the Novo Nordisk Foundation (2025–2029). This initiative, the largest of its kind globally, aims to place Denmark at the forefront of university science education research.

13.5.3 Creative Pragmatics—Connecting STEM with Transdisciplinary Fields

Let us now turn to Michael's effort to connect STEM learning with a broader curricular context through the competencies that feature in Creative Pragmatics (for a detailed pragmatist ethnography of the transdisciplinary field of archaeology see Rathje et al., 2012).

As chair of a department of archaeology at University of Wales, Michael had led a faculty team to design and implement a modular curriculum in archaeology that embraced competency development through the transdisciplinary field—integrating lab sciences, social sciences, arts and humanities, from STEM to the creative arts applied to heritage management and the cultural politics of historical identity. Heritage and identity politics remain key concerns in the challenges of contemporary globalist politics, while STEM-oriented funding has come to dominate archaeology.

This project in what can, with hindsight, be described as Creative Pragmatics, was made feasible and viable by three key factors. First, the department faculty shared a future-oriented vision of a discipline that was united in its diversity by a grounding in key concepts in critical and reflexive social theory. These were particularly agency and power-knowledge, with fields of knowledge building such as archaeology, and embracing STEM manifestations, conceived as contemporary cultural construction—archaeologists working with the remains of the past to contemporary ends. Second, the management of the department was as flat as it might be, given the necessary institutional requirements for decision-making—the faculty were a collaborating team with a concern for a progressive pedagogy that would prepare students for actionable insight through archaeology into matters of contemporary cultural concern. Third, the "resource allocation model" at the University of Wales was, at the time, unusual in that it devolved all financial responsibility and management to the level of department. This meant the faculty team led by the chair had complete control over funding decisions. Without having the middle management of a dean's office, for example, the department faculty could strategise without external interference. They aimed to promote an archaeological skill set that mobilised the archaeological imagination in the service of contemporary concerns in a world now explicitly acknowledged as belonging to the Anthropocene, an era characterised by human impacts that threaten the integrity and sustainability of our lifeworld.

The just-described administrative system of University of Wales failed as Michael left in the late 1990s because there were no means of coordinating strategic planning across departments. The departmental structures Michael encountered when he moved to Stanford were, and remain, quite different. Dominated by STEM and ICT (information and communication technology) oriented programs, the institution remains wedded to quite traditional models of learning that foreground bodies of knowledge and disciplinary skill sets (this comment is based upon a direct survey of 2500 undergraduates and their majors in academic years 2021–2023 as well as the official specification of majors to be found in the university bulletin). Departments remain committed to conventional disciplinary boundaries, and certainly in the recruitment of faculty. Large initiatives such as the new (as of 2024) Doerr School of Sustainability do represent a commitment to transdisciplinary integration, though this is mainly focused on technology-based solutions to contemporary environmental challenges. The extent to which the mission and vision of this $1.2 billion investment will embrace the kind of competencies at the heart of the European strategic plans discussed in Sects. 1.4 and 3.6 remains unclear.

Notable in relation to implementing Creative Pragmatics however, were four key opportunities. The first was an interdepartmental program, Science Technology Society (STS). This remains popular because it complements the STEM bias in undergraduate majors (about 75% of undergraduates follow STEM majors). Over two decades Michael has taught a project-based class, *Ten Things: an archaeology of design*, that foregrounds key principles of Creative Pragmatics with regard to building competence in critical contextualization of science and technology. The course was voted by students as one of Stanford's top ten: surveys and course evaluation confirm student interest in acquiring competencies that complement conventional STEM learning. Many see such competence in "soft skills" as a means of enhancing their standing and savviness in the job market.

The appeal of skills in problem-solving in complex real-world scenarios applies especially to programs in design. This is the second opportunity Michael took up. Design programs are a great way to ground STEM in the competencies of Creative Pragmatics. As mentioned in the previous section, Stanford has pioneered over more than five decades competence-based learning under the heading, since the 2000s, of design thinking. Michael has collaborated with engineers and professional designers in team-taught studio-based courses, introducing broad historical and cultural context to design challenges through archaeology and anthropology. This led directly to an experiment in transdisciplinary learning, the Revs Program (2009–2015), which centred research and learning upon the automobile ("phenomenon-based learning"), connecting the history and archaeology of the automobile with current agendas in the design and engineering of sustainable carbon-neutral mobility. The program involved working directly with the automobile industry as well as communities of car enthusiasts, modelling the creative competencies needed in the necessary evolution of mobility design as well as in the cultural curation of the archaeological automobile.

The possibilities and potential of the transdisciplinary outreach just described has been the third opportunity. A cultural feature of Silicon Valley business community, within which Stanford has long played a significant role, is ease of networking. It is relatively easy to share and collaborate within and beyond the university on matters of common concern, and these typically centre upon STEM topics and applications. A vital focus is upon competencies needed for future-oriented innovation. In collaboration with Tamara, William Cockayne, and Larry Leifer in the Center for Design Research, Michael established an interdisciplinary research group in strategic foresight and innovation that extended active learning practices underway within Stanford's design community (Leifer, 2005; Dym et al., 2013). This has been a platform for projects with businesses and organisations to help them acquire and develop competencies in Creative Pragmatics. One such ongoing project is with the Japanese tech corporation Aisin; the objective is to prompt Aisin's cultural evolution into a learning community that embraces Creative Pragmatics, expanding their capacity to innovate in science and tech while centring on social values and sustainable development goals.

The fourth opportunity involves a mindset oriented on experiment and risk. Under the leadership of Jeffrey Schnapp (Italian and comparative literature), Tim Lenoir (science studies), Henry Lowood (science and tech curation), and Michael, Stanford Humanities Lab (2003–2009) fostered project-based research-learning (Schnapp & Shanks, 2009). The lab modelled collaborative participatory learning in designing exhibitions, publications, IT applications that showcased the intersection of STEM–social science–humanities–arts skill sets and methodologies. Projects included scientific instrumentation, IT simulation, expert systems, and collaborative authoring software as well as many more that can be located within the humanities and social sciences. The experimental and entrepreneurial ethos of the Humanities Lab lives on in Michael's personal lab (https://archaeolog.stanford.edu).

What might one take away from these experiences? They illustrate succinctly what might be involved in the implementation of programs in Creative Pragmatics: matters of leadership, translation and reconciliation of different disciplinary and stakeholder interests, of institutional management and administration in relation to one's agency in being able to effect change, spotting and grabbing opportunities, collaboration and establishing common ground, clarifying and sharing vision for the future.

It will be clear that all these can be related to the competencies of Creative Pragmatics. It is those competencies that can facilitate their implementation. One does well to practise what one preaches.

13.5.4 Creative Pragmatics—The Student Experience

For further inspiration, let us now offer an example of an implementation of Creative Pragmatics with a focus on student learner experience, showing great value across multiple student cohorts.

We have discussed high level institutional concerns, policy and curriculum. In their case studies, the book chapters have placed a strong emphasis on the science and math components in the STEM mix and less on the engineering component, which defines the E in STEM. While engineering is often lightly considered, if at all, at primary education levels, let alone higher education, we certainly do not mean to overlook engineering, so, as complement to the outline in the previous section of opportunities in design and engineering, we highlight another engineering example, another real-world example of *personal* implementation of a program that highlights the principles of Creative Pragmatics. It follows a student journey from suspicion to deep engagement.

William Cockayne, a colleague of book editors Carleton and Shanks, (re)invented a STEM course that embodies the tenets of Creative Pragmatics. Cockayne adopted a set of required courses for third-year aspiring engineering students at a technical university in Sweden. Cockayne has taught the course several times with similar outcomes. At the course start, students had to be pushed to break their passivity and be coaxed into taking initiative. They were used to learning to a standard model of lectures and exams. Several students were visibly uncomfortable with the high levels of ambiguity evident in the course, which they even communicated to the university's department head. In lieu of lecture mode, classroom time was instead spent in open-ended and self-reflective discussions, where the teaching team led students through a set of exploratory journeys that in fact had been road-mapped and rehearsed beforehand by the faculty.

Outside of class, students worked in teams to build increasingly advanced prototypes through continuous exploration (e.g., "what do they want us to do?"), self-directed learning (e.g., "does this YouTube video show us how to load the code onto the circuit board?"), deep collaboration (e.g., "why don't we try using the new sensor to play a game?"), and tangible building (e.g., "what type of prototype is due next?"). Teams presented their work in the following week of class, which mixed their informal demos with class reflections that, in turn, led to the next set of questions. The hands-on course format also allowed the teaching team to support a returning student whose lack of childcare support would have limited her success in a more traditional course based on lectures, problem sets, and exams. Instead, this student served as an amazing junior coach to the other student teams because she brought valuable industry experience and insight, plus built an exceptional set of prototypes on her own time after working with the other teams.

In the second time teaching the course, Cockayne aimed to address the discomfort that some students experienced when being asked to make their own decisions on what to build. The course activity was to develop a toy concept as student teams. This toy focus helped retain a sense of fun in STEM learning, while also drawing on the students' prior personal experiences being children. Cockayne connected the toy activity to the real world in multiple ways, which caused the students' learning to feel less abstract and more relevant as career preparation. For example, students received the initial assignment in the form of a Market Requirements Document (MRD), a well-known industry tool, which was based on an actual MRD from a top tech company. Like real life, students also had to justify their concept in context, so Cockayne had them consider early pricing by benchmarking similar toys on Amazon.com because toy pricing affects choices in material, production costs, and user perception. In addition, students read recent news articles that reinforced key engineering concepts, such as an interview with a Pokémon executive about the company's growth strategy in relation to technology decisions. Subsequently, students saw how their in-school education matched expectations outside school, plus why reading widely matters to learning. Fourth, students were frequently doing math—now with a clear reason—so they saw the value of their equations in their assignments.

The course's final exam setup is worth highlighting too. Student teams had to demo functional prototypes to a visiting group of students brought by the local gymnasium STEM teacher. These gymnasium students served as representative toy users and were younger than the university students by several years. The exam setup changed the course testing experience in several ways. First, university students were delighted to be tested—when was the last time readers here remember students grinning after their science or engineering exams? Moreover, the university students took pride in showing off their concepts, even taking extra care in telling their product stories and creating colourful mock packaging to match their toy concepts and enhance appeal. Second, instead of being judged by faculty, university students were judged by a younger generation. This changed the power dynamic, and both groups became empowered to ask one another questions and direct their own learning. Third, the younger students witnessed a positive learning experience, which inspired some to consider STEM in their own futures. In fact, one teenage girl expected that the visit would be a lecture, which made our editorial team wonder why does she already assume this of university courses? Afterwards, this teen told the STEM teacher at her school: "I thought it would be the same as other visits where we just have to listen to someone talk, but this time it was fun and so inspiring to be able to participate in judging models and ask questions." This comment surprised her teacher, who told Cockayne: "This is exactly what I am motivated to do as a teacher: get students to become curious and interested to learn!"

In our informal observations, we saw how impactful the teaching model was to the potential long-term success of these students. Most of the STEM-heavy courses that these students had taken in their first 2 years at university relied mainly on thick engineering textbooks filled with math equations, backed by copious presentation slides. Frankly, most students couldn't recall key concepts or tools described in earlier courses, showing low retention of foundational material. More concerning yet unsurprising, the students didn't question the efficacy of this pedagogical approach because they assumed this was how university courses were supposed to be! Even student peers at other universities where our editorial team teaches globally—across the United States, Japan, Mexico, Switzerland, Italy, Austria, and more—have expressed similar views. The majority of these students default to a reactive mode, expecting to be spoon-fed content and crunch through mathematical problem sets. At this particular school, the overall dropout rate was historically high, and students who "quit" the university's engineering track were seen by most faculty as weak and fearful of mathematics. Yet in truth, when this math was placed in context, the students understood why and how numbers were necessary to advance the project efforts and their own knowledge and they even became motivated to shape their own learning process.

13.6 Final Reflections

By adopting Creative Pragmatics as a stance on active learning, the field of education can move beyond the transmission of facts and theories to foster a deeper understanding of scientific and STEM principles and their practical applications. Creative Pragmatics can help bridge the gap between traditional curriculum-based teaching and the demands for active, hands-on learning experiences by helping teachers design active learning environments and by foregrounding student engagement.

Within and through Creative Pragmatics, we take a standpoint in relation to how knowledge is developed, shifting more towards knowing as a process than knowledge as units. This is how we best teach and learn in higher education, giving priority to student agency, highlighting and trying to make visible learning processes, and adopting active methods. In accordance with pragmatist philosophy, interaction between knowledge institutions and publics is based on negotiation, mediation, and process. Ultimately, Creative Pragmatic approaches to STEM education aim to support the development of competent, empowered, and empathic individuals who can and want to contribute meaningfully to society with confidence and creativity. Isn't this the real point of education?

References

Andersen, A. S., & Heilesen, S. B. (Eds.). (2015). *The Roskilde model: Problem-oriented learning and project work*. Springer.

Carleton, T., West, S., & Cockayne, W. R. (Eds.). (2023). *Inventing the almost impossible: Creating, teaching, funding, and leading radical innovation*. Springer.

Davis, B., & Sumara. (2006). *Complexity and education: Inquiries into learning, teaching, and research*. Erlbaum.

Dondi, M., Klier, J., Panier, F., & Schubert, J. (2021). *Defining the skills citizens will need in the future world of work*. McKinsey & Company. Retrieved January 10, 2024, from https://www.mckinsey.com/industries/public-sector/our-insights/defining-the-skills-citizens-will-need-in-the-future-world-of-work

Dym, C. L., Agogino, A. M., Eris, O., Frey, D. D., & Leifer, L. J. (2013). Engineering design thinking, teaching, and learning. *Journal of Engineering Education, 94*(1), 103–120.
Elken, M. (2016). 'EU-on-demand': Developing national qualifications frameworks in a multilevel context. *European Educational Research Journal, 15*(6), 628–643.
European Qualifications Framework. (2017). Official Journal of the European Union, C 189/15. Retrieved January 9, 2024, from https://eur-lex.europa.eu/legal-content/EN/TXT/HTML/?uri=CELEX:32017H0615(01)&from=EN
Högberg, B. (2023). Academic performance, performance culture, and mental health: An exploration of non-linear relationships using Swedish PISA data. *Scandinavian Journal of Educational Research*. https://www.tandfonline.com/doi/full/10.1080/00313831.2023.2192752
Kennedy, T. J., & Sundberg, C. W. (2020). 21st century skills. In B. Akpan & T. J. Kennedy (Eds.), *Science education in theory and practice* (Springer texts in education) (pp. 479–496). Springer.
Leifer, L. (2005). Center for design research, Stanford University. In J. Clarkson & C. Eckert (Eds.), *Design process improvement* (pp. 522–525). Springer. https://doi.org/10.1007/978-1-84628-061-0_35
Méhaut, P., & Winch, C. (2012). The European qualification framework: Skills, competences or knowledge? *European Educational Research Journal, 11*(3), 369–381.
Mikulec, B. (2017). Impact of the Europeanisation of education: Qualifications frameworks in Europe. *European Educational Research Journal, 16*(4), 455–473.
Miller, R. (2015). Learning, the future, and complexity. An essay on the emergence of futures literacy. *European Journal of Education, 50*(4), 513–523.
Miller, R. (Ed.). (2018). *Transforming the future: Anticipation in the 21st century*. Routledge. Available at https://www.taylorfrancis.com/books/e/9781351047999
OECD. (2018). *The future of education and skills. OECD learning framework 2030*. https://www.oecd.org/education/2030/E2030%20Position%20Paper%20(05.04.2018).pdf
Pearson, M., & Shanks, M. (2001). *Theatre/archaeology*. Routledge.
Rathje, W. L., Shanks, M., & Witmore, C. (Eds.). (2012). *Archaeology in the making. Conversations through a discipline*. Routledge. https://doi.org/10.4324/9780203083475.
Schilpp, P. A., & Einstein, A. (1949). *Albert Einstein philosopher-scientist* (1st ed.). Library of Living Philosophers.
Schnapp, J. T., & Shanks, M. (2009). Artereality (rethinking craft in a knowledge economy). In S. H. Madoff (Ed.), *Art school: Propositions for the 21st century*. The MIT Press. https://doi.org/10.7551/mitpress/7811.003.0013
Smith, S., & Ashby, M. (2020). *How to future: Leading and sense-making in an age of hyperchange*. Kogan Page Publishers.
Tan, O. S. (2021). *Problem-based learning innovation: Using problems to power learning in the 21st century*. Gale Cengage Learning.
UNESCO. (n.d.). *Futures literacy*. UNESCO. Retrieved August 29, 2024, from https://www.unesco.org/en/futures-literacy
Wamsler, C. (2020). Education for sustainability: Fostering a more conscious society and transformation towards sustainability. *International Journal of Sustainability in Higher Education, 21*(1), 112–130.
Zhou, C. (2012). Fostering creative engineers: A key to face the complexity of engineering practice. *European Journal of Engineering Education, 37*(4), 343–353.
Zhou, C. (2016). Fostering creative problem solvers in higher education: A response to complexity of societies. In C. Zhou (Ed.), *Handbook of research on creative problem-solving skill development in higher education* (1st ed., pp. 1–23). IGI Global.

Tamara Carleton PhD, Universitetslektor in Mechanical Engineering, Blekinge Institute of Technology.

Tamara is an award-winning professor, industry advisor, and expert in radical innovation and strategic foresight. Recognised as one of the Top 50 Women Leaders in Education (in 2023 and

2024) by Women We Admire, she teaches at multiple business schools and technical universities around the world. In addition, she holds the UNESCO Chair in Anticipatory Leadership for Innovative and Better Futures at Tecnológico de Monterrey in Mexico. Tamara is also the founder and CEO of the Innovation Leadership Group, which helps leaders and teams to build innovation capacity and map bold futures. She has written several seminal books in innovation, including *Building Moonshots: 50+ Ways to Turn Radical Ideas into Reality* and the *Playbook for Strategic Foresight and Innovation*. She holds a doctorate in mechanical engineering from Stanford University.

Connie Svabo PhD, Professor of STEM Education and Science Communication, Director of STEM Center for Research in Science Education and Communication (FNUG), Department of Mathematics and Computer Science, University of Southern Denmark.

Connie is a transdisciplinary scholar and creative practitioner with expertise in learning environment design and experience. She is the founding Director of Center for Research in Science Education and Communication (FNUG), Section Leader and member of the management team of the Department of Mathematics and Computer Science, chair of the SDU research programme in Science and Mathematics Education and Communication. She promotes Creative Pragmatics as a model for knowing in a complex world and underpins this stance with a practice that ranges from art over education to design and consultancy. Her portfolio includes geological UNESCO heritage sites Stevns Klint and the South Fyn Archipelago, as well as the Danish National Gallery Open Air Museum and the museum of natural history, Naturama. Previously a Continuing Studies Instructor at Stanford University and Faculty at Roskilde University, where she directed the research centre Experience Lab, chaired the study programmes in Communication Performance Design 2013–2016 and served on the academic board. She has published books on STEM education, sustainability, competence development, design pedagogy and digital life. She holds a PhD in Social Science from Roskilde University's multidisciplinary programme in Society, Space and Technology.

Michael Shanks PhD Doc, Professor of Classics and Archaeology, Stanford University.

Acknowledged as one of the most original and influential of contemporary archaeologists, Michael is a specialist in long-term humanistic views of design and creativity, innovation and social change, while exploring connections across the sciences and arts in a hybrid field of theatre/archaeology. He is senior faculty in Stanford University's programs in Science Technology Society, Urban Studies, Writing and Rhetoric, Archaeology, Classics, and in the Center for Design Research, part of Stanford's d.school in the School of Engineering. He directed Stanford Humanities Lab, pioneering innovative project-based pedagogy in the arts and humanities, and helped found the Revs Program at Stanford, connecting the archaeological automobile with contemporary car design. Founding faculty in Stanford's Archaeology Center, Michael is a director of Stanford's research group Foresight and Innovation, and, for two decades until its retirement in 2023, he worked with mediaX, the affiliates' program that connected Stanford with industry through technology and innovation. He served on the city of Rotterdam's International Advisory Board from 2008 to 2019 and works with many companies and organisations in developing cultures of innovation. In another mobilisation of Creative Pragmatics, Michael continues to pursue archaeological fieldwork in the northern borders of the Roman empire and beyond.

Chunfang Zhou PhD, Associate Professor of STEM Education and Science Communication, STEM Center for Research in Science Education and Communication (FNUG), Department of Mathematics and Computer Science, University of Southern Denmark.

A researcher and practitioner working on creativity, STEM education, and teacher education. Chunfang has an interdisciplinary and cross-cultural educational background: she finished a Bachelors Degree (Industry Automation and Information Science) and Masters Degree (Philosophy of Science and Technology) in China; during 2008–2012, Chunfang finished her doctoral research on *Group Creativity in Engineering Education in a Problem and Project-Based Learning (PBL) environment* at Aalborg University, Denmark. She is a board member of Danish Development

Research Network (DDRN), and works as a senior consultant in the PBL (Problem-Based-Learning) Teaching Innovation and Research Center at Northeastern University (NEU), China. Chunfang locates her research in Science, Technology and Society (STS), with a focus on creativity and STEM education. Since 2013, Chunfang has authored and co-authored over 110 peer-reviewed publications; she is the editor of *Handbook of Research on Creative Problem-Solving Skill Development in Higher Education* (IGI Global, 2016), the author of *Introducing Problem-Based Learning for Creativity and Innovation in Chinese Universities* (IGI Global, 2020), and an editor in the book series *Arts, Creativities and Learning Environments in Global Perspective* (Brill/Sense). Chunfang is an editorial board member of *Discovery Education* and the *European Journal of Education and Pedagogy*, as well as advisory board member for the *International Journal of Engineering Education.*

MIX
Papier aus verantwortungsvollen Quellen
Paper from responsible sources
FSC® C105338

FSC
www.fsc.org

If you have any concerns about our products,
you can contact us on
ProductSafety@springernature.com

In case Publisher is established outside the EU,
the EU authorized representative is:
Springer Nature Customer Service Center GmbH
Europaplatz 3, 69115 Heidelberg, Germany

Printed by Libri Plureos GmbH
in Hamburg, Germany